Agricultural Privatisation, Land Reform and Farm Restructuring in Central and Eastern Europe

Edited by
JOHAN F.M. SWINNEN
ALLAN BUCKWELL
ERIK MATHIJS

Routledge
Taylor & Francis Group
LONDON AND NEW YORK

First published 1997 by Ashgate Publishing

Reissued 2018 by Routledge
2 Park Square, Milton Park, Abingdon, Oxon, OX14 4RN
711 Third Avenue, New York, NY 10017, USA

Routledge is an imprint of the Taylor & Francis Group, an informa business

A Library of Congress record exists under LC control number: 97074820

ISBN 13: 978-1-138-61014-9 (hbk)
ISBN 13: 978-0-429-45022-8 (ebk)

Contents

Tables

Figures

Contributors

Volker Beckmann is preparing a PhD in agricultural economics on agricultural transformation in Eastern Germany from the University of Göttingen. He is researcher at the Institute of Agricultural Policy, Marketing and Agricultural Development, Humboldt-University of Berlin, Germany.

Dr Štefan Bojnec is Professor of Agricultural Economics at the University of Ljubljana, Slovenia, and advisor to the Government with the Ministry of Agriculture and Forestry in Ljubljana, Slovenia. Previously, he was post-doctoral researcher at the Leuven Institute for Central and East European Studies, visiting research fellow at the University of California at Berkeley and at the University of Manchester. He holds a PhD from the University of Zagreb.

Allan Buckwell is Professor of Agricultural Economics at the Department of Agricultural Economics, Wye College, University of London, United Kingdom. He was the past president of the European Association of Agricultural Economics and served as senior adviser on agricultural policy reform for several Central European governments and for the European Commission.

Azeta Cungu has an MSc in agricultural economics from Wye College, University of London. She is currently working on her PhD research at the Department of Agricultural Economics, Katholieke Universiteit Leuven, Belgium.

Dr Sophia Davidova is lecturer in food policy for economies in transition at the Department of Agricultural Economics, Wye College, University of London,

United Kingdom. She holds a PhD from Sofia University, Bulgaria, and has been senior adviser to the Bulgarian government.

Dr Dinu Gavrilescu is Minister of Agriculture of Romania. Formerly, he coordinated several studies on Romanian agriculture as deputy director of the National Institute of Economic Research, Bucharest, Romania.

Dr Konrad Hagedorn is Professor at the Institute of Agricultural Policy, Marketing and Agricultural Development, College of Agriculture and Horticulture, Humboldt University of Berlin, Germany. He holds a Ph.D of the University of Göttingen and has written extensively on agricultural reform in Central and Eastern Europe.

Dr Ladislav Kabat is head of the FAO Statistics Department, Rome. Formerly, he was Rector of the University of Nitra, Slovakia, and professor at the Department of Operations Research.

Dr Diana Kopeva is Assistant Professor at the Department of Agribusiness, University of National and World Economics, Sofia, Bulgaria, and consultant at the Policy Advisory Unit, Ministry of Agriculture and Food Industry, Sofia, Bulgaria.

Dr Zvi Lerman is Professor at the Department of Agricultural Economics and Management, The Hebrew University, Rehovot, Israel. As consultant with the World Bank, he has published extensively on land reform and farm restructuring in Central and Eastern Europe and the Former Soviet Union.

Erik Mathijs is an agricultural economist and a researcher at the Department of Agricultural Economics, Katholieke Universiteit Leuven, Belgium. He is currenly working on his Ph.D research, the subject of which is farm restructuring in Central and Eastern Europe.

Dr Sándor Mészáros is senior researcher at the Research and Information Institute for Agricultural Economics, Budapest, Hungary. He published extensively on Hungarian agriculture and agricultural policy.

Dr Ewa Rabinowicz is Associate Professor at The Swedish University of Agricultural Sciences, Alnarp, Sweden. She has written extensively on agricultural sector modeling, agricultural policy, political economy and agricultural transition. She is currenly member of the Economic Council of Sweden and President of the European Association of Agricultural Economists.

Dr Tomáš Ratinger holds a PhD in Econometrics from the Charles University, Prague, and a MSc in agricultural economics from Wye College, University of London. He is a senrior researcher at the Research Institute of Agricultural Economics (VUZE), Prague, Czech Republic.

Dr Alexander H. Sarris is Professor at the Department of Economics, University of Athens, Greece. He holds a PhD in economics from MIT, Boston and was an assistant professor at the University of California at Berkeley. He is a senior advisor on agricultural sector policy and rural development with several governments including the European Commission.

Dr Johan F. M. Swinnen is Professor in Agricultural Economics and Food Policy and director of the Policy Research Group at the Department of Agricultural Economics, Katholieke Universiteit Leuven, Belgium. He has a PhD in agricultural economics from Cornell University. Previously, he was senior research economist at the Leuven Institute for Central and East European Studies.

Preface and Acknowledgements

This volume results from a research network on 'Policy and Institutional Reform in Central and Eastern European Agriculture and its Likely Impacts on Domestic and International Markets', sponsored by the COST Programme of the EU Commission and coordinated by Jo Swinnen. The project started with a workshop in Leuven, Belgium on 9-11 June 1994. A second workshop was held in Sofia, Bulgaria on May 26-30, 1995 and the concluding conference was in Sinaia, Romania, on June 14-16, 1996.

This volume includes edited research papers resulting from collaboration between the participants of the network and additional experts who later became involved. The project has focused on three aspects of the reforms in Central and Eastern European agriculture: privatisation, land reform and farm restructuring. A key objective of the project was to use a similar approach to study these aspects of the reforms in all countries in order to obtain results which would allow comparisons between countries. The analysis focused (a) on collection of detailed data on the reforms, (b) on collection of information on factors that influenced the progress and outcome of the reforms, and (c) on explaining the results of the privatisation, land reform and farm restructuring.

The project team put special emphasis on (a) and (b) because we encountered large difficulties in analysis because of inconsistencies and uncertainties about the real state of the reforms because of the absence of even the most basic data. Therefore, one of the main contributions of the papers in this volume is in producing information on the progress and process of the reforms in the various countries. We believe that such information is extremely important for researchers, policy analysts and policy makers and we hope that this volume can make a contribution in this area, as well as by indicating a series of factors which have been important determinants of reform progress.

We acknowledge the administrative support of the Leuven Institute for Central and East European Studies (LICOS) in managing the network. Katrien Verhelst has been a careful and competent manager of the budget, and has been largely responsible for organizing the first workshop in Leuven. Marvin Jackson, Director of LICOS, has been very supportive of the network. We thank Josee Verlaenen, Odette Moria and Ria Uyttebroeck for assitance in the revision and formatting of the manuscript. In addition, we like to thank Sophia Davidova and her Bulgarian colleagues for organising the Sofia workshop, and Dinu Gavrilescu and his colleagues for organising the closing conference in Sinaia.

Jo Swinnen
Allan Buckwell
Erik Mathijs

Leuven and Wye
March 1997

Abbreviations

Chapter 1

APC	Agricultural producer co-operative
CEECs	Central and Eastern European countries
DI	Decollectivisation index
DLDC	District Land Distribution Council
GAO	Gross agricultural output
NLDC	National Land Distribution Council
PFA	Private farmers association
SF	State farm
VLDC	Village Land Distribution Council

Chapter 2

AIC	Agro-industrial complex
APC	Agricultural producer co-operative
BSP	Bulgarian Socialist Party
BUL	Bulgarian Leva
CEECs	Central and Eastern European countries
FSU	Former Soviet Union
LOUAL	Law for ownership and use of agricultural land
MLC	Municipal Land Commission
MTS	Machine and tractor station
NSI	National Statistical Institute
RPK	Regional consumer co-operative
SAF	State agricultural farm
TKZS	Labour agricultural co-operative
UDF	Union of Democratic Forces

Chapter 3

CEECs	Central and Eastern European countries
CZK	Czech Krown
RIF	Restitution Investment Fund
SAP	State agricultural assets allocated for privatisation
SGFAF	Support and Guarantee Fund for Agriculture and Forestry
TAC	Total agricultural capital
TAL	Total agricultural land

Chapter 4

APC	Agricultural producer co-operative
BML	Bundesministerium für Ernährung, Landwirtschaft und Forsten
BVVG	Bodenverwaltungs- und -verwertungsgesellschaft (Land Utilisation and Administration Company)
CAP	Common Agricultural Policy
CEECs	Central and East European countries
DM	Deutsche Mark (German Mark)
EALG	Entschädigungs- und Ausgleichsleistungsgesetz (Indemnification and Compensation Act)
EU	European Union
GDR	German Democratic Republic
FRG	Federal Republic of Germany
LPG	Landwirtschaftliche Produktionsgenossenschaft (agricultural producer co-operative)
THA	Treuhandanstalt
VEG	Volkseigenes Gut (state farm)

Chapter 5

CEECs	Central and Eastern European countries
HUF	Hungarian Forint
MOSZ	National Federation of Agricultural Co-operators and Producers
TAL	Total agricultural land
TLA	Total land area
ISP	Independent Smallholders' Party

Chapter 6

Agromec	State Machinery Companies
AS	Agricultural society
CAP	Former agricultural co-operative
CC	Commercial company

CEE	Central and Eastern Europe
FA	Family association
IAE	Institute of Agrarian Economy
IAS	State Agricultural Enterprises

Chapter 7

CEE	Central and Eastern European
LPG	Landwirtschaftlichen Produktionsgenossenschaft
SLK, Sk	Slovakian krona

Chapter 8

FALF	State Fund for Agricultural Land and Forests
SY SLO	Statistical Yearbook of Slovenia

Chapter 9

ECE	East Central European
FSU	Former Soviet Union
IFC	Internal Finance Corporation

Chapter 10

AIC	Agro-industrial complex
BSP	Bulgarian Socialist Party
CEECs	Central and Eastern European countries
FSU	Former Soviet Union
IFI	Individual farming index
LC	Liquidation Council
LSO	Large-scale successor organisation
TAL	Total agricultural land

1 Agricultural privatisation, land reform and farm restructuring in Albania

Azeta Cungu and Johan F.M. Swinnen

Introduction

The Albanian agricultural reform differs from reforms in other Central and Eastern European Countries (CEECs) in several aspects. The starting point for the reform differed because, first the share of agriculture in total employment and output was much larger than in any of the other CEECs. Second, all agricultural land and other assets were nationalised during the Communist regime, unlike most other CEECs.

The reform process contrasted reforms pursued in other CEECs because Albania was the only country which decided to distribute all land to the rural families and farm workers. Furthermore, the implementation of the land reform and privatisation process was extremely rapid compared to any other CEECs. Moreover, the Albanian reform has resulted in a complete break-up of the collective and state farms. By 1995, all land was used by small-scale family farms – unlike any other CEECs.

This chapter describes the reform process and its outcome and analyses the factors which have influenced the reform policy choice and caused the radical restructuring of the agricultural economy in Albania. The paper is organised as follows: the first section gives an overview of the traditional land tenure rights and farm structures in Albania. Next, we discuss the post-Communist reforms in agriculture, the 1991 privatisation process, land reform and the emerging agrarian structures. In the last section, we analyse the determinants of the reform policy choices and factors which have affected the decollectivisation process.

Historical perspective of land ownership and farm organisation in Albania

The pre-transition structures of ownership and production in the Albanian agriculture are closely related to the Communist reforms which were the most radical reforms in the history of the country's property rights. These reforms were themselves affected by the traditional farming relations and the pre-1945 structure of ownership in the Albanian countryside. In fact, one might argue that the Communist reforms which violently expropriated all landholders of more than 5 ha of land, would have been more difficult to implement had the pattern of land ownership been less polarised. With the bulk of the rural population owning little or no land at all, this radical land reform received large popular support. It is therefore interesting to have a brief look at the recent history and evolution of the agrarian relations in Albania.

Land ownership before 1945

In 1912, only five families owned most of the arable land and 36.7 percent of all agricultural land (including forests and pastures) in the coastal plain and hilly region of the country (Table 1.1). After the independence from the Turks (November 1912), all land owned by the Turkish state and the Sultan was confiscated by the Albanian state, but private landowners were not affected. A land reform in 1924 tried to give 4 ha of agricultural land to each of those tilling the land but failed because of strong resistance from large landlords. Instead, the royal regime, which came into power shortly after the democratic government failed in its reform attempts, claimed that landholders' property is 'unviolable'. The King's government allowed large landholders and high-ranking officials to acquire even more land by manipulation of a law for the distribution of agricultural land to Albanian emigrants returning from abroad and settling in rural areas, and by forcefully removing peasants from their lands. The King's government distributed a few thousands of hectares of land, mostly state owned, to small and landless farmers in 1930. This reform only marginally affected the big landowners. Other factors reduced the concentration of land ownership between 1925 and 1945. The most important were land sales and estate divisions through inheritance (Civici, 1994). Table 1.2 shows how the land ownership distribution had become somewhat less unequal by 1945. However, these data do not account for differences in land quality: the inequality of land ownership was much stronger on the best quality land and peasant farms were mostly situated in the hills and mountainous areas. Furthermore, 13.8 percent of the farming population was still landless.

Table 1.1
Land ownership pattern in Albania, 1912

Ownership	*Total area* *ha*	*%*
Large estates owned by the State or the Sultan	56,287	14.7
Private estates	140,000	36.7
Religious institutions' ownership	13,700	3.6
Peasant ownership	171,789	45
Total	381,776	100

Source: Xhamara, 1995

Table 1.2
Land ownership pattern in Albania, 1945

Landlords category	*Number*	*%*	*ha*	*%*	*Average per family*
Big landlords	7	0.006	14,554	3.7	2,070
Rich farmers	4,713	3.034	91,133	23.16	19
State property	-	-	50,000	12.71	-
Small and medium size landowners	128,961	83.080	237,668	60.43	1.8
Landless	21,544	13.080	0	0	0

Source: Xhamara, 1995

Communist reform and farm restructuring

The Communist economic reform started with reforms in agriculture. Soon after taking over government, the Communists launched an agrarian reform, distributing at least three hectares of average quality land to every rural family of two persons. For any additional members, 0.5 additional ha of land were granted. All large landowners and religious institutions were to be expropriated and financially compensated. This approach was not considered radical enough and an amendment ruled out compensation for all those who were not themselves farming their land. Land transactions and hiring non-family labour were banned.

Owners of large and medium sized holdings were allowed to retain 5 ha of arable land; land above this amount was redistributed to small and landless peasants (Sjoberg, 1991 quoted in Stanfield et al., 1992).

After only a few months of individual farming, the Albanian farmers were required to pool their recently acquired land into collective farms. The process was initiated on a voluntary basis in one region (Lushnja) but expanded very quickly into other areas where farmers were often forced to join the collectives. Those who did not join were labeled 'enemies' and became targets in the 'class struggle'.

Collectivisation was carried out in three stages, but it was complete and radical. It was first initiated in the lowlands and later extended into hilly and mountainous areas. Alongside the process of creating the Agricultural Producer Co-operatives (APCs), State Farms (SFs) were established on the best land of the country on estates belonging to the previous state, religious institutions or big landholders. The Communist state invested substantially in modernising the state farms, which were intended to become the superior model of the socialist production structures. High quality inputs, imported machinery and other equipment, greenhouses, large storage facilities, big animal houses and selected breeding livestock were all made available to the SFs. The APCs were intended as a first step towards total nationalisation of production as their means of production were still, in theory, collectively owned. With time, it was expected that all land and assets would be state owned and APCs would therefore be transformed into SFs.

In SFs and in APCs, cropping patterns, input supplies, output marketing and sales and the structure of activities were imposed from above. The process was driven entirely by political objectives and carried out through a centralised approach. Economic considerations of profitability and incentives, efficient resource allocation and comparative advantage as well as farm specific location and climatic conditions, were less important. A characteristic feature of the period was the general trend of enlarging the collective and state farms' scale of operations by extending the collectivisation process and reducing individual property, as well as by merging two or even more of the existing APCs together (Table 1.3 and 1.4).

Collectivisation was completed in 1967. The next step was the total abolition of private farming and nationalisation of all land. This occurred in 1976.

The pre-transition farm structure

In late 1989, some modest reforms were introduced by giving APC members small plots for individual farming: 0.2 ha in the plains and 0.3 ha in the hills and mountains (Pell and Gorgoni, 1992). However, these small plots quickly produced an important share of output: Christensen (1993) estimated that

21 percent of total agricultural output was produced by the small plots which operated on only 3 percent of the total arable land. For comparison, 160 SFs produced 29 percent of total agricultural production and owned 170,000 ha (23 percent) of arable land. Their size ranged from 500 to 2000 ha with an average size of 1,070 ha. 50 percent of production was supplied by the 492 APCs which cultivated over 74 percent of arable land. Although all land continued to be publicly owned[1], most APCs were being subdivided and, by March 1991, the 492 units had increased to 1,000.

Table 1.3
Restrictions on the size of private farms, Albania 1945-1956

Farm size (ha)	*1945* 000 plots	*%*	*1950* 000 plots	*1956* 000 plots
< 1.00	69.4	44.36	39.4	37.6
1.01 - 5.00	69.9	44.66	82.9	91.2
5.01-10.00	12.5	7.98	21.0	16.9
> 10.00-	4.7	3.00	2.2	1.0
Total	156.5	100.00	145.5	146.7

Source: Albanian Institute of Statistics, Yearbook 1960

Table 1.4
Changes in land ownership in Albania, 1950-1995 (in %)

Pattern of ownership	*1950*	*1970*	*1989*	*1991*	*1995*
Private sector	91.30	4.21	3	21	94.4
Collective sector	5.37	90.98	74	55	-
State sector*	3.33	4.81	23	24	5.6

* Land which is being individually farmed by Joint Venture workers is, here, considered as state owned although recent legislation grants most of this land in property to workers.

Source: MAF, 1995; Pell and Gorgoni, 1992; MAF, 1996

Despite these changes, the diversion of members' efforts onto their private plots was detrimental for the production on land which remained with the collectives. In fact, production in 1990 and 1991 fell dramatically. The threat of food shortages and widespread economic difficulties indicated the need for much deeper reforms which would ensure a recovery of production, and a reduction of the inefficiency of the collectivised sector.

Agricultural privatisation, land reform and farm restructuring

This section summarises the main features of the current privatisation and restructuring reform in agriculture. A detailed discussion of the process and the legal framework for agricultural privatisation in Albania, is provided in Cungu and Swinnen (1997).

Privatisation and initial restructuring in agriculture

Privatisation in agriculture started spontaneously in the beginning of 1991. The *Land Law* (June 1991) recognised the right of APC members to receive land, in the framework of a distribution program, and leave the APC if they preferred so. This law was followed by a number of Government decrees which regulated the distribution of APC assets (other than land) as well as the complete liquidation of these structures. A number of regulations set out the procedures for the implementation of the legislation. The most important regulation was that the APCs' land was to be divided up amongst the member families on an equal per capita basis. Specialised bodies were created to organise and monitor the distribution process at national level, the National Land Distribution Council (NLDC); at district level, the District Land Distribution Councils (DLDCs) and village level, the Village Land Distribution Councils (VLDCs). The land distribution process was formalised by issuing land ownership titles to the APC members. Besides land, other assets were also distributed amongst members once all APC debts were paid off.[2] Also, Land Distribution Councils were authorised to correct the improper possession of assets which took place when villagers spontaneously broke-up the collectives.

The only exception to the above rules was witnessed in some mountain areas of the country where land distribution was not carried out according to the normal criteria of per capita allocation, but rather restituted to ex-owners in their old boundaries. The functions of the formal VLDCs were, for this purpose, performed by the so-called 'Councils of Elderly' which were composed of old village people known for their 'fairness and traditional wisdom'. Families who did not own land before 1945 but who were members of the APCs and would benefit from the application of the Land Law, were thus excluded from the process. Decree No. 230, did provide some room for a *de jure* accommodation of what *de facto* had already happened in that it specified that the quantity, quality and location of land for distribution are decided by the village councils which are to be democratically elected by the village members. In reality, the government accepted the proposals of the Councils of the Elderly as it seemed to be very much the will of the villagers themselves. Therefore, it continued granting the ownership titles to the beneficiaries.

The privatisation of SFs was initiated later. Those SFs which were created by merging collectives were split up into their component APCs and then privatised according to the criteria used in the collective farms. Those SFs which were estimated as the most viable were pooled into joint venture partnerships with the participation of foreign capital. Decision No.452 of October 1992, provided for the distribution of the remaining SFs' land user rights to workers and specialists. Other assets were either directly sold to workers at very low prices or otherwise auctioned. In either case, the current workers of the farm received preferential treatment. When joint ventures failed, as many of them did, their land was also distributed for use by the workers.

Compensation of ex-owners. It was initially decided that eligible former owners, i.e. the pre-1945 holders of agricultural land[3] and their heirs, were to receive financial compensation for part of the property they owned pre-1945. The former owners however demanded full, physical restitution of properties. The government responded to the growing pressure of the ex-owners by passing a law which provides for the physical compensation of ex-landowners with land in tourist areas (July 1995). However, the criteria for deciding who is eligible to financial compensation and who will get tourist land is still unclear. Similarly, although the legal provisions for financial compensation of former owners do exist, the process itself is far from implemented. Compensation claims have been evaluated but the actual compensation is yet to start.

Shortcomings and new policy initiatives. The process of re-distribution of property rights in Albania was initiated with the physical distribution of the property and was carried out separately from the formal distribution of rights. The latter process lagged behind the physical distribution. This created a situation in which many farmers were farming land without an official ownership title (Table 1.5). Moreover, many types of land transactions including renting and selling were, initially, legally banned. Under such circumstances, the resulted fragmentation of agricultural holdings and *uncertain property rights* on land were the most critical shortcomings of the reform. The new structure of farms which has emerged in Albania is unique amongst CEECs for its high degree of fragmentation. Land privatisation created more than 490,000 new private farms fragmented into 1.9 million parcels which range from 1-10 parcels per farm with an average of about 3.3 separately located parcels per family farm. The average farm size is 1.05 ha ranging between 1.29 ha in valley and hillfoot regions to 0.55 ha in the mountains (Grace, 1995 and IFDC, 1994). For some regions with limited agricultural land, the average farm size is less than 1 ha e.g., in Dibra it is 0.94 ha, in Kukes 0.88 ha and in Puka 0.55 ha (MAF, July 1994).[4]

Table 1.5
Land distribution in Albania, end 1995 (in ha)

	Total	*SFs*	*APCs*
Initial total arable land	**702,005**	**154,083**	**547,922**
Refused land[a]	64,052	8,487	55,565
Actual arable land	**637,953**	**145,596**	**492,357**
State owned land	18,548	12,524	6,024
Land retained by the state[b]	16,012	9,988	6,024
Existing Joint Ventures	2,536	2,536	-
Land for distribution of which:	619,405	133,072	486,333
- distribution completed	602,022	128,508	473,514
- in the process of distribution	17,383	4,564	12,819
Land distributed			
- as a % of total arable land	94.4[c]	88.26	96.2
- as a % of land for distribution	97.2	96.57	97.36
% of distributed land for which the titling is complete	86.4		

[a] As of 1993, there were more than 120,000 ha of refused land of which 60,202 ha had actually been distributed by the end of 1995. The remaining 64,000 ha will be used for developing private forestry. As such, this latter is not considered as arable land in the above table.

[b] Includes land retained for research and experimental purposes; land used to accommodate for out-movements of rural population from marginal areas, and land used by units of the Albanian army for self-sustaining purposes.

[c] This figure does not take account of the joint-venture land which was distributed. In addition, some 8,200 ha of land have been used to reduce land pressures from people moving out of the marginalised regions. This indicates that the ratio of the actual distributed versus total arable land is higher than what shown above.

Source: MAF, 1996

The legislation of land tenure (Law No. 7715 of 02/06/1993) as well as the creation of a market for real estate (Law No. 7983 of 27/07/1995) is expected to contribute to the reduction of fragmentation and uncertainty. However, this is unlikely to result in a substantial and quick consolidation of holdings as long as employment possibilities outside agriculture are limited and the issue of compensating ex-landowners is not settled.[5] The rate of consolidation will also depend on growth in overall living standards in the rural areas, because poor farmers are less likely to give up their land as it represents security for their families.

Although Law No. 7983 recognises and allows land sales, the official land market is not yet operational. This is because the law stipulates that no land transactions can take place until the land has been appropriately registered. The registration process itself is well-underway with the mapping and updating of field boundaries already completed. However, the registration is not expected to be complete until the end of 1998. In addition, the implementation of the programme requires that registration is organised in all regions of the country, but (as of 1996) a number of district offices is still to be set up and equipped.

In the meantime, the delayed issue of titles, restrictions in the transaction rights of the new landowners and increasing restitution pressures from ex-owners all contribute to uncertainty of property rights in agriculture. This has affected farm investment decisions. In an environment of underdeveloped markets and institutions, fragmentation and the associated high transaction costs have contributed to the creation of a predominantly subsistence farming.

The reform outcomes and emerging structures

In the APCs, 486,333 ha of arable land were made available for distribution, of which 473,514 ha were distributed by the end of 1995. Also by then, 97 percent of the SFs area available for distribution was already allocated (Table 1.5). Altogether, some 462,730 families benefited from the process.

At the same time, livestock, fruit and olive trees, farm implements and buildings were privatised. APC assets were rapidly distributed free of charge to farm workers as, in principle, these were the collective property of a group of people, i.e., co-operative members.[6] In SFs, these assets were sold to users. According to a World Bank survey[7] of early 1994, 90 percent of private farmers possess stables and 44 percent own draught horses. This reflects the high dependence on livestock and the low level of mechanisation. The World Bank survey further shows that some of the privatised assets, especially machinery and buildings, are not suitable for use on small farms. Small capacity tractors which would be more suitable for the size of the farms are scarce and are not well suited to the hard soil in many agricultural areas of Albania (Table 1.6 and 1.7).

The legal basis which regulates commercial activities in Albania is provided by the Civil Code, the Commercial Law and the Company Law. This framework however applies very little to agriculture as the emerging farms are mostly subsistence farms with limited activities operated on a commercial basis. Table 1.8 summarises some of the main features of farms in Albania and distinguishes four types of farm organisation:

Table 1.6
Possession of machinery and buildings in the WB (1994) survey area*

Type of asset	*Average number or capacity per farm*
Draught horses	0.6
Tractor >50 HP	0.1
Stables (no. of places livestock units)	7.1
Grain storage (t)	1.1
Milktank (hl)	11.3
Milk processing (hl/day)	57.6
Meat canning equipment (t/day)	0.1
Greenhouses (ha)	0.1

* See note 7.

Source: van Bekkum, 1994

Table 1.7
Possession of machinery and buildings in the WB (1994) survey area

Type of asset	*Percentage of private farmers owning:*
Draught horses	44.3
Tractor 1-15 HP	1.0
Tractor 15-50 HP	0.3
Tractor >50 HP	2.6
Grain harvester (ha/day)	0.3
Stables (no. of places livestock units)	90.2
Grain dryer (t/h)	1.0
Grain storage (t)	17.1
Milktank (hl)	8.9
Cold storage (m3)	1.0
Milk processing (hl/day)	10.2
Meat canning equipment (t/day)	3.9
Fruit caning equipment (t/day)	4.3
Greenhouses (ha)	0.3

Source: van Bekkum, 1994

Individual family farms. The big pre-transition agricultural structures have been almost completely abolished and in 1996 more than 95 percent of the arable area was being farmed by small-scale farmers who operate individual production units. Data from the 1994 World Bank survey indicate the existence of four main land tenure types within the individual family farmers' category: ownership, leasing, sharecropping, and informal ownership. According to this source, most commonly, farmers own the land (in 96 percent of the total land area surveyed)[8] whereas informal ownership (in 0.8 percent of the total area) is a transitional form of tenure to be phased out with the completion of the titling process.

Predominantly, small farms' activities combine crop and animal production to ensure adequate household consumption although the pattern of the activities varies according to climatic conditions, availability of resources and land quality. Small farms typically use family labour and are managed jointly by family members. Decisions are made by the head of the household who, in most cases, is the husband (Table 1.8).

Agricultural joint ventures. Joint ventures were created as limited liability companies with the participation of the Albanian state and a foreign partner. The foreign partner almost always participated with technology and finance.[9]

The land used by joint ventures is publicly owned as it consists of the ex-SFs land which is part of the Albanian capital in the foundation of the company. The average size of land operated by joint ventures is 150-200 ha of land. Labour is mostly hired and the company is administered by a hired manager or otherwise one of the partners, normally the one who has the biggest share in the enterprise.

Private farmers associations (PFAs). PFAs were born as a natural solution to the problem of fragmentation, restricted access to inputs especially credit, and difficult marketing of outputs. For almost three years, the PFAs existed as informally established structures based on personal trust and relationships as there was no legal basis for their existence and functioning. In May 1996, the Co-operative Law was passed which formalised their existence. In 1994, some 60 PFAs were officially registered, but only few of them still operate in reality. Land is cultivated jointly although it is individually owned by each of the members who also provide their own and, sometimes, family labour. The manager is democratically elected by the members and decisions are made by majority approval.

Research and experimental farms. In these farms ownership is retained by the state and land is used for research and teaching purposes by the University of Agriculture in Tirana, Research Institutes and agricultural vocational schools in districts. Normally the farms are operated by hired staff and students and the manager is appointed by a governing body.

Table 1.8
The characteristics of farms in Albania, 1994

Farms	*Legal basis*	*Land Ownership*	*Number*	*Average land area (ha)*	*Management*	*Labour*
Individual family farms	Law No. 7501, June 1991 Decree No. 266, Aug. 1991 Decree No. 255, Aug.1991 Decree No. 452, Oct. 1992	Formally and informally privately owned	490,000	1.05	Family head	Family labour
Agricultural Joint ventures	Law No. 7664, Nov.1993 Law No. 7683, Nov. 1992 Law No. 7632, Nov.1992 Law No. 7667, Jan.1993	Publicly owned (Albanian share)	13	150-200	Hired manager or one of the partners	Self-employed and hired
Private Farmers' Associations (PFAs)	Co-operative Law, May 1996	Privately owned/leased	60	15	Elected member	Members & families
Research and experimental farms		Publicly owned	n.a.	n.a.	Nominated	Hired

n.a. stands for not available.

Source: IFDC, 1994; MAF, 1995; Caputo and Christy, 1994

Factors affecting farm restructuring in Albania

This section examines the reasons for the choice of a distribution reform and the radical decollectivisation of the collective farms by summarising some of the main findings discussed more extensively in Cungu and Swinnen (1997).

We address the question of why the agricultural reform in Albania was so different from similar reforms pursued in other CEECs? Why was agricultural land in Albania distributed to farm workers unlike in most other CEECs and why was decollectivisation so radical?

An explanation of the choice of the land reform process

A key difference with other Central European countries where the land remained formally in private hands throughout the collectivisation period, was that in Albania agricultural land was formally state-owned on the eve of the 1991 reforms. With land in state ownership, the Albanian government had more options than the other Central European governments, because all CEECs (former Soviet Union countries not included) restituted assets which were still privately owned to former owners (Swinnen, 1997).

Second, under the conditions of virtual collapse in agriculture and widespread spontaneous privatisation, it was important to act quickly and also ensure some continuation of production by allocating land to those who were actually farming it. The restitution option would have required a decision regarding the reference date for restitution because there were more than one 'former owner' for most of the agricultural land.[10] Another consideration might have been that land distribution is technically much easier to organise and to implement than restitution to former owners in the absence of land maps indicating the previous borders.

Another factor has to do with the relative roles and influence of each of the would-be beneficiary parties (i.e., farmers, pre-1945 and 1946-1948 owners) which induce the policy-makers to design and implement a land reform able to generate the highest political returns to those supporting this reform. In the Albanian debate, those who received land by the Communist reforms during 1946-1948 have been much less vocal (unlike in many other CEECs), and much less opposed to distribution, than the pre-1945 owners. Why is this?

The first reason is that restitution based on the 1948 ownership situation would have quite similar results as the redistribution option. Most 1948 owners were to receive land under the current distribution program because the structure of the APCs in terms of member families did not change much over the years due to restrictions in labour mobility. Moreover, in most cases, the area of land assigned to families under the land distribution program was larger than what they would have received under restitution. This is partly due to increases in the arable land area and improvements in land quality resulting from state investments, and partly because restituted land would have to be divided amongst heirs many of whom were no longer involved in farming.[11]

A second reason might have been that the 1946-1948 land reform was considered a Communist action and a first step towards collectivisation. As such, those who received land in the framework of these Communist reforms could feel that they had no legitimate claim to land. These considerations might have played a role in the anti-Communist sentiments in the early 1990s.

The third reason is that for the small landowners the marginal costs of lobbying and costs of submitting valid claims per hectare of land to be restituted (or to be compensated for) was high compared to the expected additional benefits.

These three reasons also explain the much more vigorous opposition of the pre-1945 owners to land distribution. They have much more to gain from restitution in comparison to both what they receive under the distribution program and the costs of lobbying and submitting restitution claims. In addition, they feel they are the legitimate owners of the land, because the land was taken away by the Communists.

The argument of political organisation costs predicts that former owners would be much more influential on decision-making than farm workers because their group has all the characteristics reducing transaction costs of political organisation. They are a much smaller group and there are highly concentrated benefits, compared to the enormous group of farm workers who have much more diluted benefits (Olson, 1965). Why then did the reformers in government choose land distribution over restitution? Why were the pre-1945 owners less influential than the peasants in determining the reform choice?

The most important explanation, we believe, has to do with the impacts of the reform and the low income level of farm workers. While the size of the income effect of the land reform choice might have been smaller for farm workers than for pre-1945 owners, the effect of being allocated a piece of land was very substantial for farm workers because of their extremely low incomes. Even a relatively small plot of land had a big welfare impact on farm workers and their families in a situation of very low opportunity costs for their labour. Moreover, land is viewed as a secure form of wealth and income security (Ellis, 1988). Poorer members of society are even more concerned than rich people in holding a secure form of wealth and retaining their traditional source of livelihood (Binswanger and Rosenzweig, 1986).

Because of the large impact on marginal welfare and on income security the land reform choice was very influential in determining peasants' political reactions. A vast number of people (50 percent of labour force, and 65 percent of the population living in the rural areas) all benefited at least to some extent from land distribution and many of them would have reacted vehemently against restitution to pre-1945 owners. In comparison, only a very small portion of the population (an estimated 3 to 4 percent) would benefit from restitution to pre-1945 owners. Thus, the important marginal utility effect of reforms for the many peasants strongly offsets the 'land dilution effect'. In combination with the vast numbers difference, this more than offsets the concentration and organisational cost advantages of the pre-1945 owners.

A related factor is the likely opposition of most people to the restoration of a feudal structure in the best agricultural areas by restituting a very large share of the land to the heirs of a handful families.

These factors combine to explain the choice of the land distribution process over the restitution option.

An explanation of the radical decollectivisation

The radicalness of the restructuring process in Albania is shown in Table 1.9 which is drawn from Mathijs and Swinnen (1996) who rank ten countries according to a decollectivisation index (DI) which measures the break-up of the collective and state farms by the change in the share of individually farmed land between 1989 and 1994. Albania has with 95 percent by far the highest DI.

Table 1.9
Decollectivisation index for Central and Eastern Europe (1994)*

Country	*Decollectivisation index (%)*
Albania	95
Latvia	80
Lithuania	60
Romania	47
Estonia	38
Bulgaria	36
Czech Republic	20
Hungary	13
East Germany	11
Slovak Republic	5

* The decollectivisation index (DI) is calculated by dividing the difference between the share of individual farms in total agricultural land in 1994 (IND94) and in 1989 (IND89) by 100 minus the share of individual farms in total agricultural land in 1989: DI = 100 * (IND94-IND89)/(100-IND89).

Source: Mathijs and Swinnen, 1996

The difference with virtually all other CEECs is remarkable. While spontaneous privatisation has occurred to some extent in virtually all CEECs, the extent to which this spontaneous process has lead to a complete break-up of the collective farms in Albania is unique. In most CEECs, most farm workers have remained with the co-operative-under-transformation during the early years of transition (Mathijs and Swinnen, 1996). What is the reason for this difference?

The main difference seems to be the desperate situation of the APC members. Albania had the lowest per capita income in Europe and the APC members' incomes were even lower than average incomes. The production structures in agriculture, being commanded within a framework of the state plan, have always been managed without special regard to economic efficiency considerations. This also reflected the economic isolation of the country and attempts to achieve food self-sufficiency. At best, the production structures were constructed in such a way that tried to take into consideration some climatic conditions. As APCs were mainly intended to meet the food demand of the members and some quantities to be sold in cities, they were planned to produce a wide range of products whether these were produced efficiently or not. The state also dictated the APCs' relationships with input suppliers and clients, therefore state enterprises could easily discriminate against the former. The result was a squeeze of APC productivity and benefits due to high input prices, highly inefficient production, and relatively low output prices for sales to third parties (state). This lead to a steady reduction of production capacities over time, lack of investment and maintenance, very low revenue generation and severe shortages of food for the members. On the other hand, the state had restricted credit ceilings for APCs as enormous debts were accumulated, and state financial resources were limited.

The virtual collapse of the agricultural sector between 1988 and 1991 worsened the economic crisis in the country especially in rural areas, apparently to the point where it induced a radical reaction from the APC farm workers in an environment of reduced social control and political instability. The fact that a complete break-up of the APCs resulted from this 'spontaneous privatisation' and the consequent officially organised land distribution programmes, indicates that farm workers preferred the alternative (i.e. small individual farms) compared to continuation of farming within a reformed APC framework. This shift to individual farms occurred despite all problems associated with credit supply and lack of machinery, which in many other CEECs discouraged individuals from leaving the co-operative framework. This emphasises the lack of confidence and the expected poor benefits Albanian peasants held from remaining in the co-operative framework.

This argument is reinforced by the remarkably strong growth in gross agricultural output (GAO) since 1992 (OECD, 1996). Figure 1.1 shows that in

Albania, GAO increased by around 70 percent between 1991 and 1995, whereas in most of the other CEECs, it has picked up much later and at a lower rate.

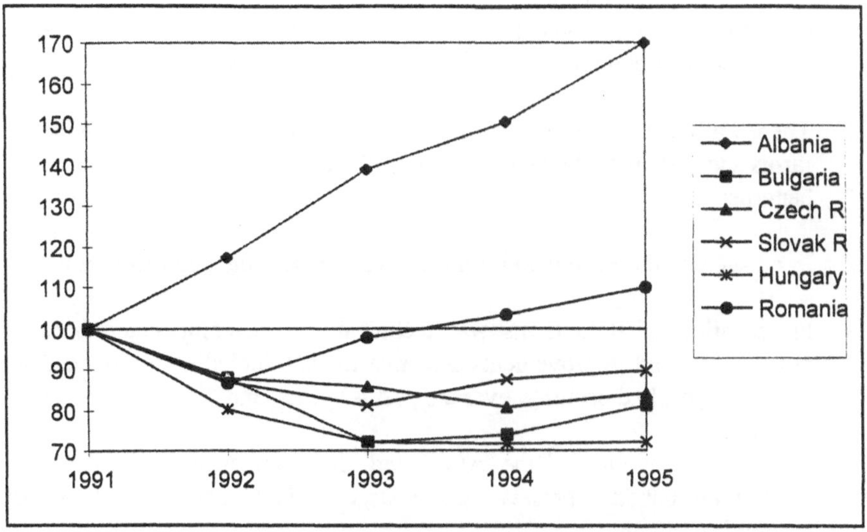

Figure 1.1 Growth in Gross Agricultural Output, Albania 1991-1995

Source: Own calculations based on OECD (1996) data

The above arguments are also consistent with the less desperate reactions of SFs workers who were in a relatively more comfortable situation and the less radical restructuring of these farms. In fact, the privatisation of SFs was done through different procedures and also more gradually than the APCs. Why was that?

To understand the rationale behind this, one has to bear in mind that the APCs' status and organisation under central planning was different from that of the SFs. At least in theory, in APCs the means of production were collectively owned, except for the land which was state property after 1976. Land used by APCs was previously privately owned by peasants and then pooled into the collectives which were created all over the country including the mountainous areas. What was produced in APCs was divided up between members, while a special quota went to the state and surpluses were sold in cities.

Unlike the APCs, SFs were operated like any other state enterprises. Everything was state owned and workers (mostly urban residents) were government employees. Their benefits were much higher than those of APC members. Some SF activities were to produce exportables. The state invested heavily in such farms by providing special facilities, including greenhouses, high quality inputs, irrigation water, and agricultural machinery. Mechanisation Stations, which provided services to the APCs on contract or rental basis, were owned by SFs. Moreover, SFs were created on big estates confiscated from the church or from the pre-Communist state and were often on the best land of the country.

While APCs were privatised (and decollectivised) quickly, the privatisation of SFs started later and proceeded slower. The reasons were:

1 There was relatively less pressure from state farm workers to privatise the farms and distribute the land because of their relatively better situation and benefits;

2 In most cases there were no former owners demanding land restitution;

3 Privatisation of state farms was technically more complicated because much more land improvements and investments, including non-land asset constructions, had occurred over the past 40 years, and

4 Given the break-up of the APCs and the fragmentation of land the government hoped to preserve SFs as large production structures to attract foreign investment.

Conclusions

The agricultural reform in Albania distributed agricultural land to rural inhabitants on an equal per capita basis. It resulted in a complete decollectivisation of collective farms and the establishment of small family farming on 95 percent of the arable area. Both these facts emphasise the difference of the Albanian case in comparison the other Central and Eastern European countries where the restitution of land was the most commonly chosen approach and decollectivisation was relatively modest in the early years of transition.

We have explained these differences with a combination of economic, socio-political and institutional factors. We argue that the choice of decollectivisation and the distribution of property was influenced by the extremely poor economic situation in rural areas where 60 percent of the population lived on the eve of the reforms. With low opportunity cost of agricultural labour and large uncertainties, the distribution of even small plots of land had large impacts on marginal welfare and income security of farm workers.

Further, the historical and legal status of the assets under privatisation and, more specifically the fact that land used by co-operative farms was state owned and the unequal pre-1945 land distribution have all affected the choice of the privatisation process.

In addition, the extremely low incomes of collective farm workers induced a radical decollectivisation of co-operative farms in the environment of a dramatic collapse of agricultural production, food shortages, reduced social control, political instability and uncertain land reform legislation that characterised the early 1990s.

Notes

1 Small plots were also formally owned by the state.
2 The state decided to write off part of the remaining APC debts.
3 See Cungu and Swinnen (1997) for an extended description of the process and the rationale behind the above choice.
4 This information was obtained from personal communication with the specialists of the Land Directorate in the Ministry of Agriculture and Food, Tirana, Albania.
5 The process of privatising urban properties was marked with several changes and sometimes reversals in the reform policy as the political power of former owners grew. Therefore, as long as the threat of the former agricultural owners is there, the farmers will be reluctant to engage in transactions or substantially invest in their farms.
6 Buildings make an exception to this general rule as these were either given to their former owners or sold out to villagers at preferential prices.
7 The survey was conducted in the districts of Lushnja, Berat and Dibra each located in a typical region reflecting Albania's three main climatic and geographical regions: lowland (Lushnja) region specialised in crop production, the hilly region (Berat) specialised in horticulture and arboriculture and the mountainous region which is livestock oriented (Dibra).
8 Note that, at national level, recent data from the MAF indicate a similar pattern of privatised versus all arable land (see Table 1.5).
9 These partnerships operate under the legacy of the Commercial Code (first dispositions, Law No. 7632 of 04/11/1992) the Company Law (Law No. 7638 of 19/11/1992), Law on Foreign Investments (Law No. 7764 of 02/11/1993) and also Law No. 7667 of 28/01/1993 on the commercial register and formalities for commercial companies. According to the Company Law, the minimum founding capital of a limited liability company amounts at 100,000 Lek and must be fully paid up on incorporation. The Contribution to the capital could be in cash or in kind or the right to use. Contribution of services is not qualified as capital contribution (Caputo and Christy, 1994).

10 There were two groups who would potentially have a claim on the same property: those who owned such property before 1945 and were expropriated by the Communists; and those who received land in the framework of the 1946-1948 reforms, who were afterwards asked to pool the property into the collective farms.

11 The population in Albania has tripled since 1945. It increased at an annual rate of 1.9 percent whereas the growth rates in rural areas were above the national average.

References

Binswanger, H. and Rosenzweig, M. (1986), 'Behavioral and Material Determinants of Production Relations in Agriculture', *Journal of Development Studies,* Vol. 22, No. 3, pp. 5-21.

Caputo, E. and Christy, L. (1994), 'Joint Ventures and Foreign Investment in Albanian Agriculture and Agri-Businesses', Report Commissioned by the FAO: Tirana-Rome.

Christensen, G. (1993), 'Review of the Subsidy and Support System for Wheat and Bread in Albania', FSG-University of Oxford (UK), Report Commissioned by the FAO: Rome.

Civici, A. (1994), 'La Question Forcière Dans La Restructuration De L'Economie Agricole En Albanie', unpublished working paper, Université de l'Agriculture: Tirana.

Cungu, A. and Swinnen, J.F.M. (1997), 'Agricultural Privatization and Decollectivization in Albania: A Political Economy Perspective', in Swinnen, J.F.M. (ed.), *Political Economy of Agrarian Reform in Central and Eastern Europe*, Ashgate: Aldershot.

Ellis, F. (1988), *Peasant Economics: Farm Households and Agrarian Development*, Cambridge University Press: Cambridge.

Grace, C. (1995), 'The Hill and Mountain Farmers in Albania', Report Commissioned by SARA/USAID, Ministry of Agriculture and Food: Tirana.

IFDC/USAID (1994), *Agricultural Production in Albania, Socio-Economic Survey, 1993-1994*, International Fertiliser Development Centre (IFDC/ USAID): Tirana.

INSTAT (1960), *Albanian Statistical Yearbook 1960*, The Albanian Institute of Statistics: Tirana.

MAF (1995), *Food and Agriculture Statistics of Albania: 1994 Annual Yearbook*, Ministry of Agriculture and Food: Tirana.

MAF (1996), 'Ecuria e Procesit te Privatizimit te Tokes Bujqesore ne Shqiperi', unpublished report, Ministry of Food and Agriculture: Tirana.

Mathijs, E. and Swinnen, J.F.M. (1996), 'The Economics of Agricultural Decollectivisation in Central and Eastern Europe', *Policy Research Group Working Paper* Nr 1, Department of Agricultural Economics, K.U. Leuven: Leuven.

OECD (1996), *Agricultural Policies, Markets and Trade in Transition Economies: Monitoring and Evaluation 1996*, OECD: Paris.

Olson, M. (1965), *The Logic of Collective Action: Public Goods and the Theory of Groups*, Harvard University Press: Cambridge.

Pell, J. and Gorgoni, M. (1992), 'The Albanian Agricultural Sector'. Report Commissioned by the FAO/UNDP: Tirana.

Stanfield, D., Lastarria-Cornhiel, S., Bruce, J. and Friedman, E. (1992), 'Consolidating Property Rights in Albania's New Private Farm Sector', Land Tenure Center, University of Wisconsin/Madison, Report Commissioned by the USAID: Tirana.

Swinnen, J.F.M. (1997), 'The Choice of Privatization and Decollectivization Processes in Central and Eastern Europe: Observations and Political Economy Hypotheses', in Swinnen, J.F.M. (ed.), *Political Economy of Agrarian Reform in Central and Eastern Europe*, Ashgate: Aldershot.

van Bekkum, O. (1994), 'A World Bank Regional Study: Farm Restructuring and Land Tenure in Reforming Socialist Economies: A Comparative Analysis of the Eastern and Central Europe, Case Study on Albania', unpublished report, World Bank: Washington DC.

Xhamara, I. (1995), 'Land Reform in Albania and its Macroeconomic and Social Impacts', unpublished report, Ministry of Agriculture and Food: Tirana.

2 Bulgaria: economics and politics of post-reform farm structures

Sophia Davidova, Allan Buckwell and Diana Kopeva

Introduction

The objective of this paper is to analyse the outcome of the land reform and farm restructuring in Bulgaria during the period 1991-1995 and to identify the factors which have brought it about.

Three questions are raised. (i) What were the changes in the farm structures during the post-reform period? (ii) What are the main characteristics of the emerging farm structures and how sustainable are they? (iii) What were the political and economic factors that influenced the choice the land reform and farm restructuring?

It is not possible to answer these questions on the basis of robust empirical evidence. After five years of reform, still no representative farm survey has been conducted in Bulgaria. Therefore the reliability of information on numbers and land area of different farm structures provided by the National Statistical Institute (NSI) and other Government institutions is not satisfactory. There is, of course, a genuine difficulty for the statistical office to catch up with such dynamic changes in the farm structures. But in the absence of completely reliable information, judgements about the nature and reasons for observed changes should be made and received with a degree of caution. It is only possible to work with the data that exist, but that may not reflect the full story 'on the ground'.

The second section of the paper describes the developments in Bulgarian farm structures during the period 1993-1995 in comparison with pre-reform reference points. The third part discusses the characteristics of each of the current farms structures and gives some arguments about their future viability. The fourth section uses a political economy framework to try and provide explanations of the choice of land reform and farm restructuring in Bulgaria.

Evolution of farm structures

Before starting discussing the evolution of farm structures, some specifics of Bulgarian agricultural statistics should be noted. Farm structures statistics do not allocate total agricultural land but arable land. Total agricultural land in Bulgaria amounts to 6.2 million hectares (Table 2.1). Since 1990, when the last land survey took place, the total agricultural land and its main components have been reported as being stable. Of these 6.2 million hectares there are nearly 4.7 million hectares of arable land made up of 4.0 million ha of annual cropped land and 0.6 million ha of permanent crops (includes orchards and vineyards) and temporary grass. The remaining 1.5 million ha is permanent grass and rough grazing much of which is common land. A portion of the area under permanent grass and rough grazing is state owned, but the common land is generally owned by the local municipalities. Local people have access to this land. Bulgarian statistics on farm structures do not allocate total agricultural land but arable land. In comparison to other countries this, to some extent, overstates the share of private ownership on agricultural land in the post-reform period.

Table 2.1
Agricultural land in Bulgaria, 1990 and 1995* (thousand hectares)

	1990	*1995*
Total agricultural land	6,203	6,203
Arable land and permanent crops	4,687	4,687
Arable land	3,900	4,042
Permanent crops	296	204
Medows	287	276
Cultivated pastures	204	164
Rough grazing	1,516	1,516

* The last land survey was conducted in July 1990. Since then it was politically accepted to give the total agricultural land and its two main components, arable land an permanent crops, and rough grazing constant.

Source: Statistical Yearbook 1994 and Statistical Book 1996, National Statistical Institute, Sofia

The second feature of the Bulgarian farm statistics is that there is no minimum requirement for a land area in order to treat an agricultural unit as a 'farm'. This is due to the great fragmentation of land under private plots pre-reform and individual farms post-reform. These millions of 'farms' occupy a significant amount of the cultivated land and contribute a significant part of total output. For these reasons it seems plausible to take them into account. Again this point should be borne in mind when making international comparisons.

In order to capture the evolution of the farm structures in Bulgaria in the post-reform period it is necessary to have as a reference the pre-reform situation. Following the collectivisation of Bulgarian agriculture in the late 1940s the initial structures created were collective farms, state farms and machine and tractor stations. In the early 1970s these were unified into a small number of Agro-Industrial Complexes (AICs). The central administration of these farming structures was complex and changed over time. The AICs were initially administered by the National Agro-Industrial Union which took over the function of the Ministry of Agriculture. The Ministry only reappeared at the time of the reform in 1989.

The pre-reform agricultural structures typical of the 1980s in Bulgaria are summarised in Table 2.2. There were principally three forms of organisation: AICs, other organisations and private plots. Eighty-one per cent of arable land was farmed in AICs. Within these organisations there were four main forms, Labour agricultural co-operatives (TKZSs), State agricultural farms (SAFs), Machine and tractor stations (MTSs) and Brigades. Two-thirds of the AICs land was farmed in TKZSs and SAFs. Brigades were sub-units within AICs with their own balance sheet, some had independent bank accounts and were relatively independent of the other organisations in the AICs. They generally specialised on a single crop or livestock production. Machine and tractor stations had no land, as the name suggests, they provided mechanisation services to TKZSs and SAFs. Towards the end of the period of the AICs they became more integrated into TKZSs and SAFs. In addition the AICs had numerous other enterprises from food processing and agricultural services to completely non-agricultural enterprises. However, in general, the state-owned food processing enterprises were outside the AICs. In AICs there were only some smaller inter-co-operative processing enterprises. During the 1980s there were many changes in the composition of AICs. For example the numbers of TKZSs and SAFs fell and their average size increased.

The second group of pre-reform farm structures is rather unhelpfully described as ‘other organisations’. These included farms attached to research stations, schools and forest enterprises, and also auxiliary farms which were part of industrial enterprises and the armed forces. The third group of farming structure were the private plots.

Because the structures were in a state of perpetual change it is very difficult to discover from the official statistic the precise numbers and areas in each of these categories for the years running up to the 1989 reforms. Table 2.2 shows the situation for 1985 for which it is felt that there is most information and the numbers seem internally consistent.

Table 2.2
Pre-reform agricultural structures in Bulgaria, 1985

	Number	*Average arable area (ha)*	*Total arable area (ha)*	*Share (%)*
AICs, comprising of:	298	12,600	3,754,800	80.7
TKZS	678	4,000	2,712,000	58.3
SAFs	196	2,100	411,600	8.8
MTS	99	0	0	0
Brigades	na	na	631,200	13.6
Other agricultural organisations	238	1215	289,200	6.2
Private plots	1,600,000	0.38	609,000	13.1
Total	1,601,509		4,653,000	100

AICs = Agro-industrial complexes, TKZS = Labour agricultural co-operatives, SAFs = State agricultural farms, MTS = Machine and tractor stations

Source: National Statistical Institute

In 1985 there were almost three hundred AICs averaging 12,600 ha. Within these were about 680 TKZSs averaging 4,000 ha and 200 SAFs with about half of that area. It is estimated that there were about 1,600,000 private plots averaging slightly less than 0.4 hectares and accounting for about one-seventh of the total arable area.

The post-reform structures are summarised in Table 2.3. This table shows data for the three years, 1993-1995. In the immediate post-reform years (1990-1992) the situation was so dynamic that the statistical services were unable to provide reliable data year by year. In fact the first thorough survey in 1993 was not published. The table defines five categories of post-reform farm organisations: state farms, organisations under liquidation, farming companies, co-operatives and individual farms. Each of these will be defined and their characteristics discussed below, but first their relative importance is summarised based on the data in Table 2.3. State farms in 1993 embrace land from the former 'state agricultural farms' and 'other agricultural organisations'. The total arable land in these two categories in 1985 was 700,000 ha and this appears to have halved by 1993 to 345,000 ha and continued to decline in the following two years. Of the 2.7 million ha on TKZSs in 1985, by 1993 about 2 million ha were in organisations under liquidation. This area diminished rapidly to 850,000 ha in 1994 and by the middle of 1995 these structures had been liquidated as planned (the official statistics show no arable land on these organisation although technically 157 of them still were not formally struck from the registers). It can be seen from Table 2.3 that most of the former AICs land is now shared between co-operatives and individual farms. The area in co-operatives has doubled from 1 to almost 2 million hectares between 1993 and 1995. The area of land on individual farms doubled from 600,000 ha on private plots pre-reform, to 1.2 million ha in 1993 and doubled again to 2.4 million ha in 1995. Land in farming companies was only 54,400 ha in 1993 and has dropped to about 35,000 ha in the following two years. This is presumably partly due to failure of some of this companies and partly because some people decided to participate in the co-operatives which were rapidly becoming established over this period. Summarising, by 1995 93 percent of arable land was farmed either on individual farms (52 percent) or on the new producer co-operatives (41 percent). Land on state farms has shrunk to 6 percent, and less than 1 percent of land is on farming companies.

The largest category of farming structures are the individual farms, whose numbers initially rose to 2.0 million by 1994 but fell to 1.78 million in 1995. The size of these units increased from 0.4 hectares pre-reform to 0.7 hectares in 1993 and doubled to 1.4 hectares in 1995. The number of co-operatives grew quickly starting in 1992, numbered 1,500 by the end of 1993, and have grown to over 2,300 by 1995. The average size has grown from 700 ha in 1993 to 815 ha in 1995.

Characteristics of farm structures

A summarised picture of business characteristics of farm structures in Bulgaria according to the Beckmann and Hagedorn classification[1] is presented in Table 2.4. It has not always been possible to find a perfect match between the Beckmann-Hagedorn classification and the characteristics of farm structures in Bulgaria. This is explained in detail when individual types of farm structures are discussed.

There are two pieces of legislation on the basis of which farm structures are registered, commercial law and co-operative law. Various types of companies and partnerships are registered on the basis of the commercial law, while co-operatives are created according to the provisions of the co-operative law. It is not required for individual farmers to have any form of legal registration. Normally one or two persons or legal entities are enough to register a company. Co-operatives need at least seven founder members. The requirement for minimum capital only applies to limited liability companies and joint stock companies. In the latter case the minimum requirement is substantial. Tradability of capital is allowed also for these two kinds of companies pending on the agreement of the General Assembly. Understandably, individual farmers and traders are fully free to sell their capital. They also normally manage the business by themselves. However, they are free to hire a professional manager. The situation with the general partnership is not substantially different. Either all partners or some of them are entrusted by the others to manage the farm. Management could be also delegated to a professional manager. The management of limited liability companies and joint stock companies is more complex. The limited liability company is managed by the General Assembly, but the every-day management functions are performed by a professional manager. The joint stock companies have as collective management bodies either a General Assembly of shareholders and Board of Directors, a one-tier system, or General Assembly, Managing Board and Supervisory Board, a two-tier system. Cancellation of participation in partnerships, companies and co-operatives is possible. In the general case financial compensation equals the share of contribution to the capital in companies and partnerships and to land and non-land assets in producer co-operatives. Given these different business characteristics, the farm structures have showed quite different development patterns post-reform.

Individual farms

Private organisations are gradually becoming dominant in Bulgarian agriculture. Some of the farms recorded as individual in Table 2.3 correspond to 'household plots' in the pre-reform language, but with a substantial difference as far as land ownership is concerned. Farmers who pooled their land into the new co-operatives[2] are eligible to get a plot for their individual use from the co-operative. The General Assembly of each co-operative votes a maximum size of these allocated plots. In the general case the plot is a fraction of the land area pooled by the owner into the co-operative. But there are cases of land-less co-operative members or smallholders who receive land for 'personal' use despite the fact that they did not pool any or they pooled less than they received. Thus the largest part of the land under plots is, in principle, owned by the users of the plots. However, the location of the plots rarely corresponds to the location of land owned by the farmer. In many cases the co-operative provides machinery services to the plot users (as in the former system), but now the plot users pay the co-operative for these services supplied.

Table 2.3
Post-reform agricultural structures in Bulgaria, 1993-1995

	Number			*Arable land area (ha)*		
	1993	*1994*	*1995*	*1993*	*1994*	*1995*
State farms	1,196	1,009	980	345,640	328,085	304,651
Organisations under liquidation	1,166	500	157*	1,980,000	849,057	0
Farming companies	144	142	122	54,461	39,946	34,590
Co-operatives	1,524	1,873	2,344	1,072,211	1,411,299	1,910,972
Individual farms	1,897,454	2,010,961	1,777,122	1,234,622	2,058,545	2,462,250
Total				4,686,934	4,686,933	4,686,933

	Share of arable land (%)			*Average size (ha)*		
	1993	*1994*	*1995*	*1993*	*1994*	*1995*
State farms	7.4	7.0	6.5	289.0	325.2	310.9
Organisations under liquidation	42.2	18.1	0.0	1698.1	1698.1	
Farming companies	1.2	0.9	0.7	378.2	281.3	283.5
Co-operatives	22.9	30.1	40.8	703.6	753.5	815.3
Individual farms	26.3	43.9	52.5	0.7	1.0	1.4
Total	100.0	100.0	100.0			

* In 1995 organisations under liquidation do not have land. They are left to allocate the remaining non-land assets.

Source: National Statistical Institute

Table 2.4
Business characteristics of Bulgarian farm structures

Characteristics	*Types of Business Organisations in Bulgarian Agriculture*					
English Term	Individual Trader	General Partnership	Limited Liability Company	Joint Stock Company	Co-operative	Individual Farmer
Bulgarian Term	Ednolichen Targovetz	Sabiratelno Drujestvo (C-e)	Drujestvo s Ogranitshena Otgovornost (OOD, EOOD)	Aktsionerno Drujestvo (AD)	Co-operatzia	Tshasten Zemedelski Stopanin
Juridical Basis for Existence	Commercial Law	Commercial Law	Commercial Law	Commercial Law	Co-operative Law	no
Registration	Necessary	Necessary	Necessary	Necessary	Necessary	no
Minimum/ Maximum of Members	Minimum 1	Minimum 2	Minimum 1	Minimum 2	Minimum 7	no
Minimum of Capital	No	No	Minimum 50,000 Leva.	By public offer - 5 mln Leva; Otherwise 1 mln leva	No	No
Individual Tradability of Capital	Yes	Yes	No transfer allowed without consent of General Assembly	No transfer allowed without consent of General Assembly	No	Yes
Liability	Full	Full	Limited, joint liability up to the amount of the contribution to the capital of the company	Limited, joint liability limited up to the value of the property	Joint liability	Full
Management		Every partner has power to	General Assembly of Members	General Assembly of shareholders Board of	General Assembly	

	manage. By partnership deed such power could be delegated to one or several partners or to a third person	Managers (the managers could be other than a member)	Directors (one-tier system) or the Supervisory Board and Managing Board (two-tier system)	Managing Board Supervisory Board
Cancellation of Membership	Possible	By a written notice three months in advance. Accounts are settled on the basis of balance sheet at the end of the month when the cancellation becomes effective	Sale of stocks and shares	Possible
Financial Compensation by Membership Cancellation		Proportion to the share of members in the capital unless otherwise agreed	Proportion to the share of members in the capital unless otherwise agreed	Share equal to pulled into the Co-operative by quantity and quality. The members receives equal quantity of land as pulled with the matching quality in the same area but not always its own parcel pulled into the co-operative

There are two main factors that contributed to the survival of the plots post-reform. These are, first, the conservatism of rural population, and second, the provision of food and animal feed to rural families, most of which have a few animals. These household plots may turn out to be transitional structures. They are rather too large to be gardens, yet not large enough to be small farms. As the market economy develops, many of these plots may disappear. Either all the land will be cultivated by the co-operative and members will be interested in money income and not income in kind, or people will get a larger land area than a plot (owned or leased) to create an individual farm.

The average size of individual farms in Bulgaria is 1.39 ha having almost doubled in size between 1992 and 1995. During this four year period the number of individual farms decreased by 9.5 percent, and the amount of land under individual farms increased by 71 percent. The dynamics of development of individual farms and farmed land in different farm groups since the beginning of the reform is shown in Table 2.5. The size groups have been classified in accordance to the situation in 1993. That is why some of the farms that increased their size from 1993 to 1995 have not been moved to the next group. This leads to the observed contradiction in Table 2.5 concerning the farm size group in the range of 5 to 10 ha in 1995. The average size of this group of farms increased to 12 ha, but the NSI did not do a new classification.

The total number of individual farms has fluctuated between 1.7 and 2 million. The smallest group (less than 1 ha) fell after 1992 and then stabilised at about 1.5 million. The next three groups (1-2 ha, 2-5 ha, 5-10 ha) all increased from 1992 to 1994 but appeared to fall in number in 1995. The only group whose numbers have consistently risen is the group over 10 ha which has increased 7 fold (from 580 to 4,007).

The generally small size, product-mix and market position of these individual farms creates many problems for their management. It is helpful classify them into four groups according to their size and market orientation (i.e. whether they are producing for self-sufficiency, for market or both); the share of their total income coming from agriculture and their dependence on co-operatives or private contractors to supply machinery or other services.

Group I comprises of 86 per cent of the number of individual farms and 26 percent of the total land area under individual farms. Farms with less than 1 ha are included in this group. These include also the private plots. Their average size is 0.41 ha. The 'farmers' in this group produce mostly for their own consumption, but they are only employed part-time. It was expected that the numbers of farms in this group may have increased as old people or unemployed moved out of towns and cities to the country to become self-sufficient. The statistics do not appear to confirm this 'de-urbanisation' or reverse migration. They show a decline in the number from 1992 to 1993, but little change in the number and average size of the farms in this group since. However, because this group of farms are largely run by old people it is expected that the number will gradually fall. This group will not play an important market role.

Group II comprises the 13 per cent of the number of individual farms from 1 to 5 ha which are farming also about 26 percent of land under individual farms. Unless blessed by high capital intensity, farms of this size are likely to be subsistence or marginal farms. In general, their activities concentrate on the production of labour intensive crops and animals. These types of farms are the main producers of ruminant livestock and small scale pig production. However the production is not very competitive and is labour consuming. This is illustrated by the average number of livestock per farm. The data from the National Statistical Institute shows that 221.1 thousand farms breed cows: 81 per cent of them have only one cow and 15 per cent only two. The same situation applies in pig and sheep production: 64 per cent of farms have only one pig; 22.3 per cent have two; 9 per cent have from 3 to 5 pigs. Forty per cent of farms have just three to five sheep, and 27 per cent have six to ten sheep.

Table 2.5
Individual farms in Bulgaria, 1992 - 1995

	Groups of farms by land area (ha)					
	0-1	*1-2*	*2-5*	*5-10*	*over 10*	*Total*
			Arable land (hectares)			
1992	na	na	na	na	na	1,438,000
1993	400,003	328,934	264,097	98,007	143,581	1,234,622
1994	652,147	371,567	407,180	160,361	467,290	2,058,545
1995	631,139	312,184	342,145	168,421	1,008,362	2,462,251
			Number of farms			
1992	1,783,808	128,874	42,520	8,608	580	1,964,390
1993	1,537,462	248,772	91,980	15,195	4,045	1,897,454
1994	1,555,090	342,340	93,568	15,762	4,201	2,010,961
1995	1,535,111	156,092	68,429	13,483	4,007	1,777,122
			Average size of the farm (hectares)			
1992	na	na	na	na	na	0.73
1993	0.26	1.32	2.87	6.45	35.5	0.65
1994	0.42	1.09	4.35	10.17	111.23	1.02
1995	0.41	2	5	12.49	251.65	1.39

Source: National Statistical Institute

One of the main problems for these farmers is the lack of marketing, servicing and credit co-operatives. Therefore new farmers are partly dependent on the producer co-operatives for services and on regional consumer co-operatives (RPKs) for marketing. In both cases the relations are not stable. The co-operatives are local monopolists and private farmers can use their services only if there is a free capacity of machinery and equipment. In the future some of these farms might be expected to move to the first group decreasing their size and the rest may switch to the groups three and four, increasing their size by buying or leasing land. These farms can not provide full-time employment for the members of the household.

As many farmers in this group tend to accept their land area and family labour as natural restrictions on their activities, these farmers, in future, may become members of co-operatives active in input purchasing, including credit, machinery use and output marketing, which are expected to grow in importance.

Group III comprises less than 1 per cent of newly established farms between 5-10 ha which occupy 7 per cent of area farmed under individual farms. It is expected that this group will grow in importance. These farms will produce mainly intensive crops and livestock products.

Group IV comprises the 0.2 per cent of individual farms over 10 ha which occupy over 41 per cent of land under individual farms. They predominantly function on leased land or land that is allocated for temporary use whilst the procedure of land division is being completed. They specialise in grain or industrial crops production. This group is expected to expand in number. It will mostly grow at the expense of farms in group two. These farms will cultivate family owned land, leased land or enter into informal associations of small groups of farmers. Generally these farms with some of those of the third group will be the main individual commercial farming structures.

State farms and farms owned by public institutions

State farms exist by virtue of Decree 56 (approved in 1987 and amended many times since). Subsequently most of them have been re-registered under the Commercial law (1992). In principle they are developed on land owned by the State and Municipalities. However, due to several reorganisations of farm structures under the central planning, including a fusion of collective and state farms into AICs, some of the land of the state farms is subject to restitution. This is land which previously was in collective farms. The number of state farms decreased slightly during 1993-1995 from 1,196 to 980. The same tendency prevailed in the area cultivated. It decreased from 350 thousand ha in 1993 to 300 thousand ha in 1995, which represented 6-7 per cent of total arable area. This decrease in land area was almost fully due to restitution claims. So far there has been no privatisation of state land. Land allocation to land-less peasants from municipal land reserves has been on a small scale, about 500 ha altogether by April 1996.

State farms are managed by a professional manager who is effectively a civil servant. They rely on hired labour. There are reasons to believe that some of these State farms will continue to exist for a relatively long time under the form of publicly owned companies. Currently some types of farming companies are in fact re-registered state farms. This applies particularly to the case of Limited Liability Companies (or, following the Beckmann-Hagedorn classification, Public Limited Companies). A brief description of the main types of these public or state farms follows.

State farms: experimental stations, experimental bases and selection centres. These are specialised farms in crop selection and animal breeding. They are treated as development organisations. They are expected to survive, but their number may decrease from the present 63. Altogether they farm 0.6 per cent of the arable land.

Municipal farms are registered under the Commercial law. The land that they cultivate is from the municipal land reserve. Their number is 231 and they occupy 1.2 per cent of total arable land. The expectations are that a significant part of this land will be allocated to the land-less farm workers under the scheme of the land settlement provided by the Law for ownership and use of agricultural land (LOUAL). This will lead to a decrease in the share of this group.

School and University farms are managed by specialised agricultural colleges and universities. Generally they operate on public land although sometimes they operate on land on which there are claims for restitution. In these latter cases the LOUAL provides procedures for compensation of eligible owners, normally with land in another location. The number of these farms is 234 and they cover 1 per cent of the arable land. They are not market oriented and they are not expected to play a significant role in commercial agriculture. Their main purpose is to help students to obtain practical experience, to be used for experimental research work, and they also provide food to university and college canteens.

Church Farms are based on land owned by the church. There are 88 of them and their land is only 0.8 per cent of total arable land of the country. Recently the church has started to lease out some of the land.

Forest farms occupy less than one percent (0.6) of arable land in 138 farms attached to forest enterprises. They are located in mountains and semi mountainous areas. The land is publicly owned.

Companies with limited liability are classified in the group of state farms because they are mainly state-owned. Their activity is mostly in intensive livestock production of pigs and poultry. In the late 1980s they had to re-register under the Decree 56 as Agro-firms. Post-reform they were required to re-register again under the commercial law as limited liability companies. This has been a necessary step towards changing the management structure and preparing the companies for future privatisation. Legally, there is nothing that defines limited liability companies as state-owned because they can be formed by one or more persons who are liable for the obligations of the company up to the amount of their contribution to the capital. However, the evidence so far in agriculture is that they are mainly state-owned. Limited liability companies cultivate about 0.8 per cent of the arable land in Bulgaria, but, given their focus on intensive livestock production, their importance for agricultural production is larger than this. These are the farms which have the largest debts in agricultural sector. Their privatisation was delayed to a point at which is doubtful that they can be privatised at all due to their debts and depreciation of assets. Most of these farms are facing bankruptcy and they will probably disappear in future.

Agro-business companies are state-owned structures. Their origins are also in the Decree 56. They had to re-registered and change their name and legal form according to the Commercial law. However, the NSI still records them under this old name. There are only 13 of them, covering 0.5 per cent of the arable land. The main distinction from the previous group is that the agro-business companies are not only engaged in farming, but also in trade with agricultural products.

Others. This group includes farms attached to various state organisations and institutions: homes for the mentally and physically disabled; homes for old people; auxiliary farms of state companies. The share of arable land farmed is 0.6 per cent.

Machine and tractor stations are pre-reform state-owned units. They were privatised in two ways. During the last pre-reform years in order to increase productivity in agriculture so-called 'groups for high productivity' were created. Collective farms gave to some groups of individuals land and machinery and a larger independence to organise their production with higher incentives to capture incomes. These groups had their own bank accounts.

Post-reform, many of these groups bought the machinery and created private companies for mechanisation services. The second privatisation route was through the liquidation of the collective farms. At the beginning of the reform the former AICs were removed and most of the machine and tractor stations became sub-units of the collective farms. When the non-land assets of collective farms were allocated amongst the eligible owners, some of them bought the machinery, while the remaining stock was pooled into the new co-operatives and owners have shares in it.

Private farming companies based on the Commercial law

All types of companies registered under the Commercial law, despite differences in their legal status, are called 'traders'. There is a contradiction in Bulgarian law in this matter. There are farming companies registered under the Commercial law, yet one of the general provisions of the law says that: 'Natural persons engaged in farming shall not be deemed Traders', (Commercial law, Part one, Chapter 1, Art. 2). However, in practice during the registration of the company the trader is required to list a range of activities, in which he expects to be engaged. This may include farming as one of the activities. The common practice in Bulgaria is to list a wide range of activities, most often not even related by technology or marketing. Thus, there are thousands of companies which mentioned farming as one of their possible activities. The small number discussed in this paper, 122, comes from NSI data, and possibly means that they are mainly involved in farming. These companies cover less than 1 per cent of arable land. However, this may be an underestimate of the involvement of private companies in agriculture due to the lack of capacity in the NSI to capture the range of the actual activities of private companies.

Beckmann and Hagedorn proposed two sub-groups of private commercial farms in their classification, partnerships and companies. In Bulgaria it is difficult to use these groups because both partnerships and companies are registered as different types of company based on the same legislation. The Beckmann-Hagedorn classification also identifies three business organisations in the group of partnerships: Civil Law Partnership, Commercial Law Partnership and Limited Commercial Law Partnership. An attempt is made to define Bulgarian companies in relation to the Beckmann-Hagedorn scheme. The types of companies which are discussed below do not exhaust all types provided by the Commercial law in Bulgaria, but they include these legal forms of companies which are typically involved in farming and are privately owned.

Voluntary Associations (Civil Law Partnership) have developed during the last three years. They are formed on a voluntary basis. Legal registration is not required in Bulgaria. Each partner has unlimited liability for his own capital, but limited liability for the joint capital. Individual tradability of the capital is possible, which is another difference from the Civil Law Partnership, as defined by Beckmann-Hagedorn. There are no requirements for minimum capital. Because there is not a legal requirement for registration these voluntary associations are fluid and they are difficult to capture and analyse without case studies.

General Partnerships (Commercial Law Partnerships) are exactly as defined the Beckmann-Hagedorn classification. General partnerships are formed by two or more physical persons under a common trade name. The partners assume joint and unlimited liability. The application for entry of the General Partnership in the commercial register has to be signed by all partners and the partnership deed should be attached. Each partner has the power to manage the affairs of the partnership or the management could be delegated to one or several of the partners, or to a third person. The consent of all partners is required for the acquisition and disposal of real estate, the appointment of a manager who is not a partner, and the conclusion of a loan contract amounting above an agreed limit.

Joint-Stock Company (Public limited company) assumes liability to its creditors up to the value of its property. A minimum number of two participants is required for the incorporation of a joint-stock company. The value of the capital and shares has to be denominated in BUL. The minimum authorised capital of a joint-stock company, if the formation is by public offering, is 5 million BUL, and it is 1 million BUL if there is not a public offering. The minimum nominal value of a share is 100 BUL. Larger nominal shares must be divisible, without fraction, by 100. The management structures of Joint-stock companies are: the General Meeting of the shareholders and the Board of Directors (a one-tier system), or the General Meeting, the Supervisory Board and the Managing Board (a two-tier system). The General Meeting is called by the Board of Directors, or by the Managing Board at least once a year. A decision of the General Meeting is adopted by affirmative vote of the majority of the shares represented. Most of joint-stock companies in Bulgarian agriculture are in the livestock sector.

Sole traders do not have an analogue in the Beckmann-Hagedorn classification. The sole proprietorship in their table corresponds more to an individual farm in Bulgaria to the extent that no registration is required. Sole traders are registered under the Commercial law (part 2 division 1). Any capable, physical person, resident in Bulgaria, may register as a Sole trader on the basis of an application in which information stipulated by the law should be stated. Agricultural Sole traders are generally farmers who are at the same time purchasers of agricultural products for resale, processing or export.

Producer co-operatives

Producer co-operatives are the second most important category of farm in Bulgaria after the private farms, measured by the share of arable land. Forty one per cent of the total arable land in the country is in such co-operatives. They produce about one half of the gross agricultural output, but their role is limited principally to three kinds of crop, cereals, oilseeds and sugar beet. Co-operatives produce about 90 per cent of these. Currently their participation in livestock farming is negligible.

Table 2.6 shows the number of the new co-operatives increased seven fold during the period 1992-1995 rising to 2,344 by the end of 1995. In 1992 there was only a small number of new co-operatives because the land reform was at its very beginning. That is why it is more sensible to analyse only the period 1993-1995. During this period the average size of a co-operative, measured by the arable land area, increased by more than 100 ha. By the end of 1995 it was about 815 ha. Thus the first characteristic of the new co-operatives in Bulgaria is that they are relatively large. They are less than a fifth of the size of the structures they replaced, the TKZSs, and 3.5 times more numerous. These facts are signs of reform, a larger number of farming units, each with a more manageable size. However, the average size of the membership, above 200 people, might be deemed large for efficient decision-making.

Table 2.6
New co-operatives in Bulgaria, 1992-1995

Year	*Number of co-operatives*	*Total arable land (ha)*	*Number of founder members ('000)*	*Average size (ha)*	*Average number of founder members*
1992	347	98,000	na	282	na
1993	1,524	1,072,221	268	704	176
1994	1,873	1,411,300	468	753	250
1995	2,344	1,910,972	571	815	244

Source: National Statistical Institute

Although the new co-operatives are more numerous than the previous TKZSs, information obtained from the Bulgarian Ministry of Agriculture and Food Industry shows that in 83 percent of the villages and towns in which new-co-operatives were established there is only one co-operative, in 13 percent there are two, and only in 4 percent of villages are there more than two. Thus in 83 percent of villages the situation of the first post-collectivisation years has been reproduced, each village has a co-operative. This design was strongly supported by Bulgarian agriculturists pre-reform, who claimed that this was the optimal model. Some of these people are still active in giving advice to the rural population about farm structures. For example, the Chairman of the Union of Co-operatives in Bulgaria, explained that his Union advises the population in the villages in which there are more than one co-operative to merge the existing co-operatives. The Chairman perceived that this advice was being followed.

The co-operatives try to maintain their place as the main, and if possible the only, significant farming unit in each locality. This serves a double objective, to avoid competition in the village and to become the only employer in agriculture, which facilitates the co-operative in setting the rules for payments for labour, land and capital. This also creates the opportunity for the co-operatives to squeeze the private farmers in the village. They can refuse to provide them with machinery services, storage space or simply create adverse feelings amongst co-operative members towards the private farmers, bringing about a social marginalisation of these farmers.

Most of the co-operatives in agriculture are producer co-operatives and they are involved almost exclusively in farming. There are also multifunctional co-operatives, e.g. agriculture and credit, producer and consumer, agriculture and trade. However, according to the NSI data these multifunctional co-operatives represent less than 1 per cent of the newly established co-operatives.

Bulgarian, producer co-operatives have the features underlined in the Beckmann-Hagedorn classification. However, currently not all of them are registered as agricultural producer co-operatives under the co-operative law. By virtue of this law a minimum number of founder members should have notarial deeds for their land. However in late 1996 many co-operatives were still operating on land allocated for temporary use because the land division has not been completed. Such co-operatives cannot register as agricultural producer co-operatives through the co-operative law, and they are registered under legal arrangements for co-operatives. Probably such co-operatives will have to re-register in the future if they would like to become an agricultural producer co-operative. How long this process will take, depends on the speed of the land reform.

Members of the co-operative have to contribute labour, land or capital. One of the basic differences with the co-operatives under the central planning is that the labour contribution is not obligatory. There are many members in the new co-operatives who do not contribute labour. These are mainly the absentee land owners. By the law it is not ruled out that a member can contribute only labour. However, such cases are exceptional. Currently the members of the new co-operatives can be sub-divided into two groups. The first group constitutes the former members of the collective farms. All people in this group have shares in the non-land assets, thus they contribute at least capital and most of them also land. The second group is everyone else. This includes the absentee land owners who contribute land and some of them also capital under the form of shares of non-land assets.

In the general case the members enter the co-operative with their land for which they know exactly the boundaries and the location. If a member subsequently wishes to withdraw from the co-operative and to withdraw his land, and if it is not convenient to release land 'in the middle of' the co-operatives property, he may receive instead, an equivalent parcel of land in another location. If the member owned the land that he used as a private plot, he has right to withdraw this land area too. Also, having joined a co-operative the member often accepts that in case of withdrawal he will receive his share of non-land assets in cash.

According to the ownership of land and payments to members for their labour, land and capital, there are two groups of producer co-operatives in Bulgaria. The difference between them is subtle and it is difficult to capture in detail because the issue is highly politicised.

In the popular parlance the first group of co-operatives is labelled 'blue co-operatives' following the colour of the non-communist political coalition, the Union of Democratic Forces (UDF). The second group is labelled 'Orsov co-operatives', following the name of a Bulgarian Socialist Party (BSP) member of the National Assembly Zlatko Orsov. The name implies that these are 'red' co-operatives, or in another words they are strongly related to the ideology of the collective farms.

It appears that the first group is closer to the requirements of the co-operative law. Most of the owners had their land with physically demarcated borders at the time of joining the co-operative, and these co-operatives try to give a fair return to labour, land and capital.

The second group, 'Orsov co-operatives', is created mainly on land for temporary use where the owners have not received their land in physically demarcated boundaries. Land ownership in these co-operatives is less clear. In fact it is a shared-ownership, each member knows that he is eligible to get some land area of specified quality from the land of the co-operative, but because the land reform has not been completed he does not know the location of his ownership. The general practice in these co-operatives is to avoid paying rents and dividends finding reasons such as a bad crop year, increase in input prices, the need for investment. Also they try to implement norms for labour payments, a method inherited from the central planning. It seems that overall the main distinction between the 'blue' and the 'red' co-operatives lays in the recognition of ownership of individual owners. The 'red' co-operatives are inclined to continue the practice from the former system when the members were owners on paper, but they were deprived of the returns of ownership. This form of co-operatives may turn out to be a temporary structure which persists only as long as the land is still not fully restituted. In future it will become more difficult for the pro-Communist managers to deprive owners of the returns to ownership, particularly as more of the first type of co-operatives appear. It will be almost impossible to do this in villages in which there are at least two co-operatives, 'blue' and 'red' and where information flows easily. This may be one of the reasons that the Union of Co-operatives, which is exclusively a union of 'red' co-operatives, works hard toward the principle 'one village - one co-operative'.

There is a contradiction between the interests of members who contribute only land and those who contribute labour as well. Several co-operatives plan to decrease the number of members by renting the land of the absentee land owners without keeping them as members. To the extent that these absentee land owners have also shares of the non-land assets, the intention is that the co-operatives will buy these shares and create a joint property. However, the evidence of this is not yet sufficient to analyse. A move in this direction could decrease substantially the membership of the co-operatives which would facilitate more streamlined decision-making as well as the future transformation of the co-operatives into farming companies. The practical implementation might be difficult because co-operatives cannot release members without their consent.

The management of the new-co-operatives comprises a General Assembly in which the principle of one-member-one-vote applies, and which elects a manager. In co-operatives with a large number of members instead of General Assembly there is an Assembly of Representatives. Full time employees are relatively small in number and include mainly administrative staff and professionals, e.g. agronomists.

Are Bulgarian new producer co-operatives characteristic of typical co-operatives? It is not easy to find a rigorous definition of co-operatives because the notion is quite broad and embraces many activities marked by voluntary participation, co-operation and will to work towards common objectives. Here some of the definitions will be discussed in order to see whether the Bulgarian producer co-operatives have the general defining characteristics.

The most widely accepted definition is the one adopted by the International Labour Conference in 1966. This states that a co-operative is 'an association of persons who have voluntarily joined together to achieve a common end through the formation of a democratically controlled organisation, making equitable contributions to the capital required and accepting a fair share of the risks and benefits of the undertaking in which the members actively participate', ILC (1966).

Bergmann (1967) defines co-operation as 'institutionalised collaboration of members with equal rights in co-operative form of enterprises. This implies an organisation with by-laws and formulation of the common aims. The main economic goal is the material promotion of the members by joint efforts, mutual aid and individual effort. All members enjoy equal rights and contribute a clearly fixed, equal minimum contribution, e.g. land, goods, labour, cash, capital, animals. The members administer the co-operative from the start or are led to self-administration, possibly by government assistance or guidance. Economic promotion of members helps in fulfilling the social and educational tasks'.

One of the fullest definitions of agricultural producer co-operatives is given by Putterman (1989):

> Agricultural producer co-operatives (APCs) are farm enterprises in which the labour of a group of farm households is pooled under an output or net-revenue sharing arrangement. Land and other non-labour means of production may be held in common, or may be privately owned by the members who contribute it to group enterprise in return for a share of output or revenue. In the co-operative proper, production has the characteristic of team production in the sense that the fraction of output attributable to the labour of any particular team member is undefined because output is a joint product of many hands. No employment relationship exists among the co-operators, and decision-making is in principle participatory or democratic, as opposed to hierarchical or authority based. Aside from the co-operative farm itself, co-operative farm communities may have some lands allocated to production on a household or individual basis.

On the basis of these definitions the following characteristics of a producer co-operatives could be deduced: voluntary organisation; democratic non-hierarchical management; common objectives; contributions from members; equitable share of benefits and risks.

Both types of producer co-operatives in Bulgaria, 'red' and 'blue' have these characteristics. The issue whether the assets are jointly or privately owned is not a defining one. Both options are acceptable if the other conditions are satisfied like individual contribution and sharing the risk and the output (net-revenue). If the issue of ownership of land and non-land assets is central in the discussion surrounding producer co-operatives in Bulgaria and in most of the CEECs, this is due to the period of transition which is characterised by a massive change in ownership. Whether assets are privately owned or jointly owned is not treated in the theory as a decisive factor for the behaviour of producer co-operatives. Much more important are factors like the trade-off between maximisation of returns per labour unit and employment of the members of the rural community. This is particularly the case in the situation where there are large problems of the lack of non-agricultural job opportunities, problems in management and supervision etc. (see Putterman, 1989).

How sustainable are producer co-operatives in Bulgaria? It is important to define the time horizon in which this sustainability is treated. It is clear that for the medium-term the producer-co-operatives have an important role in Bulgarian agriculture. In a longer term, 15 to 20 years, there are some reasons to suggest they might decrease substantially in importance. Some of the factors that could contribute to this process are first, generational turnover in the country side and the arrival of younger more individualistic people without memory of work under collectivised agriculture. Second, stabilisation of the macroeconomic environment and decrease in risk to a degree bearable by entrepreneurs. Third, an increase in profitability of agriculture from the current low level. Fourth, a higher degree of development of upstream and downstream markets. Fifth, a better development of agricultural services. Under these conditions the 'shelter' currently provided, or at least perceived as being provided, by the co-operative might be of a far smaller influence than at present. The removal of these 'negatives' which it is suggested create much of the current strong desire to stay within the familiar folds of the co-operative organisation, would then allow the positive attraction of the relative freedom for decision making under other 'capitalistic forms of organisation' to become more apparent. If these ideas are correct, what will succeed the co-operatives will be mainly farming companies rather than family farms. With an increase in profitability and a decrease in risk of agricultural production, groups of co-operative managers and/or members will discover the incentive to buy up non-land assets and buy or lease land on a long term basis and register a farming company. Of course, in the meantime, some people are expected to leave the co-operatives and to create family farms. However, on current form, the bulk of producer co-operatives will try to avoid this fragmentation by creating impediments to a mass exit. This will tend to preserve the relatively large size of the co-operative, which could also contribute to the decision to transform the co-operative into a farming company rather than into family farms.

Political economy of land reform and farm restructuring

Some general remarks

Swinnen (1995, 1997a) proposes a political economy framework to explain the choice of privatisation and de-collectivisation policies pursued in the agriculture of the CEECs. This provides an excellent basis for comparing the policies chosen and the progress in implementation across countries in relation to a relatively small number of 'explanatory variables'. The core idea is that privatisation and de-collectivisation each have both economic and political distributional effects and it is the anticipation of these effects which, together with the initial conditions, determines the decisions on reforms and the implementation. The two steps involved in applying the political economy framework are, first, a search for correlations between countries' reform policy choice and a number of factors affecting the political decision making, and second, an attempt to flesh out the correlations with fuller explanations of the political forces bearing on agriculture.

The factors identified as affecting the policy choice about the land reform include: post-collectivisation legal ownership of land, pre-collectivisation distribution of land ownership and ethnicity. The factors which seemed to be correlated with choices concerning the new farm structures include: the political composition of the government, the ways in which de-collectivisation can affect the future political power distribution, and the degree of security of political and economic reforms.

The results of Swinnen's comparative study stimulated closer examination of the particularities of each country's case. This section therefore examines the extent to which the 'cross-country' analysis stands up to closer scrutiny for Bulgarian. It will be argued that the framework does indeed steer us in the right direction in trying to understand the choice of reform instruments and the energy applied and effectiveness of the implementation of the policies chosen. However there are a number of issues for which other explanations and different interpretation are helpful in gaining a fuller understanding of the Bulgarian case. The discussion deals first with land reform and then with the farm restructuring. These are distinctly different processes, although they are also inter-linked. There is also some attempt in the discussion to distinguish policy choice from policy implementation. This distinction can be useful in understanding events, although it, too, can be taken too far. Policy choice can be viewed as concerning the strategic decisions and policy implementation concerns the tactics of putting the strategy into effect. All too often it suits politicians to blur this distinction and to effect strategic changes in policy through legislation on implementation, or indeed through the non-implementation of 'agreed' strategy.

Terminology

The terminology in the area of land reform is a minefield. There is clearly no universal acceptance of which terms should be used for which situations or policies. It is impossible for one author or group to dictate terminology; all that can be done is to explain as precisely as possible the way terms are being used. This is done in the Table 2.7 and the text which follows.

Table 2.7
Terminology for land reform in transition economies

Property Right	*Pre-reform states*	*Reform choice 1*	*Reform choice 2*
Land Ownership	Private	Continuation	Distribution
	Public/state	Continuation	Privatise
	Common/municipal	Continuation	Privatise
	Church	Continuation	Sale
Land use	Central planning	Continuation	Restitution or privatisation
	Private decision	Continuation	
Income from land	State control	Continuation	Restitution or privatisation
	Private control	Continuation	
Alienation	State control	Restitution for private and church land Continuation for public or state land	

The table serves to show that it is important to distinguish four aspects of rights in land; (i) ownership, (ii) use, (iii) usufruct, or the right to the income from land, and (iv) alienation, or the right to transact the land. There were different pre-reform states for each of these rights. As far as ownership was concerned, there were four pre-reform states.

The largest portion of land in Bulgaria and in most Central European countries was still, formally and legally, owned privately by whosoever owned it pre-collectivisation. Second, there was some publicly or state owned land. Third, there was some common, communal or municipal land. Fourth, the church, an institution which is considered neither the state, nor a private institution, owned land. However the use rights over land during the Communist period was simpler: most land was subject to central planning or state control, thus the owners, whoever they were, had little or no ability to decide for themselves whether and how to use their own land. Part of the land in each country was allocated as 'personal plots', the ownership of this land could have been in any of the four categories listed. The use rights on these plots were mostly unrestricted and left to private decision of the plot operator. Likewise, as far as the rights to the fruits of the land, or the income rights, were concerned, most were controlled by the state through the central planning system. Land users had to deliver prescribed volumes of produce at prices which were centrally determined. However, generally, the output of the personal plots accrued to the operator who could use it for personal consumption or to sell it. The alienation rights on all land pre-reform were essentially nullified. Agricultural land could not be, and was not, transacted in any market. There was no market either for selling or leasing land.

The theoretical options for reallocating these property rights are listed in the right half of the table. There really were not many options for most of these categories. It is asserted here that for land which was formally in private ownership there was practically no choice but to recognise this ownership and allow it to continue, this is abbreviated as 'continuation' in the table. However, a large amount of time had elapsed since the former legal ownership was effectively exercised, during the interim the physical boundaries had often been changed or removed, the owners had often moved or died and the rights of ownership 'passed' to heirs. Indeed, in some cases, the use of the land had fundamentally changed and it was no longer in agricultural production. These changes meant that 'continuation' of the private ownership required a new legal process of recognition of ownership rights, redefinition of boundaries and issuance and registration of new deeds of ownership. These processes are generally referred to as 'restitution'. Strictly, this is an inappropriate word as the ownership had never been wrested away from the owner, hence a preferred verb would be 'continuation or re-registration' of the private ownership. The one word which cannot be used here is privatisation. What was not in public ownership cannot be privatised.

The alternative policy for this privately owned land is to distribute it. It makes most sense to distribute it to those who are currently farming it or wishing to farm it. If this is done voluntarily with the consent of, and with acceptable compensation of, the owners, then this is, *de facto*, sale by the owners to the new farmers or to the authority which expedites the redistribution. But if it is done without a process to gain the consent of the owners and without compensation, then this can only be viewed as a post-communist expropriation of ownership rights.

For land owned by the state (central or local) there are clearly two options: to continue to keep this land in public or common ownership or to privatise it. Of course there are a number of methods of privatisation: by auction, by voucher schemes or by distribution. The details of these choices are important, but they are not the central issue of the table. Church land can presumably stay with the Church or it could be sold to individuals, the State or Municipalities.

There seem to be no choices of policy option concerning the use rights of land which are already in private hands. Unless it is proposed that some individuals should be deemed unfit to operate land, these rights must continue. The label to attach to the options for reallocating use rights in land which was under state control is less obvious. The candidates are restitution or privatisation. The case for calling it restitution is that the use rights of privately owned land had been expropriated from the owners during the Communist era, so returning these rights to the owners (or their heirs) – which is usually done at the same time as the ownership deeds are re-registered, is exactly what the word restitution means. The argument for calling it privatisation is that the use right had indeed been held publicly, therefore passing it to private hands is indeed privatisation. However, our preference is to call it restitution because this explicitly refers to the fact that the use rights had been expropriated and underlines the political imperative of returning these rights to the land owners.

Turning to the reform choices as far as the disposition of income from operating land, these are identical to the choices for using land. The state could, in principle, hang on to its control over the earnings from the use of land. This can be done for example by the central government insisting on compulsory purchase of produce at politically determined prices. The other option is that the state restitutes these rights to the private owners. Where the disposition of income is already controlled by the user of the land, there seems little alternative but to allow this to continue.

The fourth category of property right, that of alienating the land and thus transacting it, had been completely sterilised under communism. For privately owned or church land there is little choice but to create the laws and institutions necessary for it to be leased or sold. There seems to be a great reluctance to do this in ex-communist countries. The very concept of people making money out of the ownership of land is itself alien, and there is a great fear of foreign ownership of land. There is, of course, enormous scope in these circumstances to make land transactions an easily accessible process or to make them very difficult. By placing restrictions on the process of land sale or leasing, the rights of alienation are themselves restricted. For land which is publicly owned, the issue of alienation rights has no real meaning. Those who are charged with the responsibility of managing state owned land continue to have no rights to sell it.

In conclusion, the CEECs, Bulgaria included, faced a relatively restricted policy choice about the design of land reforms. The main constraint came from the post-collectivisation pattern of ownership. Where land-ownership was still in private hands the continuation of this private ownership, otherwise referred to as restitution, seems the only option which has legality. As stressed by Brooks et al. (1991) in surveying the options, few economists outside the region would have chosen physical restitution of rights to prior owners as the preferred solution. Despite this, by analysing the initial conditions in the region, these authors came to the conclusion that the restitution approach does have some redeeming features to complement its apparent political appeal, which counter some of the economic problems it raises. They particularly emphasise that if the land had been distributed to agricultural workers without payment this will be a strong argument for the Governments to exclude them from further distribution of state-owned assets.

It is a curious thing that some Western observers, who are generally associated as being strongly in favour of private asset ownership and legal transactions, seemed enthusiastic for the distribution option. Curious, because this option is essentially an involuntary process even if former owners are compensated by whatever the distributors consider is a reasonable amount.[3] The proponents of distribution profess to be concerned principally about what they consider is the most efficient solution. For this reason they seem prepared to overlook the inconvenient fact that the land that recommended be distributed was owned by someone. The idea even emerged that choosing restitution (i.e. what is termed here 'continuation') over distribution indicated favouring 'historic justice' over 'efficiency'. Distribution of land which was formally in private ownership creates a class of people who may feel deprived of their rights to land and in most countries and through most history this turns out to be an unstable situation. So much human conflict in history has taken place because people are denied the land they consider their own that this hardly needs stating. In such circumstances, the lack of security in property rights felt by the recipients of distribution programmes may undermine the efficiency expected from distributing the land to people engaged in agriculture and 'removing' the class of absentee landowners.

Land reform design: politics and potential implications for efficiency

In the Bulgarian case, pre-reform, land was mostly in private ownership. In this situation the reform option as far as ownership was concerned was straightforward, land had to be restituted. There was no significant Bulgarian voice which argued for the distribution option. The same applied to the rights of use and usufruct. Concerning alienation rights there was a hot political debate whether to attenuate these rights by regulating to whom land should be offered first in case of sale or by defining a maximum size of land than can be leased in.

In general the debate in Bulgaria was not whether restitution was the right option to choose, but how the restitution should be accomplished. One of the most important aspects of the land restitution concerns the boundaries in which the land should be restituted. This is a critical decision, because it is related to two questions, whether the restitution confers effective land ownership property rights and whether it could potentially assist improvement in efficiency.

The two main options for restitution which have been the subject of hot political debate are either to restitute the land ownership in physically identified and demarcated boundaries, or to restitute ownership on an area with specified quality of land within a bigger field (or within the land cultivated by a co-operative) without identifying a particular location. The first option has two sub-options, to restitute the land in historical boundaries, or in comparable boundaries. The latter means a physical location which is not necessarily in exactly the same location as the pre-collectivisation ownership, but with maximum possible correspondence in the area and quality of land. Both sub-options, in historical or comparable boundaries, introduce full property ownership rights and thus neither, in principle, involve any additional transaction costs as far as sale or leasing the land is concerned. There are therefore no impediments to efficiency from these forms of land reform *per se*. The very fact that the new ownership has well-defined and mapped physical boundaries creates one of the basic conditions for developing a land market and future market land reallocation which is believed to bring efficiency increases.

The second option is in fact closer to the concept of shared ownership. If the 'owner' of the land does not have a deed showing the boundaries of his land, there are bound to be higher transactions costs associated with any attempt to sell or lease such land. These additional costs take the form of the negotiations which have to take place between the land owner and the co-operative management or others who have the authority to allocate land in physical boundaries if the owner wishes to withdraw it from the co-operative. Such an allocation process provides the agents involved with a degree of power. This form of restitution of ownership therefore could be said to bring with it full rights of ownership and income, but less than full use and alienation rights. It therefore impedes the market reallocation of land. That is why it could reasonably be said that there is some efficiency penalty with this form of land restitution.

The main argument that the political design of the Bulgarian restitution process may induce inefficiency is based on the fact that it brings about fragmented ownership. This obviously depends on the degree of fragmentation of land ownership during the pre-collectivisation period. In the case of Bulgaria pre-collectivisation ownership was indeed highly fragmented, and thus restitution was bound to perpetuate this fragmented ownership. Brooks and Meurs (1994) show that the same fears of introducing inefficiency due to land fragmentation as a result of the reform exist in Romania. But as the authors stress these fears are based on the assumption that farm ownership and operation will coincide, that therefore smallholders with fragmented holdings sacrifice significant economies of scale. However, efficiency depends much more on the functioning of factor markets than on the initial distribution of assets. If there are economies of scale to capture, farm sizes will gradually adjust through purchase and lease of additional land and assets (Boyd 1991; Paarlberg, 1992).

Restitution in historical boundaries is only one of the sub-options approved in Bulgaria and it applies to land for which these historical boundaries are easily defined. Normally this is in mountainous and semi-mountainous areas, where collectivisation did not involve reshaping of fields so natural boundaries still exist (they are usually physical features such as trees and streams). In these areas the application of principle of comparable boundaries would have only increased the transaction costs of land division and of negotiating an approval by the owners. In the main agricultural areas, restitution is based on the so called 'new real boundaries', which are comparable boundaries. They are identified on the basis of land allocation plans drawn up by Municipal Land Councils. By April 1996 out of 2.84 million hectares restituted to a stage at which notarial deeds could be issued, 66 per cent was defined with comparable boundaries. It is worth noting that Bulgaria was not an exception amongst the CEECs in introducing the sub-option of land restitution in historical boundaries. For example, the same is true in Romania. 'When possible, claimants are given land in its original boundaries. When this is not feasible, a piece of equivalent size and quality is to be returned' (Brooks and Meurs, 1994).

Two questions which arise from the option approved in Bulgaria. First, what was the political economy of this decision? And second, was there an efficiency penalty paid for this choice, and if so, why?

The issue of the boundaries in which the land should be restituted showed clearly the policy differences towards land reform between the ex-Communists, the BSP, and the non-Communist coalition, the UDF. In 1991 it was under ex-Communist jurisdiction that the original land law was passed. In this law two options were provided for, namely that eligible owners could receive land back either in physical boundaries or, if they wish so, a bigger area is to be allocated to a group of people in which every owner has a share, but the physical location is not demarcated. The second option was to facilitate the creation of co-operatives through a simple re-registration of the former structures. This was the land reform path desired by the BSP. When the UDF came into power and amended the land law in 1992 it ruled out shared ownership. In this way it tried to facilitate the implementation of an exit clause from the co-operatives if owners wish to do so and to decrease the transaction costs of such exit. Ex-Communists came into power again in 1995 and proposed new amendments to the Land Law, one of which was to restore the possibility of shared-ownership. The Constitutional Court ruled out this amendment as a non-constitutional.

What is the importance of restitution in physically demarcated boundaries for this to be such a 'hot' political issue? In fact what it hides, is the political battle over the fate of the collective farms. Even with shared ownership in place, there are no legal obstacles for people to withdraw from the collectives with their land. But, practically, it becomes more difficult. The negotiating power of an individual against the interests of the collective is small. That is why the most probable outcome in cases where the co-operative management wish to discourage people withdrawing land from the co-op, is that the individual will be offered a parcel with a location or quality with which he will not be satisfied. In some countries there are provisions for a legal procedure in such cases, but this is a high cost option. (There are numerous examples in the Czech Republic which show the length and difficulties involves in such procedures). In Bulgaria, due to the relatively small size of individual ownership and a big portion of absentee owners, it is most probable that the transaction costs would have discouraged owners from withdrawing their land and would therefore guarantee the continuation of the collective as the main farming unit in each village. This suits the political interests of the former communists because it means a continuation of their influence. It cements the collective as the main employer in each village.

It is reasonable to ask if Bulgaria sacrificed a potential increase in efficiency due to the political adoption of only one option, restitution in physical boundaries? It is claimed here that Bulgaria might potentially gain efficiency in the long run by rejecting shared ownership. The latter creates more impediments to the development of land market, which is crucial for the efficient allocation of land in future and requires further procedures to delineate the ownership boundaries in order to allow a market to develop.

It could be claimed that purely from the point of view of creating efficient farm structures, there was a better option than the restitution in physical boundaries and it was proposed by an American team of experts led by Richard Ran very early in the reform process, in 1990. The substance of this option was to cut the relationship between the former and future legal ownership, thus to reject land restitution, and to sell agricultural land in auctions to the highest bidder in whatever sizes are demanded. The former owners would then be compensated in cash. This proposal did not rule out the participation of the former owners in the auctions if they wish to do so. No political party supported this proposal. The voter-politician model provides a simple explanation. This option would not satisfy the former owners who had legal proof of their ownership. They viewed land as a security against economic uncertainty, a role which would not be provided by cash compensation given the experience of high inflation and every expectation that it would continue. Thus politicians would have lost the support of the former land owners. On the other hand, politicians would have gained support of the people who would benefit from the land auctions. The latter group was expected to contain a much smaller number than the (approximately) five million Bulgarians who benefited from land restitution. The rejection of land auctions by all parties supports Swinnen (1997b) that pure efficiency considerations could not explain the political design of the reforms.

Another issue which is normally raised when the political economy of land reforms in the CEECs is discussed is the choice of date used as the legal basis for land restitution. On this matter, Bulgaria was no exception to the general rule for the CEECs. Some of the post-World War II land reforms were accepted as legitimate despite the fact that they took place under a strong communist and Soviet pressure. Bulgaria accepted as legitimate the land reform of 1946, and based its restitution law on the ownership status following the 1946 reform, but pre-collectivisation. This policy choice in Bulgaria was relatively easy without much confrontation between the ex-communists and the UDF and it has not been challenged in the subsequent proposals for amendments to the land law. In 1946 only the so called 'large land owners' had land expropriated, this happened for holdings of land above 30 ha in grain producing areas and 20 ha elsewhere. This expropriation affected mainly the Christian Orthodox Church, schools and a relatively small number of individual farmers. The nationalised land was about 300,000 ha and 120,000 ha of these were allocated to a similar number of land-less peasants (Syulemezov, 1976). Thus the political gains of the decision to acknowledge the legality of 1946 land reform were the support of all original owners or their heirs who benefited from 1946 land reform. The approval of the outcome of 1946 land reform also decreased the technical problems that would have arisen if claims had to be checked for a more distant historical date. The political loss could come not so much through the individual owners whose land was nationalised, because they were a small number, but only if the Church was an important actor in the political process. In Bulgaria the Church traditionally has not had a strong influence. Therefore it was clear that the political gains of accepting the outcome of 1946 land reform would be greater than the political loss, and thus there was not a big political struggle around this issue, although this factor was present in the debate.

Land reform implementation

If, from the point of view of political design the Bulgarian land reform seems radical and long-term sighted, the implementation process has been slow and contradictory. In fact even by late 1996 it was still far from complete. The factors contributing to this could be classified into three groups: technical, economic and political. The sheer size of the restitution process combined with insufficient budget funding for land surveys and mapping created a substantial impediment to its implementation. The requirement that all land should be restituted in physically delineated boundaries increased the technical problems in short to medium term.

At the same time there was little real pressure from land owners on the Government institutions to speed up the process. The lack of pressure was mainly due to economic factors. First of all land owners saw land mainly as wealth, offering shelter from inflation. For these reason only a small fraction of owners was keen to sell land and therefore the majority did not urgently need notarial deeds. To the extent that leasing is concerned, at a practical level land owners made use of two options which, even though they are not fully satisfactory from the point of view of legal security, gave them some returns on ownership. All potential owners have a certificate from the Municipal Land Commissions (MLCs) showing the land area, the quality of land and the approximate location of ownership, which is normally in the territory belonging to the village On the basis of these certificates, although the physical boundaries were not mapped, they provided the land to the local co-operative and received some rent. Other owners who were in regions where the land reform was at a more advanced stage, knew exactly where their ownership was located dispite that the whole legal procedure was not completed and they did not have deeds in land. They were able to lease out land not only to co-operatives but also to individuals through informal transactions. For owners who had the ambition to farm, a temporary arrangement was introduced. Land was distributed in larger parcels to groups of such people for so-called 'temporary use' and they shared the output according to agreed rules based mostly on their land shares. These latter were authenticated by certificates for temporary use issued by the MLCs. This option for temporary use was made even more acceptable by allocating land in the approximate locations corresponding to the ownership of individual farmers. These arrangements had the benefit of ensuring some income from farming to the recipients of the temporary use land, even though the informal nature of the arrangements may have created some disincentive to investment. However, a greater disincentive to farming was the decrease in real output prices, and the corresponding low profitability of agriculture during transition. This ensured that farmers did not have a strong incentive to hurry the implementation of restitution in order to start investing in the new business. Land owners seemed satisfied with the temporary arrangements because they expected that one day their ownership would be legally re-registered, and in the meantime they were able to capture rent or extract income from farming. As a result they did not exert strong political pressure for more active Government actions to complete the land reform. In reality, the individual decisions of many land owners to provide their land to the new co-operatives undermined, to a great extent, the radical nature of the land reform and its potential contribution to the development of the land market. In most cases as land owners 'joined' the new co-operatives they agreed to the condition that if they wish to withdraw from the co-op, they would receive exactly their own land only if there were no problems of access to the land in question. In other words the owner must be able to get to the land from public roads without having to cross co-op land. In all other cases they would receive an equivalent parcel to which there is a direct access. In this second case the transaction costs of withdrawing from the co-op will be the same as in the option of shared ownership.

This behaviour of land owners was well suited to the predominant political climate which for the most of the post-reform period in Bulgaria has been in the hands of parties related to the former Communists. These parties are in favour of collective forms of farming and therefore tend to down-play the role of clear and enforceable property rights. They delayed putting into place a Code for implementation of the land law, they defined complicated technical procedures with high requirements for precision in land delineation, and they made many changes in the Code in order to slow down the process of reforms.

In summary, reflecting on the first five years of the process of land reform in Bulgaria shows a radical political design, but an erosion of the radical features during implementation.

Political economy of farm restructuring

For the first five years of transition, farm restructuring has provided a field for intense political battle between the BSP and the UDF. The battle centred around the issue whether the re-registration of the former collective farms on the basis of the new legislation should take place before or after the restructuring of the farm and how deep the restructuring should be. In the original land law all collectives were given a deadline to re-register as new private co-operatives under the new co-operative law, or to register as a farming company. Physical allocation of the non-land assets amongst eligible owners was not required. The presumption was that each owner will have a defined share in these jointly owned assets on the basis of which he or she will receive dividends. When the UDF came to power and amended the law in 1992, it ruled out the simple re-registration, and provided for 'liquidation' of collective farms. What is the substance of this 'liquidation'? First the UDF removed the option of simple re-registration of the former collectives. What was required was a much deeper restructuring of the existing organisation before a new one could be registered. This deeper restructuring included two major elements, one concerning the non-land assets, and one the management.

The main requirement in this 'liquidation' was that non-land assets should be physically allocated amongst eligible owners prior to a re-registration of the farm under the new legislation. In fact, this had the same objective as land restitution in demarcated boundaries: to make it easy for people to obtain their assets and leave the collectives by decreasing the transaction costs of withdrawal.

The second element of the 'liquidation' concerned the change in management. The political decision in 1992, when the land law was amended during the UDF government, was to replace the former management with 'liquidation councils'. The 'liquidation councils' had two main functions, inherently inconsistent. They had to manage the activity of the farm, and thus to contribute to the survival of the organisation, and to allocate the non-land assets amongst the eligible owners, and in this way to contribute to the disappearance of the organisation. The replacement of the management by the 'liquidation councils' was valid only for the period up to the completion of the allocation of the non-land assets. Nothing has been decided for the management of the newly registered co-operatives once all non-land assets were distributed. The option that the management of the old collective could take over the running of the new co-operative was not ruled out. However, the UDF did not want to leave the responsibility for the deep restructuring to the former managers. Such restructuring would be adverse to the interests of these managers who wished to maintain the co-operative as the only agricultural unit in the village and thus to keep those employed in agriculture in a condition of strong economic and political dependence.

When the former Communists returned to Office in 1995 they changed once again the rules for liquidation of the collectives. They approved implicitly that the collectives had been transformed, and accepted the status which existed of a partial allocation of non-land assets as sufficient for re-registration under the new legislation. By virtue of the amendments to the land law in 1995, the liquidation councils were removed and replaced by three-member committees responsible for the completion of the allocation of the non-land assets. These committees were not vested with legal powers to collect debts owed to the co-operative or to represent the co-operative in the Court during disputes. Thus by definition they could not play the role of liquidators. Likewise, they were not supposed to manage the co-operatives. Since the 1995 amendments to the land law, the co-operatives have been managed by elected managers. In some cases they have been included in the three member committees, but sometimes not.

Given this battle raging, it is hardly surprising that the liquidation was not completed as politically designed. Some of the assets that were easy to distribute, e.g. livestock, were distributed quickly. The indivisible assets, e.g. buildings and machinery, needed sound procedures for their allocation. As such procedures were not in place, most of these assets were not distributed. In addition to the technical difficulties, the lack of motivation of the liquidators and of owners to follow the political decision was an impediment to the completion of the process. This lack of motivation was due to the strong influence of ideas from the left, in particular the wish to continue farming in co-operatives. It was also due to the risk averse behaviour of people engaged in agriculture. This is understandable given the huge economic uncertainty during transition and the chronic lack of capital, insufficient even to maintain non-land assets received through the process of liquidation. However, the strongest factor in our opinion has been the wish to preserve the co-operatives with the assets, necessary for farming.

As the new co-operatives began to register in 1995 the first requirement was for the members to vote a statute of the co-operative. At this stage almost all co-operatives in Bulgaria approved rules that brought to an end the physical allocation of the non-land assets. All eligible owners have a value share in the non-land assets on the basis of which they are paid dividends. In case of withdrawal from the co-operative they are eligible to receive cash, of value equal to their share, updated by inflation and depreciation of the assets.

Two questions usually raised about the Bulgarian 'liquidation' are whether Bulgaria was an extreme case amongst the CEECs in the political design of farm restructuring and whether the liquidation brought about a larger depreciation of agricultural capital than in other CEECs. It is beyond doubt that the transformation of the collective farms required in Bulgaria in order to register a new farm was much deeper than in the cases of re-registration in the Czech and Slovak Republics, or Hungary. However, the Bulgarian case is not the most extreme amongst the group of CEECs. In Albania collective farming was completely dismantled. The policy of UDF did not rule-out producer co-operatives. All post-reform political forces, including the UDF, foresaw that in future a large portion of farming would take place under a co-operative form of management. Indeed, co-operative ideas have been strongly rooted in Bulgarian society since the 1920s. What the UDF tried to achieve, was to increase the share of individual farming, facilitating the exit from the collective farm; to change the old management which was strongly related to the former Communists; and to have as members of the co-operatives, owners who knew exactly which were their assets. Their hope was that co-operative members in such circumstances would feel less vulnerable politically, than members who are fully dependent on the decisions of co-operatives. The latter could easily constitute a fertile ground for political pressure coming from the ex-Communists.

Concerning the disruption created by the 'liquidation' it is generally claimed that the political design of farm restructuring in Bulgaria was more disruptive and led to a greater depreciation of the existing capital than in the other CEECs. It is difficult to prove this claim due to the lack of market value of assets pre- and post-reform. Other indicators such as land and livestock productivity can help only indirectly to estimate the consequences of the liquidation process. Productivity in Bulgaria has exhibited the same downward tendency as most of the CEECs since reform. Thus it cannot be of great use to the specific issue of the capital depreciation. The rates of growth, or decline, of gross agricultural output are not very useful in answering the question raised. Many other factors (for example macro-economic conditions and price and trade policy) have had a profound effect on output. However, according to OECD data even these output trends have not shown substantially poorer performance in Bulgaria compared to the remaining CEECs (OECD, 1996). There is no doubt that even without the liquidation process Bulgarian agriculture would have shown slower recovery and development than in the 'good reformers' simply because the macro-economy has been one of the least stable amongst the CEECs and privatisation in the input supply and food processing sectors has been one of the least advanced. All this said, it still seems reasonable to claim that the process of liquidating the collectives brought about a larger depreciation of capital than would have been the case without this political decision. The implementation process ensured that neither the collectives who once managed the assets nor the individuals who might have managed them had an incentives to preserve the capital. The former because they were dismissed, the latter because they did not know exactly which assets would be their own.

Why was the land reform and farm restructuring policy design so radical?

It is beyond doubt that the political design of land reform and de-collectivisation in Bulgaria was radical, particularly in comparison with the Central European countries, such as the Czech and Slovak Republics and Hungary (Poland and Slovenia are not mentioned because they managed to avoid most of the collectivisation and therefore had much less de-collectivisation to achieve). What are the factors that have contributed to this? Swinnen (1995, 1997a) suggests two factors, the lack of political centre in the government ruling the radical reforms, and the lower security of the political reform. Here, in addition, three more factors will be discussed: demography of the rural areas; the party structure in rural areas particularly the influence of the peasants' parties; and the attitude to the Former Soviet Union (FSU).

In Bulgaria the BSP has strong support in rural areas. This situation induced the UDF to make sub-optimal decisions in order to undermine the basis for the BSP support.[4] This was important because of the significant proportion of the population engaged in, or dependent upon, agriculture. About one-fifth of those officially employed in the economy in Bulgaria is in agriculture, while in 1995 the share of agriculture in the employment was 4.7 percent in the Czech Republic, 7.1 percent in the Slovak Republic and 8.5 percent in Hungary (OECD, 1996). This could explain the fact why agricultural reforms had such a big weight in the political economy of the reform process in Bulgaria, but cannot say very much why the reform designed by the UDF was more radical than in the other three countries. One of the explanations of the latter is that the average age of the population in rural Bulgaria is high, which brings a more conservative mentality and a fear of large or rapid change. Table 2.8 shows that in 1992 when the UDF amended the land law and promoted radical reform decisions, 44 per cent of the rural population was above 50 years old, while the average for the country was 33 per cent. One third of the Bulgarian population lived in the rural areas, but over 40 per cent of the population between 50 and 69 years and more than 50 per cent of the population over 70 years old were rural dwellers.

Table 2.8
Distribution of population by age groups in Bulgaria, 1992

Age groups	*Total population number*	*%*	*Rural population number*	*%*	*Rural as percent of total population*
20 - 29	1,124,146	13	316,644	11	28
30 - 49	2,347,620	28	610,091	22	26
50 - 69	2,037,076	24	840,732	30	41
over 70	725,480	9	378,476	14	52
Total	8,484,863	100	2,780,040	100	33

Source: Statistical Yearbook of Bulgaria, 1994, NSI, Sofia

This age structure did not give strong expectations to the UDF government that it could change the behaviour of the rural population by a smooth reform providing for a simple re-registration of the collective farms. The UDF perceived that a more radical option was necessary in order to undermine the former collectives and to push the people engaged in agriculture to create a variety of forms of farming associations, and not just remain in the pre-existing situation of one producer co-operative in each village. Thus the UDF motive was to push (through legal means) the agricultural population to change the organisation of farming despite their innate conservatism and risk aversion related to some certain extent to their age.

The second factor creating strong electorate support for the BSP in rural areas is that the agricultural population does not view the peasant parties as supporters of their preference for security. The Bulgarian Peasant movement split into several parties, some in parliament some not. This deprived the rural population of a political alternative to the two main players, the BSP and the UDF. This is close to Swinnen's argument that the radical nature of the reforms was due to the lack of a political centre. But the explanation is more specific. It is the lack of a significant political force which is clearly related to the interests of farmers. That is why, within the polarised political model in Bulgaria, the rural areas with an aged and risk averse population have been strongholds of the BSP, because BSP policy has been closer to their preferences for minimal changes.

The third specific factor for Bulgaria, that does not apply in other Central European countries is the lack of a negative attitude towards the former Soviet Union, and particularly towards Russia. This stems from history since the second half of 19th century when the Bulgarians were liberated from the Ottoman Empire following the Russia-Turkey war. During the communist period Bulgaria did not have the experience seen in Hungary in 1956 and Czechoslovakia in 1968. Bulgarians therefore harbour no deep animosity towards Russia. Consequently, phenomena related to Russia, e.g. collectivisation of agriculture, do not create such strong adverse feelings as they do in some other parts of the region. This again helps to explain why socialist ideas of the BSP have not been undermined in the countryside post-reform.

In an attempt to erode the political power of the BSP in the countryside the UDF appear to have erred in their calculation of the political costs and benefits of their policies. This is particularly so with the policy of 'liquidation' the collective farms. By insisting on restitution in physical boundaries the UDF expected to gain the support of land owners. The presumption was that owners would like to be certain where their land was located because they treated it as wealth. On the other hand, eligible owners of non-land assets were not very interested in getting shares of assets, because they were often without real market value. Also most of them lived in the cities, and a big portion of the rural population did not intend to farm individually. Therefore the outcome of the liquidation planned by the UDF did not command much political support of the owners of non-land assets. At the same time it lost the support of a large group of urban consumers hesitating between the BSP and the UDF. This group believed, that the decreased supply and high price of food were directly related to the disruption created by the liquidation. Whatever its long term consequences for efficiency of land markets and farm structures, it could be claimed that through its decision to liquidate the collectives UDF incurred more political costs than benefits.

In conclusion, the land reform and the transformation of farming structures in Bulgaria were strongly politicised due to the overall importance of the sector in the economy, and particularly due to the large role of the rural population as voters. As a result, the political design of these processes was more radical than in some of the CEECs. However, implementation fell well short of the political design due mainly to the covert resistance of the political forces of the left and a mistaken calculation of voters preferences under conditions of high economic uncertainty and low profitability of agriculture.

Notes

1 The participants in this COST network agreed to work to a common set of definitions of farming and business structures. This is defined in Beckmann and Hagedorn (1997), this volume.
2 Whilst it is dangerous to apply the word 'new' to anything, it is used in order to stress that these co-operatives were created post-reform, because the collective structures that existed pre-reform are often also refer to in Bulgaria as co-operatives.
3 In the absence of a rental and purchase land market, there is little option except for the authorities to set what they consider to be a 'fair' price for land. It is inevitable that this will be contested as it has no market basis.
4 This argument is well developed in Swinnen (1997a).

References

Beckmann, V. and Hagedorn, K. (1997), this volume.

Bergmann, T. and Augur, T. (eds), (1989), *Co-operation in World Agriculture: Experiences, Problems and Perspectives,* Food and Agriculture Policy Research Centre: Tokyo.

Boyd, M. (1991), *Organization, Performance and System Choice: East European Agricultural Development*, Westview Press: Boulder.

Brooks, K., Guasch, L., Braverman, A. and Csaki, C. (1991), 'Agriculture and the Transition to the Market', *Journal of Economic Perspectives*, Vol. 5, No. 4, pp. 149-161.

Brooks, K. and Meurs, M. (1994), 'Romanian Land Reform: 1991-1993', *Comparative Economic Studies*, Vol. XXXVI, No. 2, pp. 17-33.

Co-operative Management and Administration (1988), International Labour Office: Geneva.

Information from the Department of Agrostatistics of the Ministry of Agriculture and Food Industries: Sofia.

OECD (1994), *Agricultural Policies, Markets and Trade in the Central and Eastern European Countries (CEECs), the New Independent States (NIS), Mongolia and China: Monitoring and Outlook*, Centre for Co-operation with the Economies in Transition, OECD: Paris.

OECD (1996), *Agricultural Policies, Markets and Trade in Transition Economies: Monitoring and Evaluation*, Centre for Co-operation with the Economies in Transition, OECD: Paris.

Oral Communication with the Chairman of the Union of Co-operatives in Bulgaria.

Paarlberg, P. (1992), 'The Evolving Farm Structure in Eastern Germany', International Agricultural Trade Research Consortium, Working Paper no 92-2, Purdue University.

Putterman, L. (1989), 'Agricultural Producer Co-operatives', in Bardhan, P. (ed.), *The Economic Theory of Agrarian Institutions*, Clarendon Press: Oxford.

Republic of Bulgaria, *Statistical Yearbook*, 1994, NSI: Sofia.

Syulemezov, S. (1976), *The Co-operative Movement in Bulgaria*, Sofia Press: Sofia.

Swinnen, J.F.M. (1995), 'Agrarian Reform Policies in Central Europe and the Baltics: A Political Economy Perspective', unpublished.

Swinnen, J.F.M. (1997a), 'The Choice of Privatization and Decollectivization Policies in Central and Eastern European Agriculture: Observations and Political Economy Hypotheses', in Swinnen, J.M.F. (ed.), *Political Economy of Agrarian Reform in Central and Eastern Europe*, Ashgate: Aldershot.

Swinnen, J.F.M. (1997b), 'On Liquidation Councils, Flying Troikas and Orsov Cooperatives: The Political Economy of Agricultural Reform in Bulgaria', in Swinnen, J.M.F. (ed.), *Political Economy of Agrarian Reform in Central and Eastern Europe*, Ashgate: Aldershot.

3 Changes in farming structures in the Czech Republic as a result of land reform and privatisation

Tomáš Ratinger and Ewa Rabinowicz

Introduction

The necessity to restructure agriculture goes hand-in-hand with the general transition process in the Czech Republic. From the outset there were both economic considerations and a moral dimension to restitution and privatisation of land which lay behind the transformation of the agrarian sector. Much agricultural land (39 percent) and assets were confiscated and almost all remaining area and assets were collectivised after World War II and in the early 1950s when the Communists took power. The moral principle of rectifying this injustice was the main legal base of the sector transformation. Act No. 229 of 1991 (Land Law)[1] covering restitution and relocation of property rights to land and other agricultural property was accompanied by the Collective Farm Transformation Law a year later. The principal objective of the transformation of collective farms into new co-operatives was to establish clear individual ownership relationships to all the property of these collective farms.

Land and property of members of collective farms were in most cases never (legally) expropriated, and therefore, the restoration of property rights to land could be more informal. Whilst the collective farm transformation law allowed co-operatives to continue in business, state farms were seriously disrupted by physical restitution of land and assets, or by the compensation of original owners and thus neither the owner (the state) nor managers had an interest to carry on the business. Remaining non-land assets engaged in the state farms were almost completely privatised under approved privatisation projects. State owned land was not included but was retained for later privatisation after restitution claims on the land were resolved. In the interim state land was transferred to the Land Fund of the Czech Republic, and leased to expected owners.

The restitution process led to the re-establishment of the fragmented ownership structure of the period before February 1948. It was commonly expected that small- and middle-scale farming mainly in the form of family farms would boom. Such farms, which dominate the farming in Western Europe, were widely believed to be more efficient. However, the large-scale operational structures remained in operation.

The main objective of this paper is to discuss the reasons why production co-operatives and large private firms still dominate the Czech farming system, and from this perspective, the extent to which the farm operational structures have reached long term equilibrium. The predominance of large-scale farming could be due to several reasons both political and economic. It can be legislation which is biased against small family farms or unbalanced subsidies giving unequal treatment to different farm organisations, or that large-scale farming is the most efficient way of creating and consolidating property rights of newly redistributed assets by the majority of owners in present circumstances. In this chapter, we argue for the latter, namely, that the main reasons for the outcome are economic and not political. More specific we raise three questions.

1 Was the objective of reform, which aimed at establishing clear individual ownership to most of the agricultural assets, achieved?

2 What political-economy factors structured the opportunity sets for recipient individuals and organisations?

3 How far is the resulting farm structure from a stable equilibrium?

In our analysis we are trying to follow the process of (re)distribution of rights to assets in agriculture in the Czech Republic and to identify the main determinants of this process starting with policy design and continuing in policy implementation.

Terminology

Concepts and definitions

Beyond legal ownership rights we are using Barzel's concept of property rights used also by Swinnen (1997) in the discussion of the choice for privatisation in Central and Eastern European countries (CEECs). The property right rests in the right, or the power, to consume, to obtain income from, and to alienate (Barzel, 1989). Applying Barzel's definition to collective farms (like all pre-reform co-operatives) we can observe that legal ownership was private but use and usufruct rights were public. Neither was alienation of property possible.

Although it is not necessarily valid generally, we suppose that in the democratic society in order to receive property rights it is necessary to get legal ownership rights.[2] This in turn requires an application for property rights.

The crucial point in Barzel's analysis is that assets have many attributes and that rights can be defined for these attributes. The costs of determining, capturing and retaining the attributes of an asset are defined as transaction costs.[3] The presence of transaction costs means asset attributes cannot be fully identified by prospective owners, and therefore the rights to assets are not fully delineated in attributes. In consequence, the characteristic of property rights of assets containing many attributes and transaction costs is twofold:

1 Divided property rights emerge, with two or more individuals having rights to distinct attributes of the same assets; and

2 The incomplete separation (separability) of property rights means that some attributes cannot be allocated to the property right of individuals (Fulton, 1995). Hence, some attributes remain part of the public domain (called by Barzel common property).

We have considered only two categories of ownership (in the property right sense): public and strictly private, this should be understood as predominantly public or predominantly private in attributes.

Privatisation is defined as a process of transferring legal ownership (which in a democratic society also implies a transfer of property rights) from a public form to a private form. It involves not only the transfer of ownership from state to individuals (or private organisations) but from any other (former) public form. Assets can be privatised using four different approaches: free distribution, sale of assets, usufruct divestment and restitution. Most commonly used in agriculture are restitution and sales.

Restitution implies that the property and legal rights are returned to former owners who were expropriated or in other ways forced to donate the assets to the state. The former owners must usually claim for their rights, and the authenticity of their claims have to be established.[4]

Sale (in a very general sense) is the most common form of transferring legal and property rights from the state to private organisations (including individuals) when the former owner does not exist or cannot be (or there is not an intention to be) relegitimated. Usually, the sale is neither random nor unconditional, but following certain schemes approved by governmental body. The methods used in privatisation of agricultural assets are, for example, direct sales (often an interest free loan) at accountancy (book) value of physical property (real estate, machinery, animals) with the extreme case of payment free transfer or sales of stocks/shares (including voucher method), or auctions.

The both concepts used above (sale and restitution) can be argued to have somewhat limited applicability in a situation when legal ownership and property rights are not overlapping as it was the case with state co-operatives (collective farms) which exercised property rights without reflecting legal ownership. In this case formal ownership does not have to be transferred. What is needed is a restoration of property rights to legal owners. We refer to this process as restructuring. Moreover, restructuring includes elements of restitution as well as redistribution of capital (common assets) created during the existence of the collective farm. The former is embedded in allocating corresponding non-land assets to those whose assets were (often involuntary) collectivised. The redistribution entails giving away the common assets according to various criteria. Since production factors – land, capital and labour – were responsible for generation of the common assets, legal ownership and value of these factors are usually a base for transformation procedures. There are large differences in the relative weight given to land, labour and capital in different countries, however.

Usually, assets are not free of burdens (debts, collateral) which are subject to conflict of interests. On the one hand, the state wishes to transfer as much as possible of these burdens to new owners, while the owners, of course, would prefer assets cleaned of such burdens. It evidently complicates the debate about privatisation procedures.

Distributional aspects of the privatisation - pressure groups and politicians

Assuming the indisputable right (enforceable claim) of humans to own assets, the obligation of democratic government to respect it is the early impulse for the property rights reform, although many authors (Begg, 1991; Brooks et al., 1991; Swinnen, 1994) would rather emphasise the economic efficiency objective behind the reform policies in CEECs. The former regimes ended with a mess of unclearly defined and mostly public forms of ownership of most of the national assets.[5] Only a minor proportion of assets was owned and governed (with significant restrictions) by individuals. Since there was a lack of experience how to (re)establish legal ownership and property rights to the assets on such a massive scale as well as uncertainty about the immediate effects, it is difficult to believe that any single factor like economic efficiency is the only engine of the ownership reform. Swinnen (1997) argues that the efficiency issue of privatisation cannot be separated from the distributional aspects of this process. We would like to go further saying that distributional aspects dominate economic efficiency in the property rights reform, while efficiency considerations become most important for the choice of business form, and hence, structural development. Moreover, we argue that the key factor is ideology in the property rights domain. Therefore, the main policy objective of the Czech land and agricultural property reform should be seen just in the establishment of clear independent individual ownership relationships to as large a proportion of national assets, following the idea that assets should be owned by individuals.

The decision or intention to redistribute property rights implies a set of options, but the policy is required to mark an appropriate way of adoption of only a few of them. Policy settlement and implementation is definitely influenced by the perceptions of opportunities by individuals. Individuals participate in political economy through organisations.[6] Individuals articulate preferences through markets and political and social transactions (Shaffer, 1980). Politicians (designers), reflecting aggregated preferences of individuals, define a set of rules which, in turn, defines the set of opportunities for individuals and organisations.

The main actors in this process are policy makers (also marked as designers in the text), interest groups (these are related to agriculture and include individuals, and old and new farming organisations with their members and former owners, also called clients) and the general public (voters).

Starting with the last mentioned group, it can be observed that the people who do not want to be active in agriculture have usually only a small interest to obtain additional agricultural assets (of doubtful value) and are more or less indifferent. A relatively small proportion (500 thousand in 1991 and 240 thousand in 1995) from more than three million relegitimated owners of land is active in agriculture. Therefore, the general public was less engaged in the debate about the future ownership and operational structure in agriculture than one would have expected at the outset. The general public may, however, be more concerned by effects of possible disruption caused by the process of privatisation on food availability as well as the general prosperity of the country.

Distributional conflicts are an important part of the politics of privatisation. Three groups are directly involved: (1) land owners, often organised in associations; (2) working members of former collective farms with small or no claims on restitution property[7] and (3) a small but powerful group of managers. The two latter groups may oppose withdrawal of the restituted assets and decollectivisation fearing to loose their position or livelihood. Even if there is an agreement on the basic principles, i.e. restitution and decollectivisation, two other important issues must be resolved. The first relates to allocation of the assets where legal ownership rights are missing or not recognised, the second to eligibility of business organisations in participation in various distribution schemes. There was not a consensus on either of these issues among reform minded politicians and the agricultural public. The distributional debate was dominated by two sorts of opinions.

1 *Preferential.* The power over assets should be distributed according to, or in proportion to, current (including restitution) legal rights (to land and non land assets) and/or the entrepreneurial ability of individuals.

2 *Equitable.* Assets should be distributed equally to those who are currently engaged (whilst also accepting those who have legal rights).

Not surprisingly, different interest groups may be expected to take their position with respect to these two opinions depending on the impact on the welfare of the group. The supporters of preferential distribution usually claim for discrimination of corporate successors to state and collective farms. Since they mostly want to farm independently, their interest is to get physical assets or cash. (The group has been represented by various land owners associations lobbying among conservative politicians.)

The supporters of more equitable distribution consist mainly of working members, wishing to continue as employees. Working members may, however, expect to get additional assets from the distribution anyhow.[8] Therefore, they are generally only weak supporters of equity: rather they concentrate on keeping farming in association and on social welfare.

Current managers, on the other hand, constitute a small subgroup with a strong voice and obvious interests to maintain their position within the organisation. This group tends usually to do well lobbying, emphasising the pragmatism of their proposals or requirements. (They were represented by the Agricultural Party which has already disappeared, and the Union of Co-operatives, which is still alive.)

Politicians are responsible for defining a feasible solution which balances the interests of competing pressure groups directly related to agriculture as well as the public (voters) which may be only indirectly affected. The objective of the sector policy is derived from the objective of national policy, but the specificity of the (agricultural) sector has to be reflected as well. It can easily happen that preferences or interests of pressure groups are fundamentally different from the preferences or interests of the public which necessarily leads to a conflict.

The political persuasion of the government can be expected to influence the attitude to privatisation and restructuring of agriculture. Left-of-centre governments tend to keep control of the food supply as a principal instrument of social policy. Large-scale industrialised farming might seem to them more viable and easier for exercising their policies. Right-of-centre governments can be anticipated to be more favourably inclined towards family farming. This may be due to several reasons. Such governments may be directly relying on the votes of land owners and be, moreover, more strongly committed to a market economy than socialist governments believing that family farming is more in line with a market oriented economy. Furthermore, re-emergence of family-farm agriculture lies also in the long-term interests of right-of-centre parties as it contributes to establishment of a solid non-socialist constituency in the country side.[9] The government may, however, approach the issue in a pragmatic rather than an ideological way aiming at a minimisation of distortion of production and degradation of the assets.

One important phenomena in land reform in CEECs is the national principle for restitution and privatisation (Swinnen, 1997). Probably the objective of the reformers to distribute the national property on behalf of current Czech citizens has its roots in historical justice and an underlying Czech nationalism. The exclusion of the former German land owners must be seen in a historic perspective where their expulsion was the last in a series of injustices on both sides during a thousand years of coexistence. Pragmatic considerations have played a role as well, however. The rights of the new Czech owners, who emerged after the German exodus and before nationalisation and collectivisation, would conflict with the rights of Germans.

Beside ideology, historical justice and simple nationalism, other reasons can be recognised for allocation of assets to individuals and only Czech citizens. If the only financial resources available for individuals emanate from wages over the last fifty years, a lack of own capital will necessarily lead to a heavy dependence on bank credits among small-scale businesses. Hence, privatisation may aim for a way to distribute national assets (capital) to make small businesses more economically viable and less dependent on the banking sector. Also the formation of Czech capital is vital for the economy to support national interests (in positive meaning). Some regulation for purchasing of the Czech assets and land was inevitable because the Czech currency has been most probably significantly undervalued.

Main legislation

The starting point for our analysis is an assumption of the ideological background of ownership reform in the Czech Republic. The emphasis on individual ownership and property rights has its roots in the rejection of the obfuscation of collective and people's property rights. Therefore, one of the major principles of economic and social reform is to establish individual property rights over as much as possible of nationalised or collectivised assets as quickly as possible.[10] Beside this principle, the agricultural reform policy was based on four postulates:

1 Legal ownership rights are the base for property rights;

2 Injustice must be undone which implies restitution of legal ownership rights to rightful (in the moral sense) owners;

3 Unclear forms of ownership and property rights (people's, collective) must disappear;

4 The state should not directly own or control production or factors of production in the economy. This implies that it is not in the interest of the state to own and to manage a large proportion of land and assets in agriculture.

Restitution

All restitution acts (for agriculture mainly the important Land Law, Act 229/91) restrict the eligibility to individuals who are Czech (Czechoslovak) citizens (originally only permanently dwelling in the country, but this restriction was eliminated by the Constitutional Court in 1995), and whose assets were expropriated after February 1948. The statutory conditions for the transfer of land operated by collective farms are defined in the Land Law as well. The three following aspects and rules are particularly important.

The current owner or user (state farm, later Land Fund or collective farm) approves the legitimate right of the claimant at first and takes responsibility for the registration and return of assets. Only if there is disagreement about the legitimacy or extent of the claim is a state body (like regional offices of the Ministry of Agriculture) or a court involved in the restitution procedure.

The current owner (obliged person) is obliged to give back the assets physically only if the eligible person is registered as a farmer or her rights are transferred to any farming organisation.

The eligible person is compensated if original assets cannot be returned (outlined in the Land Law and specified by the Decree of the Czech government from 4.12.1991).

Transformation of collective farms

Since the land is not a part of collective farm assets the transformation procedure focuses only on non-land assets. The original collectivised assets are allocated to the original owners or their heirs. The net asset value (representing collective assets) is calculated by subtracting from all assets all debts, the value of holdings in joint agricultural companies, investment liabilities, provision for outstanding property restitution and original collectivised assets. The net assets value is then distributed to two categories of eligible persons:

1 Members of co-operatives if they were members before 30 November 1991,

2 Owners of land and non-land property (with a value exceeding CZK 30,000) operated by the collective farm who applied before the end of April 1992.

A quarter of the net asset value in the form of shares is offered to eligible persons for sale at nominal value during a period of one month. What is not sold is returned to the share pool. The share pool is distributed to individuals under the criteria of 50 percent in proportion to the area originally put into the collective farm, 30 percent in proportion to assets originally brought and 20 percent in proportion to the number of years worked on the collective farm by eligible persons.

The distribution of shares – the most important part of the transformation project – must be approved by the assembly of eligible persons. The current management is responsible for the preparation of the transformation project. However, any eligible person can suggest their own transformation project. The transformation project must include a proposal of business forms of successors, and identify new organisations which will carry outstanding debts.

The individual recipient has the right of disposal of the shares. If the shareholder wants to farm independently from the new enterprise, the withdrawal of property has to be settled within a period of 90 days. If the withdrawal is for a non-agricultural purpose, the settlement of the property share could be delayed up to seven years i.e. till the end of 1999. The shareholder who becomes a member of the new co-operative has the right to withdraw assets or receive compensation in cash at the nominal value of the share in accordance with the internal rules governing the new entity, which are in turns subject to the Commercial Code. The rights of the shareholder of the farming company are defined by the Commercial Code.

Two further aspects are worth noting. First, there is an obvious disparity between the nominal value of the share and the market (real) value of the physical assets. Second, neither land nor other assets are collectively owned after transformation. Although the Czech statistical office continues to classify the ownership relationships in co-operatives as collective/co-operative, in fact such ownership does not exist any more.

Privatisation of state assets

In principle, state assets engaged in agriculture could be privatised in the frame of the large-scale privatisation (including the voucher method). However, due to the very large number and small scale of restitution claims only a few state farms were privatised under the voucher scheme. The dominating form of privatisation of non-land property was based on individual privatisation projects elaborated by managers of state farms or those who were interested in getting state farm assets. The total book value of assets in question was first reduced by the estimated compensation to restituents and further reduced by the assets inseparable from the land. Then the assets were transferred to new owners under the scheme approved by the Ministry of Privatisation. Most of the remaining state land and assets inseparable from the land were gathered under the supervision of a special state body – the Land Fund. Half a million hectares of state land is proposed by the government to be privatised (sold) between 1997 and 2000. No procedure (including legal framework) has been specified as yet.

The Land Fund of the Czech Republic was established in 1991 (Law No. 569) to watch over the land and other property before these are privatised. The Land Fund is not responsible for the approval of a privatisation project, but it represents the state in contracts for the return of property (restitution), leasing or selling. The Land Fund in practice has leased most of the property awaiting privatisation and in so doing has relieved some of the pressure for those who wish to continue farming activities.

Implementation of the policy and other policies affecting privatisation

Most senior politicians avoided getting closely engaged in the battle of lobbies and therefore the outcome of the land reform policy design was finally very pragmatic. The distribution and reform process were designed in such a way as to minimise degradation of the value of non-land assets and in certain respect of land as well. In consequence, the preference was given to legitimate owners, but decollectivisation[11] was avoided and all business forms, except state farms, were equally supported (legally, institutionally and administratively) by government. Generally, government followed the principle of non-discrimination of any business recognised by the Commercial Code and other Laws defining operational options for agriculture. It could be argued, however, that government should have instituted ‘positive discrimination’ in favour of individual farmers and that such a policy would in truth be non-discriminatory since it would compensate for past discrimination. To some extent, the Czech government has supported individual farmers through both policy design and implementation. Sole farmers were supported in four ways:

1 By the distribution of collective farm property: 50 percent according to land, 30 percent according to original assets, and only 20 percent according to years working in the collective farm, (see also below). In this way farmers who were forced to participate in the process of collectivisation were restituted not only for the resources they lost but were also allocated a lions share of assets generated in the socialist period, which can be accounted as obvious support to sole farming;[12]

2 By obliging current users of farm capital to transfer required assets to farming recipient within 2 (3) months;

3 By the large investment support which was in place during the first three years when the Czech government offered interest free loans and no repayment loans aiming to stimulate private activities in the sector. The major criteria for giving away investment support was the recipient's registration as a sole farmer. Less attention was paid to his business plan and ability to survive. Because of the low transparency and efficiency of this system, the government switched to a guarantee and interest subsidy scheme supervised by the Support and Guarantee Fund for Agriculture and Forestry (SGFAF). This scheme includes three programmes linked to privatisation, support to starting young farmers, support to buyers of state property and above mentioned support to restitution/return of assets. The access to all other programmes of SGFAF is limited to those organisations which have no obligation to farming eligible persons (restituents from Land Law and claimants from Transformation Law) but no specific type of business organisations is discriminated against or excluded;

4 By preventing sole proprietorships to go bankrupt until the end of 1994 (Fiser, 1993).

In some respect, the exemption of state organisations from bankruptcy in 1992 (Fiser, 1993) should also be included. The early issue of the bankruptcy law conflicted with the privatisation intention (state interest) and interests of creditors, since bankruptcy stopped the privatisation process. As a consequence, the state farms were effectively protected against bankruptcy in the most important period of their privatisation, 1992-1993. It may appear to be inconsistent with the aim of the establishment of sole farms, but the prevention of bankruptcies of the state organisation enabled the value of agricultural assets to be kept up. This was the main resource for restitution. For some reasons, which are not clear enough to the authors, government did not want to pay financial compensations to restituents, although it repaid the debts of state farms to banks later on.

Beside the support to sole farmers government has introduced a scheme for non-farming restituents by offering them shares of the Restitution Investment Fund (RIF) which when sold might help to satisfy their desire for financial compensation (cash). Farming successors of collective farms, carrying obligations to return property to those who are out of agriculture, and have not accepted a membership in the new organisation, are protected temporarily. From 1999 the situation will be the opposite, however, and such shares on the property must be compensated (see the next section on the transformation of collective farms).

The transition procedures were implemented by current managers of state and collective farms, who were responsible for the success of the reform. Giving managers this role should in principle assist the speedy completion of the reform.[13] Nevertheless, the opportunity was given to every one interested in privatised (transformed) assets to come with his own privatisation project. It assured that the best solution could be chosen.

However, timing (one year for transformation of collective farms and three years for privatisation of state farms) and divisibility of physical assets often conflicted with the interests of clients. That conflict is characterised by the schizophrenia of managers obliged to give out assets of claimants wishing to farm independently, to protect the interests of the rest of owners, and to continue the production process; and by the disappointment of independent farmers waiting for needed equipment (or cash) for running their own business, on the other hand.

The government remained rather neutral in this conflict. The government measures consisted only of a threat of imposition of a personal penalty on managers in the case of misconduct, and a restriction on the access of their organisations to state (credit) support until all claims of farming eligible persons are resolved. Later, the government introduced a special support programme to organisations obliged to return assets.[14]

Looking at the politics of implementation of the land reform it can be concluded that the pragmatic approach dominated. If policy was discriminatory it was in favour of family farms rather than otherwise. Nevertheless, advocates of independent family farms often express their disappointment at the slow pace of change in collectivised or co-operative farm structures. This is claimed to result from the lack of willingness of managers of co-operatives to give back land and assets and the low degree of governmental concern about this issue.

Implementation of the transformation of collective and state farms

After the shares of the collective farm assets have been settled or the property of state farms has been offered for privatisation or restituted, the interest of individuals turns from property acquisition to the way their assets will be utilised. The concerns of four pairs of groups are pictured in Table 3.1.

It can be observed that all participants in each pair of groups (shown as one row in the table) have similar interests, despite the former business organisation. The interests can be summarised as follows. Group 1 wish to get agricultural assets. Group 2 focus on keeping their job and maintaining their welfare (to use their assets in a way which will allow them to participate as employees). Group 3 wish to maintain the value of capital (to participate in such an organisation which will appreciate (or not degrade) the value of capital and land). Group 4 will focus on getting financial compensation (sell the assets or shares at least at nominal value). Clearly, the first decision an individual has to take is the choice between farming in association or farming independently or staying outside agriculture.

Transformation of collective farms

There were two options for this transformation:

1 To register the old collective farm as a new co-operative or farming company and to carry all the old debts including obligation to return assets to farming non-members and to compensate non-farming non-members, or

2 To create a new farming corporation and to transfer only the assets according to member shares into the collective farm and to put the old collective farm carrying all debts into liquidation.

The first option is the simpler and almost always feasible, but the successor organisation will be threatened by substantial capital outflow, and will be in a disadvantageous position for state support and therefore, less stable. (Only such farming entities, which are free from obligations to eligible persons are entitled to get state support such as direct production subsidies, direct payments or credit guarantees.) It is evident that the key role in the decision making was played by the managers of original collective farms. They had more information about the feasibility of the second option and could encourage owners to form a more viable organisation. Lack of skill, the vested personal interests of managers and time stress led often to the choice of the simpler first option, although presumably the second option could be more suitable for owners wishing to participate/undertake in agriculture in association. About 90 percent of collective farm assets were transferred to production co-operatives. Since only one-fifth of the original collective farms has been liquidated, we estimate that about 80 percent of them went along with the first option. The rest following the second option, chose half and half, co-operatives and farming companies. Tradition and familiarity with the name (co-operative) were certainly important factors in the decision of business organisations.

Table 3.1
Various actors involved in the transformation of collective and state farms

	Collective farms	*State farms*
1	Owners considering starting sole farming	Farming restituents
2	Farming owners considering to continue in association (working shareholders)	Agricultural workers and managers
3	Owners considering to leave the share in agricultural businesses	Investors outside agriculture
4	Non-farming owners waiting for financial compensation	Non-farming restituents

Privatisation of state farms

A decision problem facing the person who is going to buy state farm assets or receive them via restitution, is different from that of the former member of the collective farm, since the original organisation (actually, any successive organisation) does not exist any more. Almost one-quarter of the state farm assets (determined for privatisation) were acquired by sole farmers, but more than a half by farming companies with a predominance of limited liability companies. Certainly, at least the following factors have played a role in the decision of land and assets acquisition:

1 The availability of land and assets which were for sale;

2 The possibility of cheap leasing of land and assets (in the early stage);

3 The direct sale of state assets through interest free loans (occasionally with delayed repayments);

4 The use of voucher privatisation bids.[15]

The original intention of the government was to privatise state farms in the framework of large-scale privatisation, which certainly reduced individual options to a purchase of shares only. Because the restitution process turned out slower than expected, an alternative (specific for agriculture) scheme was eventually followed. This scheme has some resemblance with the small-scale privatisation, but several important differences too. The government did not attempt to raise prices of assets through auctions. Instead, after initial leasing, direct sales were preferred (often for interest free delayed repayments) at book value to the current operator submitting a privatisation project. The implication of the scheme applied was that the government paid-off the need for an accumulation of capital. Thus, the individual choice of business organisation depended to a large extent on two things. First, the entrepreneurial ability of the individual, who could get a sufficiently large piece of land matters. Second, the willingness to expose personal property (such as a house) to risky business in the depressed agricultural market.

The creation of a limited liability company protects the personal assets of entrepreneurs (shareholders or members). The minimum of total equity is determined by the Commercial Code at only CZK 100,000, with the individual minimum of CZK 20,000. There are sums money certainly affordable for an average Czech citizen. Hence, just five partners can easy found a secure business (in the sense of personal property) with a turnover of millions. Beside the advantage of low minimum equity (comparing to CZK 1 million requirement for joint stock companies), limited liability farms can be simple from the organisational and managerial perspective. Also, limited liability companies are considered to be more private, while joint stock companies and co-operatives are more public.[16] On the other hand, low equity organisations have already faced problems in getting bank credit necessary for the purchase of state assets because of the lack of a guarantee. But since 1994 such guarantees can be obtained from the support and guarantee fund (SGFAF).

Joint stock companies have emerged usually as an outcome of the voucher privatisation scheme. There is not direct relationship between the interests of owners and the business organisation created. Unless the shareholders join an investment fund their influence on the company performance and management is negligible. Therefore the future of the company is often in the hands of hired (old) managers, who are only partly controlled and whose interests are not necessarily congruent with the interests of individual shareholders.

Outcome of the process - Structural changes

Land ownership

In 1989 25.4 percent of the total agricultural land (TAL) was in state farms, 61 percent in collective farms (state co-operatives). Only 0.3 percent was in hands of family farms. By 1995 share of state farms has fallen to 1.2 percent of TAL while the share of family farms has risen to 22.3 percent (see Table 3.2 and 3.3).

Table 3.2
Ownership/operation farming structure development in the Czech Republic, 1989-1995, in hectares

Operation structure	*No.*	*Total area*	*Average size*	*Number of employees*	*Total labour*	*Men per hectare*
1989						
State farms	174	1,089,414	6,261	127,865	127,865	0.12
State co-operative	1,024	2,622,464	2,561	403,192	403,192	0.15
Private co-operative	-	-		-	-	
Farming company	-	-		-	-	
Partnership	-	-		-	-	
Family farm	3,205	12,820	4		2,000	0.16
Other		571,302				
Total	4,403	4,296,000	846	531,057	533,057	0.12
1991						
State farms	308	1,095,864	3,558	94,967	94,967	0.09
State co-operative	1,187	2,600,717	2,191	307,929	307,929	0.12
Private co-operative	-	-		-	-	
Farming company	-	-		-	-	
Partnership	-	-		-	-	
Family farm	14,000	140,000	10	3,815	8,015	0.06
Other		448,419				
Total	15,495	4,285,000	248	406,711	410,911	0.10

continued next page

Table 3.2 (continued)

Operation structure	*No.*	*Total area*	*Average size*	*Number of employees*	*Total labour*	*Men per hectare*
1993						
State farms	240	552,000	2,300	25,157	25,157	0.05
Private co-operative	1,334	2,117,058	1,587	18,496	18,496	0.09
State co-operative	-	-		-	-	
Farming company	1,088	623,424	573	51,744	51,744	0.08
Family farm	52,003	572,033	11	8,493	17,807	0.03
Other		418,485				
Total	54,665	4,283,000	71	270,359	279,673	0.07
1994						
State farms	239	119,022	498	5,300	5,300	0.04
Private co-operative	1,468	2,043,470	1,392	14,645	14,645	0.07
State co-operative	-	-		-	-	
Farming company	1,461	876,600	600	62,195	62,195	0.07
Partnership	47	14,570	310	697	697	0.05
Family farm	60,666	871,669	14	13,360	31,217	0.04
Other		355,669				
Total	63,881	4,281,000	61	228,002	245,859	0.06
1995						
State farms	87	52,797	607	9,817	9,817	0.19
State co-operative	-	-		-	-	
Private co-operative	1,266	1,832,101	1,447	12,828	12,828	0.07
Farming company	1,573	1,070,819	681	81,663	81,663	0.08
Partnership	33	13,509	409	381	414	0.03
Family farm	66,201	955,682	14	14,258	41,021	0.04
Other		355,092				
Total	69,160	4,280,000	57	234,408	261,204	0.06

Source: MA CR, 1994, 1995, 1996, CSO, 1996

Table 3.3
Ownership/operation farming structure development in the Czech Republic, 1989-1995, percentage in columns

Operation structure	*Total area*	*Number of employees*	*Total labour*
1989			
State farms	25.4	24.1	24.0
State co-operative		75.9	75.6
Private co-operative			
Farming company			
Partnership			
Family farm	0.3		0.4
Other	13.3		
Total	100.0	100.0	100.0
1991			
State farms	25.6	23.3	23.1
State co-operative	60.7	75.7	74.9
Private co-operative			
Farming company			
Partnership			
Family farm	3.3	0.9	2.0
Other	10.5		
Total	100.0	100.0	100.0
1993			
State farms	12.9	9.3	9.0
State co-operative			
Private co-operative	49.4	68.4	66.1
Farming company	14.6	19.1	18.5
Partnership			
Family farm	13.4	3.1	6.4
Other	9.8		
Total	100.0	100.0	100.0

continued next page

Table 3.3 (continued)

Operation structure	*Total area*	*Number of employees*	*Total labour*
1994			
State farms	2.8	2.3	2.2
Private co-operative	47.4	64.2	59.6
Farming company	20.5	27.3	25.3
Partnership	0.3	0.3	0.3
Family farm	20.4	5.9	12.7
Other	8.3		
Total	100.0	100.0	100.0
1995			
State farms	1.2	4.2	3.8
Private co-operative	42.8	54.7	49.1
Farming company	20.5	34.8	31.3
Partnership	0.3	0.2	0.2
Family farm	22.3	6.1	15.7
Other	8.3		
Total	100.0	100.0	100.0

Source: Own calculations

In Table 3.4 the changes in ownership and operation structure are given. In 1989 39.4 percent of TAL was publicly owned and operated (this figure includes land operated by state farms and by other state organisations including non-agricultural enterprises). Less than 6 percent was operated by private owners themselves. The largest category (54.9 percent) consisted of privately owned but collectively operated land. By 1995 the land owned and operated by public has shrunk to 4.7 percent or to 202,220 hectares. (2.7 percent of TAL in the state organisations and a residual of 2 percent of TAL is in the possession and operation of municipalities, and public associations for which ownership and utilisation are not well surveyed statistically). In 1995 most of land (78.8 percent) was rented to private organisations. More than a quarter of this, 937,293 hectares, consists of state land subjected to privatisation, which was transferred to the Land Fund and leased to expected future owners: restituents, family farms and farming companies, with the companies being the largest leasers. By 1995 68.1 percent of state land was rented to farming companies (see Table 3.5). The total land rented out by the Land Fund accounted for about 18 percent of TAL in 1993 and 22 percent of TAL in 1995 (Table 3.5 and 3.6).

Table 3.4
Ownership/operation farming structure development in the Czech Republic, 1989-1995, percentage in rows

Year	*Publicly owned and operated*	*Private but collectivised*	*Privately owned by operator*	*Rented to private organisations*	*Total area*
1989	39.4	54.9	5.6		100.0
1991	36.3	55.1	8.6		100.0
1993	17.3		13.3	69.4	100.0
1994	6.3		16.1	77.6	100.0
1995	4.7		16.5	78.8	100.0

Source: Own calculations

Table 3.5
Land leasing scheme of the Land Fund (1995)

	Hectares	*Share in Land Fund area*	*Share in business area*	*Share in business rented area*
State land in Land Fund	937,293			
rented to restituents	53,836	5.7%		
rented to family farms	245,264	26.2%	25.7%	41.4%
rented to farming companies	638,193	68.1%	59.6%	62.7%

Source: MA CR, 1996 and own calculations

Most of the restituted land was not physically taken by owners, but directly transferred for renting to successor organisations. In this way the 89 percent of land used by the former collective farms was transferred to new co-operatives between 1992 and 1993. All parcels are described and registered (computerised). They are also somehow identified on cadastral maps of varying quality. The main problem is with the practical implementation at the local level. All owners have the right and possibility to hire a private surveyor who will identify the parcel physically. Another problem is validation of the result of the survey because of bad quality of maps. Nevertheless, 700 thousand hectares of agricultural land (16.4 percent of TAL) was physically returned to former owners by the end of 1994 including land reclaimed from new co-operatives. Three quarters of this land was in small plots of less than 20 hectares (Table 3.5).

Table 3.6
Land distribution 1994 (summary)

	Area in hectare	*Percentage of TAL*
Reclaimed land		
under 2 hectares	88,853	2.1%
2 to 5	100,170	2.3%
5 to 20	344,881	8.1%
20 to 50	93,504	2.2%
over 50 hectares	72,626	1.7%
Subtotal	700,034	16.4%
Land Fund	767,675	17.9%
Direct transfer to farming companies and partnerships, rented	623,790	14.6%
Subtotal	2,091,499	48.9%
Private co-operatives	2,043,470	47.7%
State farms	114,042	2.7%
Other	31,989	0.7%
Total	4,281,000	100.0%

Source: MA CR (1994, 1995), own calculation and estimates

Three million hectares of agricultural land was returned to three million owners. Thus the average size of this ownership is round a 1 hectare. Almost one quarter of TAL is still owned by the state. Consequently, land is almost four fifths rented and not used by owner alone.

Operation structure

Most collective farms were turned to new private production co-operatives in 1992 (until 28 January 1993). Land owners, whether they became co-operative members or not, usually leased the land and assets to new co-operatives. At that time only 11 percent of former collectivised land was withdrawn from the use of new co-ops. This land was used by succeeding farming joint stock companies, limited liability companies and sole farmers (family farms). Since land continued to be withdrawn from the co-ops, consequently the area used by co-operatives dropped from 53 percent of TAL in January 1993 to 43 percent of TAL in 1995.

Most of state farms terminated their agricultural activities by the end of 1994. Some of them ceased to exist at all (160 from 308 original state farms, which existed in 1991) were liquidated by the end of 1995). An area of more than one million hectares used by state farms in 1989 dropped to only 52,797 hectares (1.2 percent of TAL) in 1995 (see Table 3.2 and 3.3). The total book value of non-land assets of state farms had amounted to CZK 48 billion (ECU 1.4 billion) before restitution and privatisation started. From this value, 44 percent was used to compensate original owners, 17 percent was inseparable from the land and will thus be privatised jointly with the land. The remaining assets (in value of CZK 21 billion, ECU) together with non-land properties of other state enterprises operating on the land (in total CZK 32 billion, ECU 0.68 billion) was transferred to the Land Fund and determined for privatisation (while the total value of state property linked to agriculture allocated for privatisation amounted to CZK 50 billion).

Farming joint stock companies, which appeared as an alternative result of transformation of collective farms, joint agricultural enterprises of collective farms or when state farms were privatised in the framework of large-scale privatisation (voucher method) received 19 percent of the state agricultural assets allocated for privatisation (SAP) in the Land Fund. So far, farming joint stock companies are a minor type of farm business. Those companies operate less than one third of the land which is used by farming companies which amounted to 25 percent TAL in 1995 (see Table 3.3) i.e. 7 percent of TAL.

Limited liability firms seem to have become a pervasive business form in agriculture. They covered about 18 percent of agricultural land in 1995. As mentioned in the previous paragraphs they originated predominantly from the process of privatisation of state farming enterprises, acquiring 35.4 percent of the state agricultural assets allocated for privatisation (SAP).

Nearly the same area, which constitutes 22.6 percent of TAL is utilised by sole proprietorships (family farms) and partnerships. Family farms in this context involve also part-time farming. Family farms and partnerships acquired 25 percent of SAP. More than one fifth of the TAL and less than one fifth of the Total Agricultural Capital (TAC) is utilised by sole farmers, the rest is managed in a corporate way.

Most of the non-agricultural activities separated from the collective farms quickly.[17] While shifts in land ownership and farm operational structures are consequences of legal changes and the administrative process, the structure of production has mostly been influenced by depressed markets for agricultural products. From 1989 to 1994, total gross agricultural output (in 1989 prices) contracted by 28 percent, (animal production by 32 percent and crop production by 21 percent) in real terms. The share of agriculture on GDP decreased from 6 percent in 1990 to 3 percent in 1994. The ratio between crop and animal gross output shifted from 41:59 to 45:55. However, 1995 seems to be a new dawn for the agricultural sector, as overall economic growth stimulated better sector performance with a 4.2 percent increase in gross agricultural output and a 10 percent rise in gross agricultural product. This improvement was projected to the shift from noticeable losses in the first four transitional years to a slight profit in 1995 (see Table 3.7).

Table 3.7
Profit/loss figures per hectare of TAL in nominal values

Operational form	*1994*		*1995*	
	Profit	*Profit + wages*	*Profit*	*Profit + wages*
Private co-operatives	–809	5,955	–8	7,706
Farming comapanies	–618	5,331	340	7,231
Sole farming proprietorships	140	1,459	1,106	2,758

The markets for animal products (milk, beef, pork, poultry) and sugar beet were affected most seriously by the price adjustment, while the market for cereals was touched only slightly. On the other hand, the oilseeds market has expanded since the beginning of the reform. The changed market situation caused a radical reduction of cattle herds and numbers of other animals. By the spring of 1995 the number of cattle fell about 40 percent compared to 1989, pigs and poultry about 18 percent. The intensity (use of variable inputs, and consequently yields) of all kind of production dropped as the new price relationships encouraged a lower input strategy during the transition period (1990-1994).

Despite these dramatic changes, the Czech Republic remains self-sufficient in basic foods. More than half of the national herd of cattle is found in co-operatives, but almost all the beef and milk production of state farms was transferred to private (non co-operative) hands. Pigs and poultry moved out of the co-operatives to a large extent. While milk and beef production has mostly been loss generating, intensive pig and poultry production has been profitable. Pig and poultry units with a high concentration of animals (usually organised as joint agricultural enterprises of collective farms) suffered from the bad performance of many former or new co-operatives. Since pig and poultry production is technologically independent of other agricultural production, the effort of managers of these units has been to separate them from co-ops. By 1995 62 percent of pigs and 80 percent of poultry were allocated in farming companies, partnerships and family farms (see Table 3.8).

Table 3.8
Ownership of animals (1989-1995)

	Cattle	*Cows*	*Pigs*	*Poultry*
1989				
State farms	30.0	30.6	22.1	29.2
Collective farms	67.2	68.0	71.4	49.1
Family farms, self supply	1.7	1.1	5.0	18.8
Other	1.0	0.4	1.5	2.9
1995				
State farms	1.4	1.2	1.0	3.1
State co-operative				
Private co-operative	53.7	53.3	36.6	16.8
Farming company	24.7	25.4	41.6	80.1
Partnership	0.2	0.2	0.0	*
Family farm	16.4	16.9	10.0	*
Other	3.5	3.0	10.7	*

* Included in farming companies.

Source: CSO, 1990; CSO, 1995; CSO, 1996

The degree of specialisation of agricultural producers was low in the pre-reform period. State farms were a bit more specialised[18] with 12 percent in crop production and 17 percent in animal production, while 96 percent of collective farms had mixed production (see Table 3.9). This phenomenon is persisting in post-reform farming. The most specialised among agricultural enterprises are farming joint-stock companies with 30 percent of farms oriented to animal production by 1994. These are mainly intensive pig and poultry units of former state farms or collective farms. Part-time farming is also partly specialised, equally (19 percent) in crop and animal production. Full-time family farms and private co-operatives adopted a mainly mixed production pattern. Diversified production may follow from the technological tradition, for locational reasons and through risk aversion of entrepreneurs, helping them to increase their resistance to market instability.

Table 3.9
Degree of specialisation of farms (1989-1994), in %*

Degree of specialisation	*Crops*	*Animal*	*Mixed*
1989			
State farms	11.5	17.2	71.3
State co-operative	2.0	2.0	96.0
Joint stock companies			
Limited liability companies			
Sole proprietorships and partnerships			
1993			
State farms	12.5	14.6	72.9
Private co-ops	3.8	4.8	91.4
Joint stock companies	3.6	40.2	56.2
Limited liability companies	9.3	10.1	80.6
Sole proprietorship and partnerships	9.8	10.2	80.0
1994			
State farms	10.5	14.0	75.5
Private co-ops	5.9	5.2	88.9
Joint stock companies	5.1	30.1	64.8
Limited liability companies	8.1	9.0	82.9
Sole proprietorship and partnerships	8.6	17.8	73.6
from it part-time farming	19.0	19.8	61.2

*** A farm is classified as specialised in crop (animal) production if more than 66% of output originates from crop (animal) production; other farms are classified as mixed.**

Source: MA CR, 1994, 1995

Scale

Since the beginning of agrarian reform the average size of farm has decreased significantly. This is most evident in state farms associated with their liquidation. The average size of a state farm contracted from 6,261 hectares in 1989 to 3,858 hectares after some splitting, mainly for managerial reasons, in 1991. In 1994 there were 229 state farms averaging 498 hectares. The majority of them faced liquidation. According to the recent Agricultural Census there were 87 surviving fully operating state farms (including school farms) with an average size of 607 hectare in 1995 (see Table 3.2).

As transformation of collective farms began in 1991 there were 1,187 farms operating on average 2,191 hectares. Over four-fifths (84.6 percent) of the collective farming area had been covered with farms larger than 1,500 hectares. At the end of compulsory transformation only 59 percent of the original area was occupied with co-operatives extending beyond 1,500 hectares. The process of withdrawal of land and assets by owners and splitting of farms has continued; by 1995 the average size of succeeding private co-operatives fell to 1,447 hectares (see Table 3.2).

The majority of farming companies are derived from the former large state and collective farms. By Western European standards they are large scale farms; in 1994 their average sizes was over 600 hectares (see Table 3.2).

The average size of family operated farm grew from 4 hectares in 1989 to 14 hectares in 1995 (see Table 3.2 and 3.10). By 1995 eighty percent of sole proprietorship were smaller than 10 hectares, of an average size of 3.8 hectare and covered only 21.8 percent of farming land occupied by family farms or 4.9 percent of TAL (see Table 3.10). On the other hand, 83 family farms with more than 1,000 hectares per farm occupied 13.7 percent of farming land used by family farms or 3.1 percent of TAL. Farms over 100 hectares amounted to 1,379 and occupy 11.4 percent of TAL. Comparing figures for 1993 and 1995 in Table 3.8 it can be observed that the land operated by sole farmers increased considerably (from 13.3 to 22.3 percent of TAL). The total number of sole farmers increased from 52,003 to 66,200 during the same period. The largest relative increase can be observed for farms between 100 and 1,000 hectares, which increased by 33 percent in numbers, followed by small farms, less than 10 hectares, which increased by 30 percent. By comparison, farms between 51 and 100 hectares became only 10 percent more in numbers. The land used by farms between 100 and 1000 hectares increased by 72 percent.

Factor mix

To this point we have followed mainly the changes in the ownership and use of land. During the transition period the use and composition of factors of production have significantly changed. In spite of the structural changes and the decline of output, more than 99 percent of total agricultural land remained in agricultural production.

Table 3.10
Area structure of family farms in 1993 and 1995

Size in hectares	*Number*	*Area*	*Average size*	*Proportion in sole farmer area*	*Proportion in TAL*
1993					
Less than 10	42,689	71,015	1.7	12.4%	1.7%
10 to 50	7,034	127,035	18.1	22.2%	3.0%
51 to 100	1,222	65,327	53.5	11.4%	1.5%
100 to 1000	977	207,125	212.0	36.2%	4.8%
more than 1000	81	101,532	1,253.5	17.7%	2.4%
Total sole farmers	52,003	572,033	11.0	100.0%	13.3%
TAL*		4,287,000			
1995					
Less than 10	55,491	208,264	3.8	21.8%	4.9%
10 to 50	7,985	168,300	21.1	17.6%	3.9%
51 to 100	1,345	91,778	68.2	9.6%	2.1%
100 to 1000	1,296	356,574	275.1	37.3%	8.3%
more than 1000	83	130,766	1,575.5	13.7%	3.1%
Total sole farmers	66,200	955,682	14.4	100.0%	22.3%
TAL*		4,280,000			

* TAL = Total agricultural land.

Source: MA CR, 1994, 1996 and own calculations

Because fixed assets and other capital has been exposed to restitution, privatisation and transformation of collective farms, the mobility of capital has been restricted by the law for the transition period. Pre-reform capital was frozen in the agrarian sector (mainly between 1991 and 1993) and investment dropped 70 percent (1989/1994).

Three million owners of land plus some others acquired nearly CZK 200 billion of state and collective farm assets, from it round CZK 120 billion of fixed assets. Own capital of agricultural organisations is under the 40 percent level. This is because a large proportion of privatised assets have not been repaid yet and also because a large proportion of assets are in the possession of individuals having no interest to leave it in agriculture.

The land lease scheme (almost 80 percent is leased, as illustrated before, the land, thus, cannot be accepted as a collateral) and the bad economic performance of the whole agricultural sector have complicated the access of farmers to bank credit. The government offered several schemes to encourage investment giving slight preference to new farming entrepreneurs (family farms, partnerships and farming companies) as discussed earlier.

Labour moved out from farms very rapidly. Labour resembled a variable input adjusting accordingly to the declining output. The total labour fully engaged in agricultural decreased by 54 percent from more than half a million to less than a quarter of a million (see Table 3.2). Much of this labour outflow is associated with the separation of non-agricultural production from farms (52 percent of the change 1989-1995) cut from excessive managerial staff (19 percent of the change 1989-1995), but a lot of the labour outflow is associated with the reduction of production (16 percent of the change for animal production and 13 percent for crop) due to the drop in demand as well (Table 3.11). The cut of agricultural labour caused problems only in a few industrially underdeveloped regions (MZe, 1994) a rising local unemployment up to 8 percent. But the national economy, in general, offered enough opportunities for employment outside agriculture so that the released agricultural labour force did not affect national unemployment figures.[19]

Consequently, the labour to land ratio dropped from 15 per 100 hectares in collective farms in 1989 to 7-8 per 100 hectares in co-operatives and farming companies today (Table 3.2). Since family farms have smaller proportion of animal production than large scale farming and owner labour is employed more than 8 hours per day, their labour intensity was around 4 people per 100 hectares in 1995. Overall, average labour productivity has doubled between 1989 and 1995 from 12 to 6 man per 100 hectares. Operational efficiency of capital decreased on the other hand. This is particularly true for original or transformed collective farms, from where capital, particularly fixed assets is moving out only slowly.

Certainly, less than one tenth of owners seek their job in agriculture (only 4.6 percent of the total employment are in agriculture). According to the sociological survey of VUZE between 60 and 70 percent of households depending on agricultural income considered this income sufficiently to cover their basic needs. in 1994. More than 80 percent of engaged in farming companies and co-operatives work less 60 hours per week while almost 70 percent of sole farmers work more than 60 hours.

To summarise, there is a clear shift in the resource mix since the beginning of reform. Employment of labour has declined substantially, capital has been reduced only modestly and the use of land has not changed significantly. This reflects the relative immobility of the resources in question (capital and land). Existing farm buildings and machinery are not easily utilised outside agriculture. Land has, by and large, no alternative uses outside agriculture. Some labour can apparently be used elsewhere.

Table 3.11
Allocation of labour

	Total labour	*Animal production*	*Crop production*	*Mangement*	*Rest*
1989					
Absolute	533,057	124,328	84,111	80,201	244,417
Share		23.3%	15.8%	15.0%	45.9%
1993					
Absolute	279,673	80,917	50,359	34,082	114,315
Share		28.9%	18.0%	12.2%	40.9%
1995					
Absolute	261,204	80,973.3	49,628.8	28,732.5	101,870
Share		31.0%	19.0%	11.0%	39.0%

Source: Horska, Spesna, 1994, CSO, 1996

Corporate farming is based on joint capital, not on joint labour or land. Nevertheless, large proportion of labour is derived from shareholders (members) and almost all shareholders (members) rented their land to their business.

Interpretation of structural changes

The previous section has documented the structural changes in Czech agriculture by facts and estimates. This section contains an attempt to identify a combination of internal (sector or within organisations) and external factors which led to the observed changes in the farming structure to the end of 1995. Future developments are explored in the final section.

Five years after the Land Law was issued the Czech farming system continues to be dominated by relatively large-scale farming with production co-operatives occupying nearly a half of the agricultural area. This seems to suggest that being the owner of land is an insufficient incentive for starting-up one's own business or family farm. Land ownership must be accompanied by other motivation such as carrying on the farming tradition, as a source of income or as a safe way of securing the future value of their capital assets. An opposite interpretation could be that 'restructuring of collective farms' as we defined it i.e. restoration of property rights of formal owners de facto failed. Practical difficulties with identifying and alienating of land, as described earlier, could be taken as an argument for this interpretation. Not denying that there is something to this argument we still believe that the difficulties mentioned above have only played a minor role.

By assessing the outcome of the process of privatisation, it should be remembered that there were obvious constraints to opportunities which could not be fully controlled by either politicians or interests groups. Four such constraints were the limited divisibility of non land assets, the limited divisibility of land, the fragmented ownership and missing efficient markets for exchange of property (land, real estate).

These constraints originated in the past when technology was biased toward large-scale operations. Small plots were merged into large fields, animals were pulled to large sheds. Consequently the rest of the assets were adjusted to this structure, roads to fields, irrigation, storage, machinery and silage containers. Technology and equipment were unified in only a few centres per enterprise. This limits the physical separability of assets. Some farm units would be impossible to operate if some important part is missing. At the same time the land ownership is very fragmented with more than 3 million owners of 3.3 million hectares.

We argue that in this sense it is inevitable that the traditional view of land and farming must change. The ownership of land should be similar to the ownership of any other real estate, like buildings or shares. (Of course, it can be disputed if land is just like as another asset.) That implies that a property right to a piece of land does not necessarily mean physical access to this piece, particularly, if this is a part of a larger field with additional infrastructure e.g. irrigation. (In such a case field splitting can take place only, if all owners agree to share the cost of that investment.) In Barzel's terms, property right is incomplete (some attributes remains in public domain), but might be acceptable, since it does not differ from other common types of ownership. Hence, our attitude is opposite to the ('common') belief that the splitting of fields to original boundaries is natural and should be done. A definition of boundaries of land may, however, be necessary to establish a land market.

This view on land property rights has been close to the perception of managers of co-operatives and probably continuing members. The Czech government in order to give the opportunity to sole farming introduced an arrangement which obliges current users (collective farms, later co-operatives or another legal successor) to offer either the original plot or compensatory land of the same quality for new farmers. In the same time, the Land Authority was established, which together with the Cartography Office is responsible for land consolidation and settlement of field boundaries.

Until the legal ownership and property rights are settled, markets for land and other real estate cannot satisfactory exist. Also a market cannot be introduced it must develop. Several factors can be responsible for slow market development: low need for exchange, improper rules, character of commodity (real estate) lack of boundary definition or outside factors preventing exchange. One can assume that during and shortly after ownership reform, especially, when it distributes assets to a large part of the population, the need for the exchange is small because the benefit from exchange is not easy to see. It is also possible that the land market is inhibited because of low profitability and uncertainty in agriculture resulting in low land prices. In such circumstances land may be kept by small land owners because they do not loose much by hanging on to land. But certainly, the character of the good which is either too large for the buyer or only a share in a large field limits the possibility for exchange.

In order to offer an opportunity to overcome the problem of indivisibility of non-land assets, it has been possible to trade informally the property shares on the original collective farm assets which were not included in the new (transformed) business.[20]

The opportunity set of an individual depends on the character of assets available to him/her, the types of organisations, their legal status and power, government support and regulations, and overall economic and political conditions during the ownership and property rights reform process. The perception of opportunities changes over time, and it may improve through learning. The reform process dramatically changed the opportunity set of individuals, but due to their lack of experience, people were not able to realise all opportunities. Evidently, new business organisations enlarge the opportunity set of owners who must choose a business organisation to be able to grasp their legal right. Sole farming proprietorships are characterised as easy to establish (it just requires registration), and a direct and transparent link between land, capital, labour and management through one owner/operator, but full liability requires the owner to bear a personal risk. Establishing farming corporations (farming companies and co-operatives) requires the fulfilment of a set of conditions (minimum of shareholders and equity, setting internal rules and a foundation document), and land, equity capital, labour and management are not necessarily linked except through the business itself (see characteristics in Table 3.12).

Table 3.12
Characteristics of different types of business organisation in the Czech Republic

Characteristics	*Types of business organisation*					
English Term	Sole Proprietorship	Commercial Law Partnership	Limited Commercial Law Partnership	Limited Liability Company	Public Limited Company	Co-operative
Czech Term	Podniky fyzickych osob	Verejna obchodni spolecnost	Komanditni spolecnost	Spolecnost s rucenim omezenym	Akciova spolecnost	Druzstvo
Juridical basis for existence	Law about Sole Farmers/ Business Licence Law	Commercial Code	Commercial Code	Commercial Code	Commercial Code	Commercial Code
Registration	Not necessary	Necessary	Necessary	Necessary	Necessary	Necessary
Minimum/Maxi-mum of members		2/-	2/-	1/ max 50	1 for legal entity, 2 otherwise/-	2 for legal entities, 5 otherwise/-
Minimum of capital				CZK 100,000	CZK 1 million	CZK 50,000
Individual tradability of capital	Yes	No	No	No, transfer allowed only with the agreement of general assembly	Yes, usually not publicly (on the stock exchange)	No, transfer allowed only with the agreement of board of management
Liability	Full	Full	At least one partner with full	Limited	Limited	Limited

Management		Unlimited partner	Chief executive, general assembly	Board of management, supervisory board, general assembly	Board of management, supervisory board, general assembly
Cancellation of membership	Not allowed	Not allowed	Application for a court decision	Sale of stocks	Application for a board of management decision + internal rules
Financial compensation by membership cancellation			Share in the total capital relative to the equity share		Share in the (total capital - indivisible fund) relative to the equity share + internal rules

Most owners of agricultural assets do not have sufficient assets to form a viable farm unit. Note that just over half of family farm land is operated by large farms (over 100 hectares), indicating that economic viability, which increases with the size, has been an important factor from the beginning. Those who contemplate starting a private farm must rent resources from others to achieve a viable farm size. Perhaps people, who do not want to farm individually, prefer to rent to a co-operative.[21] Current low profitability is also important. Willingness to venture a new enterprise in such circumstances must be low. Improved profitability may induce more individuals to take the risk. The possibility of off-farm employment with higher paid, more secure and comfortable jobs and to leave farming altogether has been important as well. Small farms in several other CEECs are driven by subsistence considerations. This can, in particular, be observed in Latvia and Lithuania where farm employment has increased in spite of falling agricultural output.

Thus, we argue that efficiency considerations were the main driving forces behind the emerging farm structure in Czech agriculture. Efficiency is defined as a minimisation of cost associated with the process of ownership transformation. The most efficient operational form for grasping the property rights to assets has appeared to be a large scale farming corporation. Although we do not deny that most sole farmers believe that their way is at least as efficient as that of other farming enterprises, the decision for starting independent business was determined often by very personal objectives: family tradition, desire for independence or, perhaps, loss of job (or necessity for extra income- part-time farming).

Future structural development

The present Czech governing coalition is satisfied with the result of the privatisation and transformation in agriculture, believing that further structural changes are only a matter of market forces.[22] As we demonstrated before, the reform policy was, in principle, not discriminating against any business form. The support to sole farmers was the minimum to give them some chance of survival. The government also created a conducive macro economic environment and fair rules instead of getting involved in owners' disputes. On the other hand, the Czech government did not take measures to correct the asymmetry in information in the transformation and privatisation procedure. There was almost no support to extension service during the transition process. That asymmetry was disadvantageous to individual owners.

The land reform distributed agricultural assets to the 'poor' Czech population. Therefore, the reform did not bring new capital in the sector, on the contrary, the non-farming owners tend to withdraw capital from agriculture.

We estimate that the capital withdrawal from the successors of collective farms amounts to about CZK 30 billion, which is one fifth of the total capital of these organisations. It will be difficult for the successors of the collective farms to recoup such capital before the end of the century. If the restitution capital remains allocated predominantly in co-operatives or other businesses with no tradable shares, then there will be only four solutions to the 1999 withdrawal problem either (1) ask for bank credit, or (2) escape from the organisation carrying restitution (debt) obligation and establish new organisation without debts, or (3) pursue the non-farming non-members to become members, or (4) transfer this problem to the government.

The third solution is cost free but unlikely, while the other solution bring additional costs. In the first case, the cost will be carried by co-operatives. The second solution will divide the cost asymmetrically between both sides, leaving compensation of non-farming restituents just to the liquidation procedure. The last option will transfer the most of the cost to taxpayers. It should be noted that the Czech government already paid CZK 7 billion for covering the cost associated with privatisation and liquidation of state farms. While the land is usually assumed to be undervalued by the administrative price, the book value of non-land assets is commonly considered to be 'quite real'. In the further process it can turn out that the non-land assets value was overestimated or that the value will be reduced by splitting farming units. This may bring a feeling of being cheated among people who participate in the process in spite of the fact that justice has been intended by the designers of the reform scenario.

The government coalition believes that the economic viability of farming organisations which have emerged during reform will be the main determinant of the future structural development. Current economic results may exhibit that farming companies are more profit oriented than co-operatives, but certainly, they show that farming corporations do not care too much about their stockholders. Although the value of fixed assets is similar, the depreciations (book, not for taxation) of farming corporations are one half higher than the one of sole farmers (who present only tax depreciation). It illustrates that managers operating with cash flow want to leave all this sum in the business instead of exposing a large proportion of it to a debate about dividends. (It is, however, also possible that co-operatives do not fully realise the need to replace old capital.) It consequently illustrates that either concern or power of the owner is not sufficient to control managers. It is often criticised that the whole Czech privatisation led to an overly fragmented private ownership, which in its consequences does not differ too much from the public.[23]

The intention of government to sell state land to current tenants and the fragmented ownership of the rest is essentially reducing the possibility to start independent farming. The physical land reform (which will probably last until the end of the century) will arrange borders of fields and environment protective requirements according to the current need of use of the land. The Land Authority, providing settlement of boundaries, must build up quickly an efficient organisational structure, since its good functioning is the necessary condition for development of the land market.

It follows from our discussion up to now that farming corporations were probably the most efficient organisations for grasping redistributed assets and that inexperienced individuals preferred to continue in such associations for lower income rather then expose themselves to entrepreneurial risk in the depressed industry. It is also obvious that the main disputes have been concerning the distribution and utilisation of capital (non-land assets). It is noticeable that all corporate farming (farming companies and new co-operatives) is based only on joint capital, while labour and land are hired at a fixed rental rate. In terms of the Barzel approach, the owners of capital become the residual claimants having obvious interest to maintain or reshape the capital in the most efficient fashion (Fulton, 1995). The capital intensity of contemporary agriculture with all its technological advantages reduced the possibility of substitution among factors. Thus in the process of distribution of assets the best labour and the best land cannot substitute for a tractor or harvester. This fact has made labour and land abundant and uniform, but capital scarce. If Fischer (1993), Fischer and Ratinger (1994) argue that capital is employed inefficiently in agricultural businesses, this is mainly because capital is wrongly evaluated by the users and not because of lack of the organisations/owners' incentive. From this perspective the further structural development depends definitely on the relationship between factors in respect of output variability.

It brings us again back to the role of government in the reform and post-reform process. It seems that the pumping of capital throughout various schemes (SGFAF since 1995) could not compensate for the cost of running the agricultural reform at the book value of assets and reform-transaction costs. This must be taken into account in formulating agricultural policy for the near future. The re-evaluation of capital is necessary if it is to be employed efficiently. This is obviously a very sensitive issue, since it is linked to the compensation of non farming eligible persons.

Summary and conclusions

Returning to the first of the questions formulated for the study, it can be stated that the objective to distribute as much as possible the agricultural assets among individual owners has been achieved for non-land assets, while substantial part of agricultural land is still awaiting privatisation. Nevertheless, the private sector has become predominant in agriculture, 95 percent of TAL is operated by private businesses.

While there was a common agreement about the recognition of former individual owners' legal right to agricultural assets, a dispute appeared concerning who will get the real power over these assets. In the end, this dispute was resolved very dogmatically in the rigorous belief in individuals' property rights on the one side, but very pragmatically that these rights can be applied through any private business organisation, on the other side.

Advocates of independent family farms often express their disappointment at the slow pace of change in collectivised or co-operative farm structures. This is claimed to result from the lack of willingness of managers of co-operatives to give back land and assets and the low degree of governmental concern about this issue.

Defenders of the concept of farming in association express the attitude of a large part of the rural population which is deeply anchored in the pre-reform co-operative experience. Building on this socialised view of the very structure of farming, they call for protection of the whole sector. Special favoured treatment of the sector is justified as it is the most sensitive industry producing basic foodstuffs, providing employment for the rural population and creating non-market values such as the landscape. The objective of this advocacy is quite explicitly to keep and raise the level of social welfare in rural society. On the contrary, the family farm advocates emphasise more individual and liberal goals through the re-establishment of the traditional farm and farmer.

The values of the association of land owners (sole farming, family farms advocates) has been seemingly congruent with the reform objectives proclaimed by the governing coalition. However, the government is obviously satisfied with the outcome of the redistribution of property rights (privatisation) in the agricultural sector and will, most likely, avoid any actions which could cause social disruption in rural areas. In this respect, the association of land owners feels right to be cheated but not by the government and the whole process, but by their unrealistic expectations.

The Agriculture Chamber, unifying all agricultural producers, which was established in autumn 1991, contributed by a large deal to this bitterness. Representatives of the Chamber have been always reluctant to enter or even mediate the debate about land reform and its implications.

Introduction of new private production co-operatives is explained by the fact that they simplify and speed-up the process of establishing property rights, and that they create a defensive option for the conservative rural population in the 'risky' agrarian business. When co-operatives offer a certain level of income and there are sufficient opportunities for jobs outside the agricultural sector, there is only a small incentive for active or passive members of co-operatives to withdraw land and assets and start their own independent farms.

Although privatisation (mostly direct sale) of state non-land assets was an exceptional opportunity to start sole farming, the dominate forms arriving from this process have been farming corporations. A fragmented structure of land ownership calls for farming in association to match the capital available and to be able to catch some scale advantages. State or collective farm capital was designed for large-scale farming and generally cannot be transformed into technology convenient for small farms.

From this point of view, the farming structure adjustment will only decelerate with the finishing of privatisation of state farms and transformation of collective farms. The land reform brought a new distribution of land among operators, and there are not evident reasons for changing it quickly. The most weighty argument for a faster farm structural change, which would diminish the role of co-operatives, follows from their non-market objective (keeping social welfare), which is deeply embedded in the concept of a co-operative and which is not coherent with the competitive market environment. On the other hand, the majority of other agricultural businesses have non-market objectives too, and ultimately this is the problem of the whole sector.

Notes

1 There were several restitution laws preceding the Land Law (e.g. two rehabilitation acts, 119/90 and 87/1991) creating the first legal base for restitution but having only a small impact on the farm sector. The Land Law brought a complete list of injustices which were to be rectified.

2 Fulton (1995) states in the accord with Macpherson's arguments (1978) that property is not thought to be a right because it is an enforceable claim of something what is thought to be a human right.

3 The transaction cost should be understood in the very broad sense; if someone is prevented to determine or capture one of the attributes of the asset in question, then the transaction cost must be infinite.

4 While usually the view is opposite, some Czech law experts concluded that once the state returns to a normal democratic legal system respecting human rights the restitution is a logical implication. The definition of eligibility of individuals to claim their assets back is necessary for protecting the rights of current owners.

5 Examples are: people's ownership, co-operative ownership in industry, co-operative ownership in agriculture and co-operative ownership of flats and houses.

6 The definitions of agricultural business organisations are scattered in the text, which we understand is difficult to follow by the reader. We consider three types of old organisations, state enterprises (dominated by state

farms), collective farms (sometimes called state co-operatives) and sole farmers of negligible importance, and five types of new organisations: sole farmers (often referred to as family farms), partnerships, limited liability companies, joint stock companies and agricultural co-operatives (see Table 3.12). Limited liability companies are referred to as farming companies and together with co-operatives as farming corporations. We use the term 'farming in associations' for any form of non-single operation (owner and family labour). We must point out that (a) there is a substantial difference between a state and collective farm, as collective farms always enjoyed more managerial independence and played an important social role in rural areas; and (b) that the new Commercial Code is very flexible in respect to co-operatives, permitting many specifications of rules for the general assembly e.g. cancellation of membership, or vote distribution.

7 In the sense to withdraw physical assets.

8 They will get (already have received) a property right to a share of the collective farm assets reduced by the originally brought assets, now given back.

9 Conservative politicians often seek for support among the conservative rural population. Their belief in the similarity of values might, however, appear as a pure illusion, since conservativism of peasants rests in keeping what they have been accustomed to which is exactly the 'collectivised village'.

10 This explains why such a large proportion of national assets was privatised by the voucher method. On the other hand, it resulted, paradoxically, in the ownership structure, where power of individual owners (not necessarily individuals) is on average very small. The critics of the voucher method argue that it led only to transformation from one kind of public ownership to another, and therefore, the economic effect of such a reform is negligible. This was particularly true in the early years of the reform, when the capital market was evidently underdeveloped.

11 As a principle.

12 Comparisons can be made with other countries, such as Latvia, where almost two thirds of assets was allocated to labour.

13 Their own, obviously biased, interest was not taken into account, rather they were considered to be some kind of state servants (as in the past).

14 However, there are doubts about the real impact of these government actions. While publicity was given to the penalty issue, but almost no administrative support, the second one (credit) was administratively backed up but without publicity and facilitation.

15 The voucher privatisation was often taken less seriously by the public. It acquired the sensation of a game such as gambling or lottery, thereby

favouring gamblers who could win for a small stake (CZK 1,000), a sum 80 times higher on average, as promised by official propaganda.

16 This is a perception.

17 Former collective farms often had various non-agricultural activities, which were not linked to agricultural production at all, like construction activities, production of computers, spar parts for cars etc.

18 Specialised production: more than 66 percent of output creates crop or animal production; Mixed: otherwise. Percentage in operational form output.

19 The low unemployment rates over the whole transition period are considered an exceptional success of the economic policy governed by liberal minded reformers. Government representatives stress that the main agents of this success are low minimum wages and fast privatisation. Critics (left or right) argue the opposite that the government in order to make the reform simple and feasible, has avoided making sufficiently deep reforms.

20 One might say, they have been allowed to be considered as informal bonds, which expire immediately for those wishing to resume farming or in the fall of 1999.

21 This might be a very important behavioural phenomenon. Government, as an institution, enjoys significant credibility among the Czech public, which is transmitted to organisations and institutions owned or managed by the state. Therefore, there may be an incorrect association with co-operatives or just with the matter of ownership reform – a belief that the government carries its deal of responsibility and the assets are, therefore, in the right hands.

22 See the presentation of the Minister of Agriculture J. Lux and the Minister of Finance I. Kocarnik at the Agricultural Chamber meeting in Lucerna, Prague on 27. February 1996

23 In terms of Barzel's approach, too many attributes remained in the public domain.

References

Bardhan, P. (ed.) (1991), *The Economic Theory of Agrarian Institutions*, Clarendon Press: Oxford.

Begg, D. (1991), *Czechoslovakia,* Political Economy 3.

Beckmann, V. and Hagedorn, K. (1997), this volume.

Brooks, K., Guasch, J.L., Braverman, A. and Csaki, C. (1991), 'Agriculture and the Transition to the Market', *Journal of Economic Perspectives*, Vol. 5, No. 4, pp. 149-161.

Buckwell, A. (1994), 'Restructuring of Agriculture in Central and Eastern Europe: Progress, Problems and Policies', Paper presented at the OECD meeting of the *ad hoc* Group of Experts in East-West Relations in Agriculture, 12-15 September, Paris.

Buckwell, A., Haynes, J., Davidova, S. and Kwiecinski, A. (1994), *Feasibility on an Agricultural Strategy to Prepare the Countries of Central and Eastern Europe for EU Accession*, Study prepared for DG I of the European Commission.

CSO (1990), *Statisticka rocenka CSFR/Statistical Yearbook*, The Czech Statistical Office: Prague.

CSO (1995), *Soupis hospodarskych zvirat k 1.3. 1995*, The Czech Statistical Office: Prague.

CSO (1996) *Agrocenzus 1995*, The Czech Statistical Office: Prague.

Davis, J. (1994), 'A Comparison of the Effects of Structural Adjustment on Agricultural Development in Developing Countries and Transition Economies', unpublished working paper.

Divila, E. (1992), *Privatizace a vlastnická participace zamestnacu*, Institute for economics, CSAV: Prague.

Divila, E. (1994), 'Otázka casu nebo koncepce', *Ekonom*, Vol.18, pp. 21-22.

Fischer, M. (1993), 'Pouzití produkcní funkce v analýzách výkonnosti zemedelských podniku a v rozborech alokace práce a kapitálu', Conference paper: Forming of Agricultural Policy and Its Implementation, University of Agriculture: Prague.

Fischer, M. and Ratinger, T. (1994), *K modifikovanému vyjádrení výkonnosti zemedelského odvetví*, Interim Report of Grant no /1994, Grant Agency of Czech Republic, Institute for National Economy: Prague.

Fiser, Z., (1993), *Zakon o konkurzu v podminkach ceskeho zemedelstvi*, Sbornik referatu o konkurzu a vyrovnani, Ustav zemedelskych a potravinarskych informaci: Prague

Fulton, M., (1995), 'The Future of Canadian Agricultural Co-operatives: A Property Rights Approach', *American Jorunal of Agricultural Economics*, Vol. 77, pp 1144-1152.

Horska, S. (1994), 'Socialni souvislosti procesu transformace zemedelstvi', *Research study* no. 9, VUZE: Prague.

Horska, S. (1996), 'Signaly zmen socialniho postaveni zemedelcu', *Research study* no. 30, VUZE: Prague.

Kraus, J., Doucha, T., Sokol, Z. and Prouza, B. (1994), 'Agricultural Reform and Transformation in the Czech Republic', in Swinnen, J. (ed.), *Policy and Institutional Change in Central European Agriculture,* Avebury: Aldershot.

Macpherson, C.B. (1978), *Mainstream and Critical positions*, University of Toronto Press: Toronto.

MA CR (1994, 1995, 1996), *Zpráva o stavu ceského zemedelství*, Ministry of Agriculture of the Czech Republic: Prague.

OECD (1995), *Review of Agricultural Policies: The Czech Republic*, OECD: Paris.

Ratinger, T. (1994), *Agri Marketing System in the Czech Republic during Economic Transition*, Grant No. 91/93, RSS-CEU: Prague.

Shaffer, J.D. (1980), 'Food System Organisation and Performance Toward a Conceptual Framework', *American Journal of Agricultural Economics,* Vol. 62, pp 310-319.

Swinnen J.F.M. (ed.) (1994), *Policy and Institutional Change in Central European Agriculture,* Avebury: Aldershot.

Swinnen, J.F.M. (1997), 'The Choice of Privatization and Decollectivization Policies in Central and East European Agriculture', in Swinnen, J.F.M. (ed.), *Political Economy of Agrarian Reform in Central and Eastern Europe*, Avebury: Aldershot.

Symon, J. (1994), *Analýza procesu transformace zemìdìlské výroby v Brnì mìstì a venkovì,* Diploma report, Mendel Agricultural University: Brno.

VUZE (1996), *Vyberove setreni hospodarskych vysledku podnikatelskych subjektu v zemedelstvi za rok 1995 v siti testovacich podniku*, VUZE: Prague.

4 Decollectivisation and privatisation policies and resulting structural changes of agriculture in Eastern Germany

Volker Beckmann and Konrad Hagedorn

Introduction

When agricultural economists in Europe talk about major changes in agricultural structures and policies which started in the early 1990s, they are probably discussing one of the following two topics: either the recent reforms of the Common Agricultural Policy (CAP) of the European Union (EU) which were induced by the GATT negotiations, or the privatisation and decollectivisation process in the Central and East European countries (CEECs) where a fundamental transformation is taking place. Germany is the only country in Europe which is immediately involved in both processes of change. Agriculture in Eastern Germany has had to adjust simultaneously to both a market-oriented economy in a pluralistic society and a supranational framework regulating many policy-making processes, particularly with regard to agriculture. As a consequence, the rural population in Eastern Germany had to cope with large economic and social changes within a rather short period of time.

This chapter will discuss just one of those two major changes, the restructuring process of Eastern German agriculture in the course of German unification. Particularly, it aims at analysing the impact of privatisation and decollectivisation policies on structural adjustment of the agricultural sector. The development of the agricultural processing sector is excluded from the analysis. The chapter includes both economic and political determinants of restructuring. The procedures of privatisation of nationalised land are more subject to political influences than decollectivisation of production cooperatives, although the latter process is also politically regulated. Accordingly, the political economy involved in the process seems to be more important in the case of nationalised land, and we give more emphasis to economic determinants in the case of collectivised land.

The chapter is organised as follows. In the following section, the initial situation in agriculture in Eastern Germany just before unification and the political circumstances are presented. The next two sections provide an overview of the policies which were adopted in Germany in order to regulate decollectivisation and privatisation respectively. This is followed by discussion of changes in agricultural structures. In this context, also some results from farm accounts are presented in order to compare different farm types which have emerged in the restructuring process. In the last section the structural adjustments and policies which we have pointed out will be discussed from a theoretical point of view, and an outlook on the future of the transformation process will be presented.

The starting point for decollectivisation and privatisation

If we consider the starting point of the transformation process of agriculture in Eastern Germany, we have to take into account the structure of agriculture shaped by the preceding policies of the GDR, as well as the structure of political forces involved in transformation of agriculture. The latter are also strongly influenced by the historical background, for example, if we consider the contrasting interests of former land owners and East German co-operative farms. For these reasons, we will first describe the initial conditions in the agricultural sector and then the initial situation regarding the political actors and interests.

The initial conditions in agriculture - requirements and opportunities

In order to understand the initial situation in agriculture, we first have to look at those agricultural policies of the GDR, by which ownership structures and organisation of agricultural production were fundamentally changed after World War II. There were two major political decisions which affected the ownership structures and organisation of agricultural production in a fundamental way:

1 *Land reform (1945-49)*: The nationalised agricultural land which has to be privatised now originates from two sources: mainly from the so-called land reform, which took place in the Soviet Occupied Territory from 1945-1949, and to a minor extent from expropriations during the first phase of collectivisation of agriculture in 1952 and 1953 (for details, see Klages, 1994, pp. 106 ff.). In the process of the so-called 'land reform', which actually consisted of confiscation procedures, 7,160 farm and forest operations whose owners were classified as 'big landlords' and 'squires'

and which were larger than 100 ha (average size: 350 ha), were expropriated without compensation. In addition, 4,537 farm operators were expropriated because they were said to be 'Nazi-leaders' and 'war criminals' (average size: 350 ha). From the land acquired through these confiscations, together with agricultural land and forests which already belonged to the state property before the end of World War II, a land reserve was established with a total area of 3.3 million ha. From this land reserve 2.2 million ha were (preliminarily) distributed to so-called 'new settlers'. Those were farmers who were refugees from the Eastern provinces of Germany which now belong to Poland and Russia, peasants with small farms, and farm workers who received this area of land as so-called 'working property' (average size: 8 ha). The rest of the expropriated land remained nationalised property and was mainly used to establish the state-owned estates ('Volkseigene Güter' – VEGs).

2 *Enforced collectivisation (1953-60)*: Collectivisation had little impact on the (formal) ownership structure of land. During the period of collectivisation, another significantly large area of land was expropriated, but only part of this became registered state property. 24,200 farms were expropriated during the first wave of collectivisation in 1952 and 1953. In this period, 700,000 ha were confiscated and held in trust by the 'county councils'. In most cases the county councils handed this land over to the agricultural production co-operatives ('Landwirtschaftliche Produktionsgenossenschaften' – LPGs), founded in those years, without charging them any rent. In contrast, collectivisation did change the organisation of agriculture by establishing and expanding the LPGs. Formally, the land still belonged to the individual owners, but the LPGs were endowed with the unrestricted right to use the land, including the right to set up buildings. The assets of the LPGs were considered collective property, so that there was no private claim upon collective assets when members were dismissed.

According to these policies, the initial ownership structure and organisation of production can be characterised as follows:

1 About 70 percent of agricultural land was still formally privately owned, 30 percent was state property or state governed private property. The main physical assets in agriculture belonged to the collective property.

2 Organisation of agriculture was dominated by LPGs (see Table 4.1). In 1989, 4,530 LPGs cultivated 82.2 percent of the agricultural land and held 74.5 percent of the animals. State farms were only of marginal importance:

580 state farms cultivated 7.5 percent of the agricultural land and held 16.4 percent of the animals. Both collective and state farms specialised in either animal or plant production. Even in 1989 only 5.4 percent of the agricultural land[1] was used for small-scale private farming.

Table 4.1
Structural characteristics of farms in the GDR in 1989

Legal form of farm	*1989*						
	Number of farms	*Workers*		*Agricultural land hectares*		*Livestock GV*[a]	
		Total 1000	*per farm*	*Total 1000*	*per farm*	*Total 1000*	*per farm*
Private farms[b]	3,558	5.5	1.5	335	94.1	165	46
Household plots[c]	(375,000)	.		297		351	
LPG (Collective farms)	4,530	694.9	153.4	5,075	1,120	4,343	959
Plant production	1,164	306.9	263.6	4,987	4,284	69	59
Animal production	2,851	343.6	120.5	73	26	4,273	1,498
Horticulture	199	27.6	138.7	15	75	1	5
VEG (State farms)	464	95.3	205.3	447	963	479	1,032
Plant production	152	46.2	303.9	408	2,684	36	236
Animal production	312	49.1	157.4	39	125	443	1,419
VEB (State farms)	116	29.5	254.3	17	147	492	4,241
Total	8,037	825.2	102.7	6,171	768	5,830	725

[a] Livestock unit
[b] Farms with a size of more than 1 ha including church farms,
[c] Farms with a size below 1 ha

Source: Agricultural Report 1991, p.141, own calculations

Accordingly, collective farms prevailed in the initial situation of the transformations process. Thus, the main challenge for reform was de-collectivisation, i.e. the restructuring of the LPGs. Privatisation of nationalised land was far less important from a sectoral point of view. In spite of this, the political economy of restructuring agriculture in Eastern Germany was predominantly shaped by discussions and struggles around the issues of privatising nationalised land. This may be due to the fact that the problems of indemnification and privatisation can only be solved by extensive political and administrative procedures. They involve, for example, a whole variety of legal questions which require extensive legislation and administration.

Reliable information on the precise quantity of nationalised land is difficult to obtain. The best sources are the statistics of the BVVG ('Bodenverwaltungs- und -verwertungsgesellschaft) on land it leased to the East German farmers. Table 4.2 shows the regional distribution of this land. With regard to absolute as well as to relative figures the largest area can be found in Mecklenburg-Vorpommern. In some counties of Mecklenburg-Vorpommern the share of nationalised land even amounts to 50 percent (without including the land cultivated by former state farms, i.e. VEGs), on the average the share is nearly 30 percent. Also in Brandenburg nationalised land is rather important, it represents nearly 20 percent of the farm land. It is less important in Sachsen-Anhalt where we find 15 percent of nationalised land on the average; but in the northern part of the Bundesland the share is higher (up to 20 percent). In the southern Bundesländer of Eastern Germany, i.e. in Sachsen and Thüringen, the shares are below 10 percent, so that nationalised land is no major issue here. The above-mentioned differences reflect the historical ownership structures in Eastern Germany. Before 1945, in Mecklenburg, large estates prevailed. In Sachsen-Anhalt mainly medium- and large-scale farms (20-100 ha) were typical, and Thüringen followed the Napoleon code for inheritance resulting in smaller farms.

Table 4.2
Nationalised land leased by the BVVG and its regional distribution in June 1994

Bundesland	*Leased area 1,000 ha*	*Share of the total agricultural area %**
Mecklenburg-Vorpommern	462	29.1
Brandenburg	290	19.1
Sachsen-Anhalt	199	14.8
Sachsen	100	9.3
Thüringen	90	9.7
Total	1,141	17.7

* Information on the agricultural area is taken from the GEMDAT of the Federal Ministry of Finance

Source: Klages (1994, p. 109); Federal Ministry of Finance

Privatisation in Eastern Germany was guided by the principle 'restitution deserves priority to indemnification', only a part of the expropriated land can be included within the privatisation schemes. Klages (1994, p. 109) provides an estimate of the area of agricultural land available for privatisation. In 1993/94, the area leased out by the 'Land Administration and Utilisation Company' (the BVVG', to which the 'Treuhandanstalt – THA', i.e. the German privatisation agency, has been assigned the task of privatising agricultural land) amounted to approximately 1.14 million ha. Furthermore, the privatisation agency possessed approximately 245,000 ha of nationalised land which formerly belonged to the state-owned estates. In total, the agricultural area which has to be restituted to former owners is estimated at approximately 500,000 - 600,000 ha, from which already 200,000 ha have been given back. As a consequence of these various 'deductions', only the remaining 1.0 to 1.1 million ha of agricultural land is expected to be available for privatisation schemes. This amounts to approximately 18-20 percent of the total agricultural area in Eastern Germany.

The initial political situation: policy networks and their interests

The privatisation process crucially depends on whether (and to what extent) the political actors involved want to keep these decisions by which institutions like land ownership are shaped *in* their own hands or, at least, to maintain direct control of these decisions, and to what extent they are ready to leave them to more pluralistic coordination mechanisms which they can only influence indirectly. This is a familiar question in institutional economics, for example it was formulated by Schenk (1982) for the purpose of systems comparison in the area of comparative economics. This approach is based on Williamson's (1975) concept of 'Markets and Hierarchies' as two main institutional choices, but within the category 'hierarchy' he introduces additional differentiation. Agencies ('Regieorgane') can employ different governance structures ('Regie-verfahren'); for example, a commercial and a regulating governance structure and a governance structure based on direct state control are distinguished. If we go a step further, we could think of a continuous scale of governance structures, in which the competence of the governing agencies changes from very strong at one end of the scale to very weak at the other end. Correspondingly, the scope for market mechanisms and pluralistic and policentric coordination mechanism would be small and large respectively.

It depends on the political economy of the privatisation process whether it allows for modifications which act against a governance structure of direct bureaucratic control and in favour of a governance structure with more elements of markets and participation. In Germany, we find a very specific political situation if we look at the forces which are relevant to policies of indemnification and privatisation. Traditionally, issues of agricultural policy are negotiated and decided on within one political subsystem, which protects itself against influences from outside (see Hagedorn and Schmitt, 1985; Hagedorn 1996). However, in this case, several systems are involved:

1 The *traditional policy network of agriculture*, mainly located in Bonn, still represents the decision-making centre in the area of agricultural policies. It has been shaped by patterns of negotiation and decision-making which were institutionalised during the 'old' Federal Republic.

2 The *policy network of finance* is also located in Bonn. The minister of finance is responsible for the area of indemnification, and since expropriation of agricultural property can involve large sums of money, he must coordinate his decisions with the processes which are going on in the agricultural network (and vice versa).

3 From these two networks in Bonn we have to distinguish the *policy network of agriculture in the New Bundesländer* which consists of the agricultural ministries of the Länder and the agricultural administration and groups at different levels. Due to the specific historical background and the geographic distances involved, this network is dominated by other interests and relationships.

To show the differences in political influence between these networks we take into account two main agricultural groups which have contradicting interests in the process of privatisation: 'former owners of land from West Germany' and 'large-scale agricultural enterprises in Eastern Germany', particularly the successors of the former agricultural production co-operatives:

1 *Imbalances or asymmetries of information*: the advocates of the former owners who live in Western Germany know how to use the channels of information and communication within the systems of government, bureaucracy, and associations in Bonn. The actors who work for the large farm enterprises in Eastern German agriculture are quite familiar with the administrative, personal, and social circumstances in the rural areas in the New Bundesländer.

2 *Different ways of influence*: the lobbyists of the former owners are able to make use of those political mechanisms (co-operative relations, mutual obligations, log-rolling agreements) which work within the political networks ('Feinsteuerung'). Those who represent the interests of the large farms in Eastern Germany can exploit other sources of power ('Grobsteuerung'). For example, they can enact pressure by influencing the polls, and create disturbances by organising farmer demonstrations.

3 *Unequal impact on implementation:* the former owners from Western Germany are in a good position during the first phase of the policy process, that is, the legislative phase. But then, during the implementation phase, those who want to help the successors of the East German production co-operatives are stronger, because they can modify or even determine the criteria and procedures which are ultimately applied. Politicians tend to exclude those problems from decision making in Parliament which cause major conflicts and could be detrimental for a consensus, preferring to leave them to the administration. Therefore, access to implementation presents a considerable source of influence. In the case of privatisation of agricultural land, an example is provided by the criteria applied to lease or sales contracts which are established in this phase of the process.

If we assume rational behaviour, political actors will try to make use of the strong elements of their position, and try to compensate for the weaknesses they have in the policy-making process. They will try to achieve a privatisation regime which provides them with opportunities to mobilise their specific potential to exert influence, and to prevent their weaknesses from becoming relevant. Whether these hypotheses can be empirically substantiated has been analysed in other papers which describe several episodes of political decision-making on the privatisation issue as well as in other studies (see Hagedorn 1993a; 1994a; 1995; 1996; Hagedorn and Klages, 1994; Klages and Hagedorn 1993; Klages 1994; Klages and Klare, 1994; 1995a,b,c). In this chapter, we will concentrate on the main results of the processes of political decision-making, institutional rearrangement and economic restructuring.

Policies for decollectivisation and privatisation of agriculture in Eastern Germany

The German government introduced a variety of instruments to regulate and to support restructuring of collective agriculture in Eastern Germany. Among others, the Agricultural Adjustment Act and special programmes for financial support of re-established and newly established family farm partnerships and transformed collective farms are important.

Regulating decollectivisation: the Agricultural Adjustment Act

The first version of the Agricultural Adjustment Act ('Landwirtschafts-anpassungsgesetz') from June 29, 1990, (GBl. I, No. 42, p. 642) was enacted by the first (and last) freely elected government of the GDR. The major elements of this law can be summarised as follows:

1 Private property in land was re-established (§ 1). Nevertheless, the owner did not receive an unrestricted claim on his original plot. In these cases, where returning of the original plot was impossible for objective economic or legal reasons, the owner should be compensated with other plots of equal value (§ 45-47; § 51,52).

2 All forms of property and business organisation were supposed to have equal opportunities in a competitive environment (§ 2).

3 Rules for the division, merger and liquidation of the LPGs were provided. By the division of an LPG, new registered co-operatives, partnerships or companies could be established. A merger of LPGs was only permitted to lead to a new LPG. Decisions on division, merger, liquidation or transformation had to be made by a qualified majority of two thirds in an assembly of members. Only if a new company with unlimited liability was created, did the decision have to be unanimous.

4 The LPG could transform directly into a registered co-operative (§ 27 ff.).

5 The members had the right to leave the co-operative, and in this case they had a claim on their share of the co-operative's assets (§ 43). The calculation of the claims of quitting members on their assets was left to the free decision of the LPG (§ 44).

Since the first version of the law was enacted, and since re-unification of Germany, the Agricultural Adjustment Act has been amended five times. Above all, distribution of the assets had mainly been based on labour input and far less on the size of the land and the amount of capital the members had contributed to the farm during collectivisation. Since this was considered inadequate, parliament decided to establish more detailed rules for the treatment of claims. The main amendment from July 3, 1991 (BGBl I, p. 410), changed the first version in many points:

1 Direct transformation of LPGs into the whole range of legal forms of business organisation provided by the West German law was no longer prohibited (§ 23 and § 23a). The decision to transform must not only be based on a two thirds majority of the members but also had to be confirmed by a simple majority of the land owners.

2 The law now laid down detailed conditions how the claims of quitting members had to be calculated (§ 44). First, the contributions members had to made to the LPGs assets during collectivisation had to be returned.

Second, if the equity capital was larger than the members' contribution to assets, rents and interest rates for their contribution to assets had to be paid. This category of claims of quitting members was limited to 80 percent of the collective's equity capital. Third, if the equity capital was higher than the sum of the first two elements, additional claims calculated according to the members' labour input into the collective farm could be submitted.

3 The board of directors was allowed to dismiss workers for reasons of structural adjustment without affecting the membership of the workers (§ 43a).

4 If an LPG did not change its legal form until January 1, 1992, it had to be liquidated.

The aim of these amendments was to simplify and accelerate the process of transformation and to regulate the distribution of collective assets in detail (see Agrarbericht 1992, p. 122). Especially the interests of land owners were supposed to be protected. Critics of the amendment felt that the regulation of the claims of dismissing members were effectively an instrument to enforce liquidation of the LPGs.

The last notable amendment of the Agricultural Adjustment Act from November 8, 1994, (BGBl I, p. 3265) allowed those co-operatives which were built out of former LPGs to transform into different forms of partnerships. Before this amendment was made, the Co-operative Act did not allow co-operatives to transform directly into a partnership. They could only transform directly into a joint-stock company. Again, the goal of this amendment was to speed up transformation of co-operatives, especially to reduce the costs of transformation caused by taxes and asset valuation (BML, 1995b, p. 54).

In August 1996, the coalition of the Christian and Free Democrats started a new initiative for a further amendment of the Agricultural Adjustment Act (Agra-Europe 1996a). The aim of this initiative is to provide rules for calculating the equity capital of the former LPG. The first amendment of the Agricultural Adjustment Act provided rules for the distribution of equity capital, but it does not lay down how the equity capital should be calculated. Critics of this procedure argued that the leaders of the former LPG calculated the equity capital very low, so that they could keep the claims of quitting members low. They argued that this procedure means a second expropriation (DER SPIEGEL, 1995). In reaction to this initiative all East German governments, the farmers union and the co-operatives union refused the amendment (Agra-Europe 1996b). It is open to question if the amendment will pass the legislative.

Financial support of agricultural adjustment

Since unification the German government has supported restructuring of agriculture in Eastern Germany by means of considerable financial resources. As Table 4.3 shows, until the end of 1995 circa 17.4 billion DM of financial support will be supplied. About 33 percent of the expenditure was used as adjustment aid or for compensation, 35 percent was channelled through the investment funds for the improvement of agricultural structures, which is a part of the joint project 'Improvement of Agricultural Structure and Coast Protection' of the German Federal Government and the Federal States ('Bundesländer').

Table 4.3
Financial support for agriculture in Eastern Germany, 1990 - 1995

Year	*Financial support billion DM*
1990	4.895
1991	4.231
1992	2.355
1993	2.066
1994 (target)	1.998
1995 (target)	1.813
1990 - 1995	17.358

Source: Bundesministerium für Ernährung, Landwirtschaft und Forsten (BML), Agrarwirtschaft in den neuen Bundesländern, 1995.

The adjustment aid is supposed to help the farms to overcome difficulties of structural adjustment in the period of transition. This programme has expired in 1995. Nearly every farm with more then 300 hours of calculated manpower requirements obtained this support, except those farms which are public property, in liquidation or successors of those LPGs which did not transform itself according to the rules of the Agricultural Adjustment Act. In 1994, 85.6 percent of all farms in Eastern Germany received adjustment aid of 10,700 DM per year on average. The transfer consisted of a lump sum and an additional digressive payment both based on calculated manpower requirements. According to this method of calculation, new and re-established family farms obtain more support per ha land than large-scale farms (see Agrarbericht 1992, p. 125).

The investment support measures for the improvement of agricultural structure mentioned above were introduced in 1991 by mainly two programmes providing investment subsidies:

1 'Principles for Supporting Re-establishment and Modernisation of Family-type Full-time Farms' and

2 'Support for Restructured and New-established Agricultural Enterprises in the Legal Forms of Companies and Partnerships'.

According to the 'Principles for Supporting Re-establishment and Modernisation of Family-type Full-time Farms', new or re-established farms are eligible for a payment of 23,500 DM to facilitate the start of farming. Furthermore, a lump sum called 'grassland subsidy' of 50,000 DM can be granted if grassland-based animal production systems are adopted. These payments are three times larger if the new established firm assumed the legal form of a civil partnership ('Gesellschaft bürgerlichen Rechts - GbR') instead of a sole proprietorship. Companies are excluded from this subsidy. In addition, all farms get a special subsidy for evacuation out of town, for new and re-established farms this payment is 70,000 DM, for transformed corporations 100,000 DM.

All farms are eligible for preferential interest rates which are up to 5 percent (in less favoured areas up to 6 percent) lower than the interest rates paid for a regular bank loan. Family farms can receive another 1 percent reduction in interest rates if they have access to the young-farmers programme. For family farms in the legal form of a sole proprietorship, the low interest rate is granted up to an amount of 400,000 DM per farm or 329,000 DM per labour unit. For civil partnerships (GbR) the amount is three times higher. Companies can claim preferential interest rates for a maximum credit amount of 3,150,000 DM per farm or 143,000 DM per labour unit.

In addition, new or re-established family farms are entitled to a government loan with 1 percent interest rate and 2 percent repayment per year with a maximum amount of 400,000 DM for sole proprietorships and 1,200,000 DM for civil partnerships. These government loans are not available to co-operatives and companies.

Summing up, the programmes provide support for all legal forms of farm organisations, but the new and re-established farms are supported by a higher amount of subsidy than the other farm types (see Forstner 1994a,b; 1995). This unequal treatment may be justified by the fact that successor organisations of former LPGs are in a far better starting position than new or re-established family firms.

The financial support measures introduced for agriculture in Eastern Germany after the unification of Germany were shaped especially for the requirements of Eastern Germany. After January 1, 1997, the programmes for investment support will be unified for Eastern and Western Germany. According to the new unified programme, investments up to a maximum of 2.5 million DM can be subsidised regardless of the legal form of the enterprise (see Agra-Europe 1996c).

Old debts, taxation and social security

In addition to these programmes which provide direct support to farmers, the German government introduced some special regulations facilitating adjustment of farms in a non-direct way. These measures refer to old debts, taxation of farms and social security of the farm population. Since these measures have an impact on the competitiveness of different legal forms of farm organisation, they will be outlined in the following paragraphs.

On the day of German unification *old debts* (i.e. bank loans which existed before July 1, 1990) in the LPGs in Eastern Germany reached an amount of 7.6 billion DM or about 1,200 DM per ha of agricultural land. In order to maintain liquidity of the agricultural enterprises concerned the German government introduced two measures for regulating old debts:

1 According to § 25 paragraph 3 of the Unification Contract ('Einigungsvertrag'), the government agency for privatisation of state property (THA) had to assume responsibility for old debts up to an amount of 1.4 billion DM. This measure aims to write off debts which the farms were forced to accept by the government of the GDR. It was carried out in two steps and is almost completed. About 1,400 farms will be relieved of an average amount of 1 million DM of liabilities.

2 On the basis of § 16 paragraph 3 of the D-Mark Balance Act ('D-Mark-Bilanz Gesetz') the government allows the farms to defer the payment of capital service for the remaining old debts. Payments on old debts are only demanded if profits are made, and in this case only up to an amount of 20 percent of current profits. Nevertheless, the old debts are incurring interest at 8 percent, which is therefore cumulating and increasing the total amount of debt.

This regulation of old debt is considered as a crucial problem for the LPG-successor organisations. According to a survey by Fiedler et al. (1994, p. 57), 68.6 percent of the agricultural enterprises dealing with the problem of old debts are waiting for a 'political solution', i.e. they expect old debts to be written off completely. Only 26.2 percent of the farms are prepared to pay the (cumulating) capital service.

Regarding *taxation* of agricultural enterprises, government introduced some special rules for Eastern Germany on July 1, 1991. The main measures are: (a) special rules on depreciation until the end of 1996 and (b) release of trade and asset taxes until the end of 1995. These measures are to maintain the liquidity of the farms in the transformation process. Unequal treatment of farms with different legal forms results from the fact that Germany has a special social security system for self-employed farms which only insures farmers and family members in natural entities, and excludes members and workers in legal entities. Since the social security contributions to the agricultural insurance system are largely subsidised, registered co-operatives and companies do not receive considerable financial transfers which are channeled to sole proprietorships and partnerships by means of this social security system (see, for details, Mehl and Hagedorn, 1994; Mehl, 1995). In economic terms this means that the cost of labour are not reduced by these particular state subsidies for legal entities as they are for natural entities.

This section only has summarised the main policies for restructuring collective agriculture. On the one hand, rules for the restructuring process are established by these policies. They certainly regulate the collective decisions of the LPG successors, never-the-less, much of the decision-making is left to the economic actors and to the competitive process itself. On the other hand, the policies try to reduce individual costs of transformation by providing financial aid or tax exemption. Government also provides pecuniary incentives for restructuring especially for the new or re-established family farms.

Privatisation policies of nationalised agricultural land

After Germany had been reunified in 1990, the majority of politicians probably thought that privatisation of nationalised agricultural land could be completed within a few years by selling the land as soon as possible. However, it soon became apparent that this quick solution was not feasible, not only for technical and administrative reasons, but also because political decision-making on this issue required a long time due to the fundamental difficulties or finding a consensus among the interest groups involved. As a consequence, it proved necessary in 1993 to introduce a pre-privatisation stage by which most of the land was not sold but leased by means of long-term lease contracts. The law which contains the rules according to which the land will be sold was only finalised in 1995, implementation of this programme has begun in 1996, and in 1997 we may expect that the first sales contracts on nationalised land based on this programme will be concluded.

Conflicts and concepts in the political decision-making process

Before the political decision-making process finally arrived at this end, it passed through some very important phases of conflict and discussion (see Hagedorn, 1996; Klages, 1994). The first rules for privatising the land were laid down by the so-called 'Treuhand Guideline' of March 3, 1992, which was modified and finally implemented on May 29, 1992 (see Top-Agrar-Spezial, 1992a; Agra-Europe, 1992a, b). In this concept, five basic principles were listed:

1 The land has to be utilised or sold primarily in accordance with economic objectives.

2 Selling the land has to be given priority to leasing the land.

3 For reasons of objectivity and transparency the buyers have to be selected by means of auctions.

4 The prices of land should, as a rule, correspond to market prices.

5 Settlement programmes or similar state-supported programmes may only serve as complementary policies.

These principles for land privatisation were complemented by a working group from the coalition of the Christian and Free Democrats chaired by a member of Parliament named Gerster in June 1992; accordingly, the concept was called the 'Gerster paper'. The reason why this paper caused controversy was that it tried to combine privatisation procedures with aspects of compensation for former expropriation. Due to an agreement between Germany and Russia which had been confirmed by the German Constitutional Court ('Bundesverfassungsgericht'), those persons whose property had been expropriated during the so-called land reform, i. e. from 1945-1949, are not eligible for restitution, but may receive compensation within certain limits. Such an element of compensation was included in the Gerster paper, i.e. it proposed a so-called 're-establisher programme' with special privileges for the former owners as regards to their access to nationalised land. Three groups were supposed to be consciously given preferented treatment:

1 Former land owners whose land had been expropriated during the so-called land reform from 1945 to 1949.

2 Other former owners who were eligible for restitution.

3 Land owners from Eastern Germany who have extracted their land from a LPG and have re-established their farm.

Correspondingly, three groups were supposed to be excluded from the programme.

1 Residents of Eastern Germany who did not own any land before, but want to establish a new farm of their own.

2 Agricultural enterprises originating from the reorganisation of LPGs.

3 Persons who wish to establish a new private farm and become local residents, but do not yet reside in Eastern Germany.

An additional stipulation suggested by the Gerster committee was a differentiation in priorities. This meant that first only the 're-establishers' may buy land, then – after a deadline has elapsed – the 'establishers' are also permitted to buy land. In contrast, the THA and the agricultural ministers of the New Bundesländer insisted on equal treatment of all groups in East German agriculture. In fact, the Gerster concept acted as a mobiliser of the 'East-West conflict' in Germany. Special opposition and criticism was caused by the exclusion of the LPG successors and the new establishers from the acquisition of reasonably priced land. In Eastern Germany, there was no sympathy for the proposal that only those already possessing property would could benefit from a highly subsidised programme, whereas landless farmers would come away empty-handed.[2]

In this situation government had to react very quickly. A working group chaired by Bohl – the Minister in the Chancellor's Office – soon presented a new model for privatisation of land – the Bohl paper of November 1992. A three-phase model was proposed (see Grund, 1993; Agra-Europe, 1993d). In the first phase all land was to be leased for twelve years by means of a bureaucratic procedure. Applicants for land must present a farm development plan, and must have sufficient qualifications to run the farm. If there happens to be more than one equally qualified applicant then the BVVG proceeds according to a priority list.

1 First were re-establishers including the heirs of these groups, and local farmers who established a new farm. If there is rivalry between applicants from these two conflicting subgroups, the so-called 'victims of the land reform' are to be taken into account in the sense of a 'balance of interests', i.e. this group is supposed to obtain some land regardless of the outcome of the application process.

2 Next are successors of LPGs.

3 Third are persons who establish a new farm, but are not residents of Eastern Germany.

In the second stage of the Bohl concept, a 'Land Acquisition and Settlement Programme' was planned in order to offer land at favourable conditions to certain groups. The first component – the 'Land Acquisition Programme' – aimed at former owners who are eligible for indemnification, i.e. for a compensation for the losses they suffered by former confiscation or expropriation of their land and other properties. Within certain limits they then could choose whether they want to be compensated by payments or by a certain amount of land. For the second component of the programme – the 'Settlement Programme' – only natural persons were supposed to be eligible, i. e. 're-establishers' or 'new establishers' who must be local residents. Companies and registered co-operatives which are often the successors of the LPGs, and also the individual members of these farm firms should, have no access to this programme. (This restriction was waved in the final version.) Other farmers who operate family farms or belong to civil partnerships also should have access to the programme. However, the applicants may only buy that land which they already have rented in the first phase of the programme, provided that they still operate it. Other persons are excluded. In this way there was a 'closed shop', market mechanisms or any sort of competition was abandoned.

In the third phase of privatisation, after these two programmes have been carried out, the rest of the land will be sold on the land market without any restrictions and without financial support from the government.

The response to this concept[3] was again a wave of criticism which came predominately from the New Bundesländer. Again, the special treatment of the former land owners by means of the above-mentioned 'interest harmonisation clause' was regarded as a privilege for this group, and this was neither accepted by the New Länder nor by the Social Democrats in Parliament.[4]

This episode of decision making reveals the following peculiarities of the East German privatisation process: first, the Gerster proposal originated from the 'Feinsteuerung' within the Western policy network in Bonn; second, this proposal tried to make the privatisation procedure more flexible in some parts by introducing small market components; third, however, it also tried to build up strong market restrictions by implementing priorities for one particular group (the former land owners); furthermore, the counter-reaction from the East German policy network arose immediately. In this case the mechanisms of 'Grobsteuerung' were employed by mobilising public protest, and this finally resulted – along with the presentation of the Bohl paper – in the near complete abolition of market mechanisms and an introduction of bureaucratic procedures.

It took nearly one and a half years until another attempt was made to replace the bureaucratic procedure of privatisation by some elements in favour of market components. A working group of the governing coalition, i.e. of the Christian and Free Democrats, whose chairman was the member of Parliament Gattermann, proposed vouchers as a means of compensation for former expropriations. In substance, the Gattermann concept proposed in November of 1993 was a response to the draft bill of the 'Indemnification and Compensation Law – EALG', presented on May 5, 1995. According to this draft bill, compensation payments for expropriations should partially be financed by a 'property levy', i.e. a levy on property in Eastern Germany that was (or will be) returned to its original owners. This turned out to be a major source of conflict, because these owners who already usually have to finance investments would be additionally burdened by the levy. Therefore the essential idea behind the Gattermann proposal was first to eliminate the financing of indemnification -the property levy– and second to offer the victims of former expropriations compensation through real property, which would be achieved by the following voucher scheme.

1 The vouchers could be used in three ways:
   a) trade the vouchers: on the market for certificates;
   b) buy public property: directly or by means of associations;
   c) return to the state: after January, 2004, to receive a money payment.
2 Public property is offered in auctions which allow certain groups to use a right of pre-emption:
   a) owners of vouchers who belong to those persons who offered the highest price would obtain a right of pre-emption, if their vouchers cover at least one quarter of the price;
   b) if there are several persons who are within the upper price margin and also fulfil condition 2(a), then a priority list which favours the former owners would have to be applied.
3 During the auctions the former owners could make use of a special right: they may join the group who offered the highest prices, even if their own first bid was lower.

The reaction to this new concept was similar to the response to the Bohl paper. First, the Gattermann proposal also was seen as a 'Western' initiative originating from the political network in Bonn. Secondly, the dominating strategy also was to open the bureaucratic privatisation scheme in a market-oriented way in order to allow access to the former land owners. Third, targeted priorities of this group were again included in the competitive model by means of an example. Last, this was the main stimulus which mobilised a strong political reaction from East German agriculture. The East German farmers, of course, were afraid that the rented land they cultivate was to be sold to West German investors.

As a consequence, a political consensus on the Gattermann model could not be achieved, mainly because of the strong protest from the East German politicians and farmers' associations (see Agra-Europe, 1993f). Therefore, the representatives of the governing coalition, the Christian democrats and the Free democrats, who participated in the working group mentioned above, agreed upon an alternative model on November 23, 1993 (see Klages and Klare, 1994). The new proposal was based partly on the draft bill of the EALG from May 10, 1993, but at the same time integrated three elements of the Gattermann model (see Grund, 1993; Ernährungsdienst, 1993a,b). These were, first the levy on restituted property proposed in the previous version to finance the indemnification payments was cancelled. Second, the assessment basis for the indemnification payments was raised considerably. Third, the digressive scale applied to the relationship between the amount of compensation paid and the area of land formerly expropriated was reinforced to counteract the raising of the assessment basis and to keep the resulting budgetary burden within the constraints agreed earlier. Simultaneously, the deadline for payments was postponed by suggesting a model according to which indemnification claims could only be made by mortgage bonds due in year 2004 and not yielding any interest in the meantime.

The outcome of the decision-making process: a bond model and a Land Acquisition Programme

All market elements of the Gattermann proposal were deleted from the records in the new 'Bond Model', leaving only the (fiscal-politically motivated) issuing of 'budget certificates' effective far into the future, nevertheless without being related to any property but rather only as transformer for indemnities. These certificates are designed to be marketable; however, they are only equal to the comparable market value under realisation of the loss of interest when claimed before the maturity date (i.e. 2004).

Finally, on the basis of this modified model a consensus was achieved to pass the Indemnification and Compensation Act (EALG) from September 24, 1994. The Bundesrat refused at first, however, to accept the draft of the new bill, because the interests of the New Bundesländer came into play. It was first after the joint mediation committee from the Bundesrat and Bundestag had presented a compromise proposal during a second attempt at mediation, in which the unequal treatment of legal persons (essentially representing the successors of the LPGs), compared with the previous owners, new establishers and re-establishers concerning their possibilities to buy nationalised land and forest property, was given up, that a settlement was achieved (see Agra-Europe, 1994a).

Two components of the the resulting law are the special interest: the assessment of compensatory payments for former expropriations, and the simultaneously passed Land Acquisition Programme (see Indemnification and Compensation Act - EALG, 1994; Agra-Europe, 1994b; Klages, 1994, p. 112f; Klages and Klare 1994; 1995a, pp. 8f; 1995b, p. 44; 1995c, p. 46). The assessment of the indemnification and compensatory payments is derived from the threefold rateable value of the expropriated operations, based on those values from 1935. However, the amount of indemnification will be graded digressively, the increases of the compensatory payments being inversely proportional to the amount of the rateable value of that property. Compensation which the recipient has already obtained – the so-called 'Lastenausgleich' – are to be credited to the result of the calculated indemnification. In addition, the amount of compensation decreases through the loss of interest arising from the fact that the indemnification is issued in the form of bonds which first mature in 2004.

Both programmes of the second phase of the Bohl paper pictured in the first section (Land Acquisition and Settlement Programme) were summed up in one Land Acquisition Programme. All buyers of nationalised land participating in this programme are to pay three times the amount of the 1935 rateable value for the purchased land and forest property. This amount, on average, 3,000 DM/ha, corresponds to, approximately, half of the market value. The land cannot be sold again within 20 years without permission from the BVVG. The Land Acquisition Programme represents, as it were, a special area of the Indemnification and Compensation Act, because such farmers who can make no claim upon indemnification nor compensation also belong to the entitled category of persons. As a matter of fact, all active farmers may take part in the programme if they can fulfil the following requirements:

1 They have to be new establishers in the locality, re-establishers, legal persons or – in respect to the latter, but of less importance – members of farms organised as legal persons, e.g., co-operatives, limited or share holding companies.

2 The land for which they strive to purchase has to already have been leased by them over a long period of time: at least for 12 years.

3 The purchase of land is subjected to specific restrictions which ensue firstly from the overall amount of soil points to be acquisitioned (maximum of 6,000), and secondly from the restriction that the share of

property, including that newly aquisitioned, not exceed 50 percent of the overall size of the farm.

What is remarkable is that the original predominant objective of the Land Acquisition Programme – that of providing those previous owners having no claim for restitution with an option to buy back their properties – lost considerable influence in the course of the political decision-making process, and ended up playing a rather insignificant role: the privileges described above are certainly available to previous owners who have leased Treuhand land for long terms. Those previous owners no longer active as farmers, however, may only purchase land under two additional restrictions.

1 Their purchase claim for agricultural properties amounts to half the amount of indemnification.

2 Further restrictions to be taken into consideration are:
   a) the second-place status of the previous owners (without lease contracts) in comparison with an equally interested buyer, who, being a lease holder of the property at hand, is more therefore entitled;
   b) the obligation to extend the lease contract for a total of 18 years to the present lease-holder, independent from his economic situation and expiration of the existing lease contract;
   c) a reduction of the soil point ceiling restriction (see above) to 3,000 instead of 6,.000 (Klages, 1994, p. 113).

A complete sterilisation of competition between the interested buyers results from this Land Acquisition Programme. The tenants of the land already belonging to the privileged groups declared in the Three-Phase Model of the Bohl paper now receive priority in comparison to their sole competitors, who are the previous owners having no claim for restitution. The development of the amended Land Acquisition Programme together with the prospective effects of this programme – as analysed by Klages and Klare – make it very clear that in the course of the political process, East German agriculture showed a growing ability to exert its influence, and that the previous owners from West Germany continually had to scale back their originally high objectives. As a result of a quantitative analysis, Klages (1994, p. 118) emphasises, for example, 'that from the purchasing possibilities provided by the EALG, it is not so much the previous owners having no claim for restitution that can greatly benefit here, but very much the BVVG lease-holders, and above all, the legal persons'. The Land Acquisition Programme is arranged first to serve the interests of the active agricultural businesses already holding a lease-agreement. 'All other competitors, especially the previous owners not entitled to restitution, could be characterised as the 'losers' of the political arena in respect to the original concept of privatisation' (Klages, 1994, p. 119; see also Klages and Klare, 1994; 1995a,b,c).

Only through the compromise made in the last steps of the legislative procedure, which administratively secured the priority of the interests of those farmers active in Eastern Germany, was a conclusion of the process attainable. Moreover, this points to the rather fast learning process of the East German farmers and their interest representatives in using the political system of Western Germany to enforce their priorities. The design of the Land Acquisition Programme shows, in the end, that even the remaining residue of competition between tenants and previous owners was cancelled.

There are two main explanations for the rejection of the market-oriented, opening of the procedure and complete elimination of competition (see Top-Agrar-Spezial, 1993, p. 12). First, the East German farmers were afraid of what they perceived to be their weaker position in a market process. They hoped to receive land by the Land Acquisition Programme, and this land would -if the Gattermann plan had been accepted- no longer be available; instead, it would have been sold in auctions. Second, the political actors were afraid that they would no longer be able to control the conflicts caused by privatisation if more market mechanisms were to be introduced. This is easy to explain, because public opinion and protest movements in Eastern Germany are very important for politicians in a period of fundamental transformation when social peace is one of the most valuable pre-requisites of successful development.

Pre-privatisation by leasing nationalised land before selling it

Since the Land Acquisition Programme has still to be implemented, agricultural structures in Eastern Germany were still in 1997 affected by leasing procedures applied by the BVVG. Until 1994, the BVVG concluded nearly 13,000 leasing contracts. Half of the contracts are on a long-term basis, i.e. lasting 12 years. In 1993/94, 1.14 million ha were leased, 0.7 million ha on a long-term and 0.445 million ha on a short-term basis. About 60 percent of the farmland was given to legal entities, 18 percent to local residents who re-established a farm, and 10 percent to local residents who established a new farm (see Table 4.4). Persons without restitution claims, i.e. the so-called victims of the land reform 1945-1949, received less than 5 percent of the land. The average area rented from the BVVG by legal entities was 176 ha, by, local residents who re-established a farm 46 ha, by local residents who established a new farm 61 ha, and by former owners expropriated during the land reform, 106 ha.

Table 4.4

Agricultural land leased by the German Land Utilisation and Administration Company in Eastern Germany

Type of applicants	*Land leased on a short-term basis* ha	%	*on a long-term basis* ha	%	*Total* ha	%
Local residents who re-establish(ed) farms	77,458	17.84	130,470	18.75	209,928	18.39
Persons who re-establish(ed) farms with restitution claims	5,461	1.23	8,477	1.22	13,938	1.22
Persons who re-establish(ed) farms without restitution claims	13,696	3.07	38,684	5.56	52,380	4.59
Local residents who established new farms	42,115	9.45	67,521	9.70	109,636	9.61
Legal entities	274,347	61.59	418,678	60.16	693,025	60.72
Persons who establish new farms and will become local residents	27,237	6.11	27,481	3.95	54,718	4.79
Other applicants	3,149	0.71	4,654	0.67	7,803	0.68
Total	443,463	100	695,965	100	1,141,428	100

Source: Agra-Europe 29/1994, p. 43f

Development of agricultural structures in Eastern Germany

In the starting situation of the transformation process the ownership rights of many land owners were re-established, and a broad distribution of property rights in farmland was achieved. The government decided on a transformation strategy which only took care of legal rules and provided (generous) financial assistance for restructuring collectivised agriculture, but refrained from determining the route or the outcome of the structural changes.

Due to the characteristics of the initial property rights structure and of agricultural policies in Germany, adjustment of agriculture structures occurred mainly as a response to the new situation induced by self-organisation. The government has defined the rules that guided the process of self-organisation and offered support to reduce the financial burden of private actors caused by the adjustment process.

A different approach is used in the case of the privatisation policies for public property which also has an impact on the structural adjustment process. These policies represent external determination (by policies) rather than self-organisation as we have seen in the last Section. According to the privatisation procedures applied to nationalised land, the THA and the BVVG act as a dominating landlord, at least in some regions. The policies of the THA and the BVVG have exerted a monopolistic influence on the land market, because access to leased land is restricted. Furthermore, the THA and BVVG have assumed the role of price leadership and, in this way, also influenced the rents of other land leased by private land owners.

The development of the agricultural production structures can be divided into two periods: the first period from 1989 until the end of 1991 was characterised by drastic changes in the political, economic and social framework. According to the Agricultural Adjustment Act LPGs were forced to change the legal form until the end of 1991. The second period covers the time from 1992 until 1997; in this period, the changes occurred in a more continuous and regulated way.

The evolution of agricultural structures will be outlined in the following paragraphs focusing first on legal forms of farm organisation[5] and then on farm sizes.

Farm organisation

Some results of the first period of transformation are shown in Table 4.5. A huge increase in the number of farms so that in 1992 there were 18,575 units. Farms were established in nearly every legal form provided by West German law, but especially sole proprietorships, registered corporations, limited corporations, civil partnerships and limited partnerships (for a description of the legal forms, see Table 4.6). In 1992, most of the agricultural land was cultivated by co-operatives, followed by limited companies, sole proprietorships, limited partnerships and unlimited partnerships.

Although these data provide some insights into the *outcome* of the first 'round' of transformation process, it does not provide any information on the transformation process itself, i.e. on the details and determinants which shape the evolution and the path of the changes as well as on the network of actors involved. Since we are primarily interested in this field of research (see also Kabat and Hagedorn, chapter 7), we will present some information on this issue.

According to a survey made by König (1992) in 1991, about 40 percent of the LPGs included in the sample were eliminated due to the unification of plant and animal production, fusion with other enterprises, liquidation and bankruptcy. About 60 percent of the LPGs transformed into one of the legal forms offered by Western German law. Until 1992, about 49 percent these transformed enterprises changed into corporations, about 39 percent into limited corporations, 9 percent into limited partnerships, 4 percent into unlimited partnerships, and 2 percent into joint-stock corporations. This means that all co-operative associations, limited corporations, limited partnerships and joint-stock corporations originated from transformed LPGs, but only one sixth of the civil partnerships did so. The majority of unlimited partnerships were established as new enterprises. Furthermore, it can be assumed that no sole proprietorship was created by transformation of a LPG.

In the second period of transformation between 1992 and 1995, the number of farms increased by 63 percent up to 30,248. Co-operatives were the only important legal entities which decreased in numbers. In contrast, limited companies were the only legal entity which are increasing in numbers. There was a very big increase in numbers of unlimited partnerships, which grew by 183.8 percent from 1992 to 1995, and the area of land cultivated these farms expanded by 120.9 percent.

Farm sizes

The development of farm sizes is illustrated in Table 4.7. From 1992 to 1995, the average size of the farms dropped from 275 to 183 ha. In nearly every category of organisational form the average farm size decreased as the number of organisational units increased. In 1995, co-operatives were the largest enterprises with 1,435 ha on average, followed by joint stock companies with 1,344 ha, limited partnerships with 1,164 ha and limited companies with an average size of 843 ha. Sole proprietorships worked on a comparatively small scale with 46 ha on average and unlimited partnerships on a medium scale with 373 ha. In comparison, sole proprietorships in Western Germany had in 1995 an average size of 22.1 ha.

Table 4.5

Development of agricultural enterprises in Eastern Germany from 1992 to 1995

Legal form of the enterprise[a]	*Year* 1992	1993	1994	1995	*Share in 1992*	*Share in 1995*	*Change 1994 to 1995*
		Number of farms				%	
Sole proprietorship	14,602	20,587	22,505	24,588	78.6	81.3	+9.3
Partnerships	1,123	1,879	2,379	2,671	6.0	8.8	+12.3
Unlimited Partnership	760	1,416	1,890	2,157	4.1	7.1	+14.1
Limited partnership	257	311	331	335	1.3	1.1	+1.2
Other	106	152	158	179	0.6	0.6	
Co-operatives	1,464	1,388	1,336	1,315	7.9	4.3	-1.6
Companies	1,386	1,514	1,563	1,674	7.5	5.5	+7.1
Limited company	1,178	1,302	1,335	1,417	6.4	4.7	+6.1
Joint stock company	63	64	64	59	0.3	0.2	-7.8
Other	145	148	164	198	0.8	0.6	+20.7
Total	18,575	25,368	27,783	30,248	100	100.0	+8.9
		Agricultural land, 1000 ha[b]				%	
Sole proprietorship	674.0	932.4	1,081.7	1,141.3	13.2	20.7	+5.5
Partnerships	706.3	959.6	1,116.4	1,199.2	13.8	21.7	+7.4
Unlimited Partnership	320.9	562.0	708.9	805.3	6.3	14.6	+13.6
Limited partnership	379.7	393.4	403.3	389.9	7.4	7.1	-3.3
Other	5.6	4.1	4.2	4.0	0.1	0.0	

Co-operatives	2,250.6	2,053.7	1,952.1	1,887.4	44.1	34.2	-3.3
Companies	1,477.6	1,351.0	1,286.2	1,292.7	28.9	23.4	+0.5
Limited company	1,314.2	1,234.8	1,178.0	1,193.9	25.7	21.6	+1.3
Joint stock company	97.4	87.3	81.6	79.3	1.9	1.4	-2.8
Other	66.0	29.5	26.6	19.5	1.2	0.4	-26.7
Total	5,108.6	5,297.3	5,436.4	5,520.6	100.0	100.0	+1.5

[a] Farms with a size of more than 1 ha agricultural land.

[b] The statistical methods of measuring agricultural land differ between the GDR and the FRG. Therefore and because of possible statistical errors in the first years of the transformation process the data in Table 4.1 and those provided in this table are not directly comparable.

Source: Agrarbericht 1993, 1994, 1995, 1996

Table 4.6
Characteristics of different legal forms of enterprises in Germany

Characteristics	*Type of legal form of enterprise in Germany (Private Law)*							
English term (GB)	Sole proprietorship	Civil Law partnership	Commercial Law partnership	Limited Commercial Law partnership	Limited Commercial Law partnership & Limited company	Limited liability company	Public limited company	Co-operative
German term	Einzel-unternehmen	Gesellschaft bürgerlichen Rechts (GbR)	Offene Handelsgesell-schaft (OHG)	Kommandit-gesellschaft (KG)	Kommandit-gesellschaft und Gesellschaft mit beschränkter Haftung (GmbH % Co. KG)	Gesellschaft mit beschränkter Haftung (GmbH)	Aktien-gesell-schaft (AG)	Einge-tragene Ge-nossen-schaft (e.G.)
Legal basis for existence		Civil Law	Commercial Law	Commercial Law	Commercial Law and Limited liability company Law	Limited liability company Law	Share Law	Co-operative Law
Registration	No	No	Yes	Yes	Yes	Yes	Yes	Yes
Minimum of members	-	2	2	2	2	1	2	7

Minimum of capital	-	-	-	-	-	50,000 DM	100,000 DM	-
Individual tradability of capital	Yes	No	No	No	No	Yes (with restrictions)	Yes	No
Liability	Unlimited	Unlimited	Unlimited	Unlimited partner and limited partner	Limited	Limited	Limited	Limited (it can be unlimited)
Management		Partners, no delegation	Each partner, delegation possible	Unlimited partner (other arrangements possible)	Many arrangements possible	- Chief executive - General meeting of members	- Board of mana-gement - Super-visory board - Stock holders' meeting	- Board of mana-gement - Super-visory board - General assembly

Source: See the line: 'legal basis for existence'

Table 4.7

Development of average farm sizes in different legal forms of the enterprises in Eastern Germany, 1992-1995

*Legal form of the enterprise**	*Year*				*Changes*	
	1992	*1993*	*1994*	*1995*	*1993 to 1992*	*1995 to 1994*
	Agricultural land per farm, ha				*%*	
Sole proprietorship	46	45	48	46	-2.2	-4.2
Partnerships	626	511	469	449	-18.4	-4.3
Unlimited partnership	422	397	375	373	-5.9	-0.5
Limited partnership	1,477	1,265	1,218	1,164	-14.4	-4.4
Other	53	27	27	22	-49.1	-18.5
Co-operatives	1,537	1,480	1,461	1,435	-3.7	-1.8
Companies	1,006	892	823	772	-16.3	-6.2
Limited company	1,116	948	882	843	-15.1	-4.4
Joint stock company	1,546	1,364	1,276	1,344	-11.8	+5.3
Other	455	199	162	98	-56.3	-39.5
Total	275	209	196	183	-24.0	-6.6

* Farms with a size of more than 1 ha agricultural land

Source: Agrarbericht 1992; 1993; 1994; 1995; 1996

In Figure 4.1 more detailed information on the distribution of organisational forms and farm sizes is presented for 1994. About 54.6 percent of the sole proprietorships cultivated less than 20 ha, 17.8 percent more than 100 ha and only 0.3 percent more than 1,000 ha. The majority of the partnerships, i.e. 54.3 percent, had between 100 and 500 ha, only 22.2 percent cultivated less than 100 ha and 7.4 percent more than 1,000 ha. Co-operatives were concentrated in the class between 1,000 and 2,500 ha, but also 11.5 percent of the co-operatives had less than 100 ha. Companies and limited partnerships cover a wide range of sizes, the distribution is triple peaked: 25.3 percent had an agricultural area of less than 20 ha, 20.7 percent were located between 100 and 500 ha and 29.5 percent were larger than 1,000 ha.

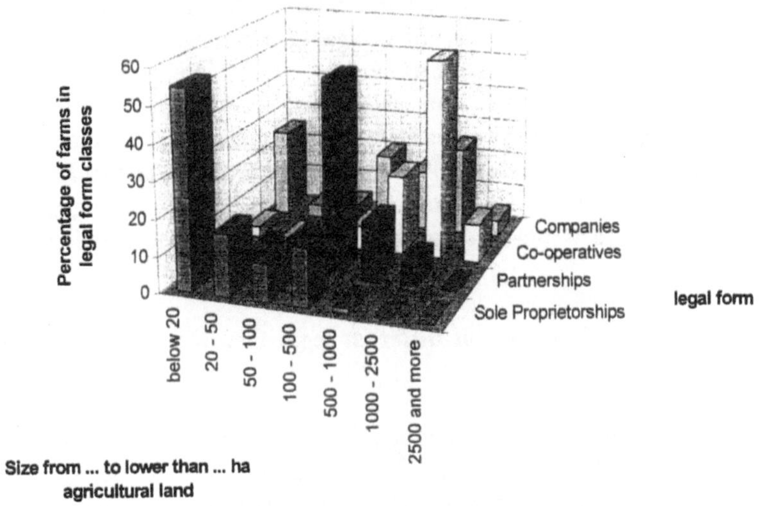

Figure 4.1 Farm size and legal form of the farms in Eastern Germany, 1994

Source: Kruse, 1995.

These data reveal that although there is a close relationship between farm size and farm organisation, sole proprietorships and partnerships can operate on a large scale as well as co-operatives and companies, and co-operatives and companies can also work on a small scale. Therefore it is imaginable that an adjustment in size does not affect the organisation of a farm as well that an adjustment with regard to organisation does not affect its size.

Farm size and land ownership

One major question of the structural adjustment in agriculture is how the ownership structure affects the structure of the operating units. In 1949, after the so-called 'land reform' 838,000 farms existed in Eastern Germany. Only 10 percent of the land was rented land, 90 percent was cultivated by the owners themselves. If every land owner would have taken the land back after the re-establishment of the property rights in land in order to rebuild their private holding, more than 800,000 farms would have been re-established. However, in 1993 only 8.26 percent of the agricultural land was used by the 1979 owners (or their heirs) themselves and most of the land was leased out. Nevertheless, a certain causal connection between the ownership of land and the size of the farm can be identified. In farms with a size of less than 50 ha 54 percent of the cultivated land is owned land, while in farms with more than 50 ha this is only true for 6.7 percent of the land. The data indicate that most of the new established small farms use their own land.

Before these changes are interpreted theoretically, some information concerning the economic performance of farms will be drawn from farm accounting results. Since the Agricultural Report 1993, annual farm accounting data are provided for three legal forms of farm organisation: sole proprietorships in full-time farming, partnerships, and corporations.

Economic performance of different legal forms: some results from farm accounting

In this section, we will present and discuss some selected economic results of farms with different legal forms[6] based on the annual Agricultural Reports of the German government from 1993 to 1995. It should be pointed out that these data lack representativeness due to a relative small sample of farms. However, until now no better data are available.

Profits and wages

It is a serious problem to compare the economic performance of different leagel forms of enterprise. This problem arise mainly because the proportion of unpaid production factors differs considerably between farms with different legal forms (see the hired labour ratio in Table 4.8. In family farms the labour force is mostly unpaid, because family members usually do not receive wages. Therefore, the Agricultural Report uses profits plus wages as an indicator to compare the economic performance of enterprises with different legal forms.

Based on this indicator, partnerships show better economic results than the other types of farms. In 1994/95, the sum of profits and wages in partnerships reached 55,542 DM per labour unit, followed by sole proprietorships with 41,314 DM per labour unit and 34,884 DM per labour unit for co-operatives and companies. However, Table 4.8 also indicates that co-operatives and companies improved their performance by 54.2 percent from 1991/92 to 1994/95. Although sole proprietorships and partnerships also achieved a small increase in this nominal measure of performance, they improved relatively less than co-operatives and companies.

It is interesting to try to explain these differences in economic success of enterprises with different legal forms.

Farm size, factor mix and performance

For an explanation of the different performance of the farm types it may be useful to start with the size of the farm measured in labour input. Figure 4.2 shows the profits and wages per labour unit depending on the total labour input. In 1994/95, sole proprietorships used 2.16 labour units (see Table 4.9) and received 41,314 DM of profits and wages per labour unit (point A), partnerships used 5.11 labour units and reached an amount of profit and wages of 55,522 DM per labour unit (point B). Co-operatives and companies used 44.9 labour units and got profits and wages of 34,884 DM per labour unit (point C). If we draw a line that covers this three points, we may get an imagination of the average 'profit and wages curve' which first increases up to 5 labour units and then slowly decreases. This 'profit and wage curve' may be interpreted as the average return to labour curve. A growing average return to labour indicates economies of size and a falling average return to labour diseconomies of size. The better economic performance of partnerships may therefore be caused by some economies of size, and the relatively weak performance of co-operatives and companies may therefore be caused by some diseconomies of size.

In addition, this stylised 'profit and wages curve' may explain some of the adjustment processes we actually observe in the last years. Table 4.9 shows, that the labour input in sole proprietorships is growing since 1991/92 which may be explained by the fact that they try to benefit from some economies of size. In principle, the adjustment made by co-operatives and companies can be understood in the same way; but co-operatives and companies decrease their size according to labour input to improve their economic performance. In 1991/92 co-operatives and companies used 67.5 labour units and received profits plus wages of 22,609 DM per labour unit (see Table 4.8 and 4.9). This is more or less represented by the point D in Figure 4.2. In the last four years co-operatives and companies moved from point D to point C. In contrast, partnerships first increase the labour input from 4.89 labour units in 1991/92 to 5.5 labour units in 1993/94 and than decreased to 5.11 labour units in 1994/95.

Table 4.8
Profits in farms with different legal forms in Eastern Germany, 1991/92 - 1994/95

Legal form of the enterprise	*Year* 91/92	92/93	93/94	94/95	*Change* 94/95 to 91/92
	*Profits per farm, 1000 DM**				
Sole proprietorship (full-time farming)	64.4	66.9	74.1	70.6	+9.6
Partnerships	179.3	293.3	193.4	174.9	-2.4
Co-operatives/ Companies	-322.1	-93.2	-82.7	-40.8	+87.6
	Hired labour ratio, %				
Sole proprietorship	18.3	23.9	29.4	28.7	+56.8
Partnerships	60.3	56.9	60.2	55.4	-8.1
Co-operatives/ Companies	100	100	100	100	-
	*Wage per hired labour unit, DM**				
Sole proprietorship	23,836	27,502	28,948	29,792	+25.0
Partnerships	30,243	31,956	34,248	38.567	+27.5
Co-operatives/ Companies	27,702	30,557	35,506	35,763	+29.1
	*Profits plus wages per farm, 1000 DM**				
Sole proprietorship (full-time farming)	72.3	79.6	92.1	89.2	+23.4
Partnerships	268.6	385.5	306.7	283.8	+5.4
Co-operatives/ Companies	1,580.5	1,552.3	1,555.6	1,619.1	+2.4

	*Profits plus wages per labour unit, DM**				
Sole proprietorship (full-time farming)	40,151	40,620	43,148	41,314	+2.9
Partnerships	54,920	75,880	55,766	55,542	+1.1
Co-operatives/ Companies	22,608	27,987	31,479	34,884	+54.3
	*Profits per unpaid labour unit, DM**				
Sole proprietorship	43,158	44,785	49,903	45,943	+6.5
Partnerships	92,413	133,945	88,302	76,705	-17.0
Co-operatives/ Companies	-	-	-	-	

* Monetary statistics are in nominal terms.

Source: Agricultural Report 1993; 1994; 1995; 1996

Table 4.9

Size of farms according to labour input, agricultural land and livestock in different legal forms in Eastern Germany, 1991/92 - 1994/95

Legal form of the enterprise	*Year*				*Change*
	91/92	*92/93*	*93/94*	*94/95*	*94/95 to 91/92*
	Labour per farm, labour units				%
Sole proprietorship	1.80	1.96	2.11	2.16	+20.0
Partnerships	4.89	5.08	5.50	5.11	+4.5
Co-operatives/ Companies	67.5	54.6	48.1	44.9	-33.9
	Agricultural land per farm, ha				%
Sole proprietorship (full-time farming)	114	141	157	160	+40.3
Partnerships	328	444	436	415	+26.5
Co-operatives/ Companies	1.754	1.786	1.736	1,721	-1,9
	Livestock per farm' Livestock units				%
Sole proprietorship (full-time farming)	28.9	36.5	43.0	46.7	+61.6
Partnerships	79.7	83.5	129.1	131.1	+64.5
Co-operatives/ Companies	1,303.2	1,225.2	1,173.5	1,128.9	-13.4
	Assets[a] per farm, 1000 DM[b]				
Sole proprietorship (full-time farming)	276.8	367.1	419.2	445.6	+60.9
Partnerships	631.7	824.1	934.7	940.3	+48.9
Co-operatives/ Companies	4,439.3	4,172.1	4,393.8	4,620.9	+0.4

		Sales per farm 1000 DM[b]			%
Sole proprietorship (full-time farming)	253.9	292.6	346.2	359.6	+41.6
Partnerships	826.8	1,070.8	1,066.0	1,047.2	+26.7
Co-operatives/ Companies	5,689.6	5,610.6	5,594.9	5,788.9	+1.7

[a] Machinery and buildings
[b] Monetary statistics are in nominal terms

Source: Agricultural Report 1992; 1993; 1994; 1995; 1996

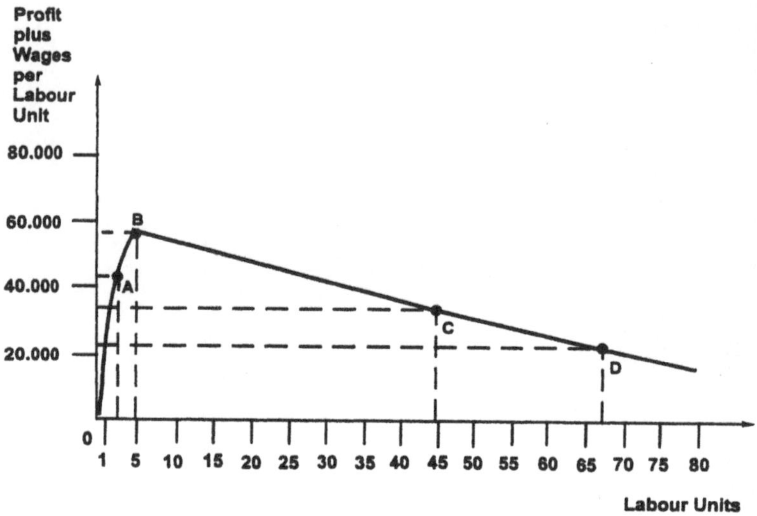

Figure 4.2 Farm size according to labour input and profits plus wages per labour unit

Although this simple model seems to have power of explanation, it has to be recognised that the different labour inputs are simultaneously connected which a different factor mix. Table 4.10 shows that in 1994/95 sole proprietorships used 74.5 ha of land and 206,300 DM of assets (machinery and buildings) per labor unit, while partnerships employed 81.2 ha and 184,100 DM of assets per labor unit. In contrast, co-operatives and companies had 38.3 ha land and 102,900 DM of assets per labour unit. These data reveal that co-operatives and companies uses too much labour in relation to land and capital. The poor performance may therefore not be caused by diseconomies of size but by allocative inefficiency. Co-operatives and companies, therefore, have to adjust the land/labour and capital/labour ratio. There may be two ways of adjustment: (1) holding the labour input constant, renting additional land and investing in assets or (2) reducing the labour force and keeping the land input and assets constant. If there exists no diseconomies of size both way of adjustment may be equally possible. However, as it can be seen in Table 4.9 co-operatives and companies are going the second way of adjustment. They tried to keep their size according to land and capital input constant and reduced there labour force. The reduction in the labour force in co-operatives and companies of 33 percent from 1991/92 to 1994/95 was simultaneously connected with a reduction in animal units of 13.1 percent and in agricultural land of 1.7 percent.

What would happen if the co-operatives and companies adjusted to the structure of partnerships? Let us assume that co-operatives and companies kept the size of the farm measured in land constant so that they would have about 1,700 ha. In order to cultivate 81.2 ha per labour unit as partnerships do, they would than have to reduce their labour input from 44.9 to 21.2 labour units, and simultaneously they have to reduce their livestock from 1,128.9 to 542.7 animal units. As it can be seen in Figure 4.2, such a reduction in labour force may lead to the same profits and wages per labour unit as the sole proprietorships are earning.

The role of government support

As shown earlier, transformation of agriculture in Eastern Germany has been encouraged and facilitated by considerable government support. Since 1993, the annual Agricultural Report provides some information on how the farms with different legal forms benefited from this financial assistance (see Table 4.11). In 1994/95, co-operatives and companies received 683 DM government support per hectare of agricultural land, sole proprietorships 678 DM, and partnerships 638 DM. However, if government support is calculated per labour unit, the picture changes fundamentally: partnerships received 51,877 DM per labour unit, sole proprietorships 50,353 DM, and co-operatives and companies 23,767 DM.

The lower government support for co-operatives and companies per labour unit can be explained by lower labour productivity of these type of farms. They cultivate less than half of the land per labour unit compared with sole proprietorships or partnerships. As a consequence, equal government support per hectare must lead to unequal government support per labour unit. From this point of view, there seems to be no direct impact of government support on co-operatives and companies responsible for their weak economic performance. The main problem of co-operatives and companies must rather be seen in low labour productivity.

From 1991/92 to 1994/95 the role of government support for the economic performance of all farm types increased drastically. This is not only an effect of the government support mentioned earlier but also a result of the agricultural reform of the EU policy. The data shows how deeply the economic performance especially of sole proprietorships and partnerships depends on the government support. In 1994/95 the government support amounts in sole proprietorships to 121 percent of profits and wages and in partnerships to 93.5 percent of profits and wages. Co-operatives and companies do not depend so deeply on government supports; the government supports amount only to 68 percent of profits and wages.

Table 4.10
Factor proportions in different legal forms in Eastern Germany, 1991/92 - 1994/95

Legal form of the enterprise	*Year*				*Change*
	91/92	*92/93*	*93/94*	*94/95*	*94/95 to 91/92*
	Land per labour unit in ha				
Sole proprietorship (full-time farming)	63.3	71.9	74.4	74.5	+17.7
Partnerships	67.1	87.4	79.3	81.2	+21.0
Co-operatives/ Companies	26.0	32.7	36.1	38.3	+47.3
	Livestock per labour unit, Livestock units				
Sole proprietorship (full-time farming)	16.1	18.4	20.4	21.7	+34.8
Partnerships	16.2	16.3	23.5	25.6	+58.0
Co-operatives/ Companies	19.3	22.4	24.4	25.1	+30.1
	Assets[a] per labour unit, 1000 DM[b]				
Sole proprietorship (full-time farming)	153.8	187.3	198.7	206.3	+34,1
Partnerships	123.2	164.8	170.0	184.1	+49.4
Co-operatives/ Companies	65.8	76.4	91.3	102.9	+56.4
	Sales per labour unit, 1000 DM[b]				
Sole proprietorship (full-time farming)	141.1	149.3	164.1	166.5	+18.0
Partnerships	169.1	210.8	193.8	204.9	+21.2
Co-operatives/ Companies	84.3	102.8	116.3	128.9	+52.9

[a] Machinery and buildings. [b] Monetary statistics are in nominal terms.

Source: Agricultural Report 1992; 1993; 1994; 1995; 1996

Table 4.11

Government support in farms with different legal forms in Eastern Germany, 1991/92 - 1994/95

Legal form of the enterprise	*91/92*	*92/93*	*93/94*	*Year 94/95*	*Change 1991/92 to 1994/95*
	Government support[a] per ha land, DM[b]				*%*
Sole proprietorship (full-time farming)	390	390	597	678	+73.8
Partnerships	255	323	577	638	+150.0
Co-operatives/ Companies	439	501	618	683	+45.3
	Government support[a] per labour unit, DM[b]				*%*
Sole proprietorship (full-time farming)	24,657	27,835	44,551	50,353	+104.2
Partnerships	17,123	28,714	45,734	51,877	+202.9
Co-operatives/ Companies	10,782	16,420	21,842	23,767	+120.4
	Government support[a] in % of profit plus wages				*%*
Sole proprietorship (full-time farming)	61.0	68.6	102.2	121.8	+99.7
Partnerships	31.0	52.3	81.9	93.5	+201.6
Co-operatives/ Companies	48,0	58.8	69.6	68.1	+40.9

[a] Including direct price compensations, animal premiums, subsidies on interest payments, investment supports, set-aside payments, and other transfers.
[b] Monetary statistics are in nominal terms.

Source: Agricultural Report 1992; 1993; 1994; 1995; 1996

The presented farm accounts are not representative for all farm types. Especially for sole proprietorships the data covers only larger full-time farms. If measured in ha agricultural land the farm observed in the farm accounts have an average size of 160 ha while the sole proprietorships in the general statistics have an average size of 46 ha (see Table 4.7). Also partnerships in the farm accounts sample are larger than in the population. The same holds for co-operatives and companies. Another problem is that the farm accounts data does not distinguish between co-operative and companies. As Table 4.7 shows, co-operatives are usually larger than companies especially limited companies.

Despite this lack of representativeness, this data may explain why sole proprietorships and especially partnerships are increasing in numbers and in agricultural area. Quite simply, partnerships and sole proprietorships perform better than co-operatives and companies. However, co-operatives and companies have increased their performance and this may explain why the speed of structural adjustment has slowed down in the mid-1990s. If the hypothesis formulated in Figure 4.2 is right, we can expect that co-operatives and companies may be able to increase their performance in the future may be up to the level of sole proprietorships or even partnerships. This will be reached by a further reduction in the labour force and may be also in livestock and land.

But still some puzzles remain:

1 If there exists some economics of size why do not all farms exploit it?

2 Why are co-operatives decreasing and limited companies increasing in numbers?

We try to answer these question in the following section.

Some theoretical interpretation of the structural adjustment in Eastern Germany

Among agricultural economists in Germany there was a very intensive discussion on what will be the outcome of transformation of agriculture in Eastern Germany in a competitive environment. The debate on the question which new structures will replace the old ones concentrated on two questions: (a) What farm size, and (b) which organisational or legal form of farms will prevail (see, for example, Henrichsmeyer 1990; Köhne 1990; Langbehn 1990; Schmitt 1990; Hagedorn 1991):

In search for an answer to these questions, several agricultural economists in Germany have arrived at two hypothesis: (1) co-operative production has only a poor chance in competition with other organisational forms, and (2) in the long run, i.e. after a long transition period with a great heterogeneity of farm sizes and organisation, family farms – and perhaps other organisational forms with similar economic features – will be the predominant outcome of the transformation process. The economic arguments leading to this suggestion are:

1 Co-operatives in agricultural production have comparatively high transaction costs due to collective decision making and inappropriate incentive structures regarding, e.g., work effort, management and investment (see Schmitt, 1990, 1991; Bonus, 1986; Beckmann, 1993).

2 Transaction costs in agricultural production are able to be minimised within the organisation of a family farm, and the main reductions in production costs can also be achieved within the potential size of a family farm (see Schmitt, 1990, 1992; Pollak, 1985; Hagedorn, 1991, 1992).

This theoretical framework is based on the assumption that the organisation of production is efficient when the sum of production and transaction costs together are minimised (see Williamson, 1985). In addition, it is assumed that efficient structures would emerge if different organisations and sizes would compete with each other and government intervention did not take place.

According to the results of the farm accounts we can conclude that although agriculture is subject to huge government intervention, there is no clear evidence that government supports influence the competitiveness of different farm size and organisations in a selective way. The different farms size and organisations compete on the product and factor markets, especially on the land market.

The outlined structural adjustment and performance of agricultural firms in Eastern Germany seem to support some of the theoretical expectations.

1 The only legal form of farms which is shrinking in numbers is the co-operative. This seems to support the hypothesis of the lack of competitiveness of co-operatives in agriculture production. The farm accounts show that co-operatives and companies perfom worse than sole propritorships and partnerships. Since co-operatives are usually larger than companies (see Table 4.7) we would expect form Figue 4.2 that they perform worse than companies. This has to be substantiated be additional empirical evidence.

2 The number of sole proprietorships and partnerships is increasing. This result corresponds with the better economic performance of sole

proprietorships and partnerships according to the farm accounts. Sole proprietorships are predominantly family farms. Even the majority of partnerships are family partnerships between spouses or father and son, or else. This seems to support the expectation that the family farm is a efficient organisation.

3 The farm accounts indicate the existence of some economies and diseconomies of size. The average profit plus wages per labour unit increase with the size of the farm and may be maximised in the size covered by partnerships. Nevertheless, on average the increasing area of cultivated land by sole proprietorships and unlimited partnerships is not caused by a growth of existing farms but by entry of new farms.

However, even if the evidence cited seems to support the hypotheses, the transformation process is slow. It must be recognised that the transformation process itself causes considerable costs. Individuals as well as society as a whole must be able to pay for huge investments in institution building, education and training, clarification of property rights, solving social conflicts, equipment and buildings, livestock and circulation capital, advisory services and infrastructure. Even if in Eastern Germany the state has been able to cover some of these transaction costs, a major proportion will be borne by individuals. There are mainly two kind of cost involved in this process: (1) the cost of entry for new farms and (2) the costs of transformation for existing farms.

Costs of entry

The structural adjustment occurs on one side by the entry of new or re-established farms. They enter from the left hand side of Figure 4.2. If a new farm to be established the operator must sign contracts for land, labour and capital. But the costs of entry may increase with the size of the farm. If a farm is to be established on a small scale it can use mainly family labour, owned land and equity capital. If a farm is to be established on a larger scale, hired labour has to be employed, land has to be rented and debt funds may have to be borrowed. As Table 4.12 shows from the farm accounts data, sole proprietorships and partnerships made huge investments in the first years. The net investment in 1991/92 amounts for sole proprietorships to 83.800 DM and for partnerships to 89,600 DM per labour unit. In contrast, co-operatives and companies showed negative investments in the 1991/92 and 1992/93.

Table 4.12

Investments in different legal forms in Eastern Germany, 1991/92 - 1994/95

Legal form of the enterprise	*Year*				*Change*
	91/92	*92/93*	*93/94*	*94/95*	*91/92 to 94/95*
	*Gross investment per ha land, DM**				*%*
Sole proprietorship (full-time farming)	1,607	1,024	949	846	-47.4
Partnerships	1,619	862	819	761	-53.0
Co-operatives/ Companies	461	465	645	866	+87.8
	*Gross investments per labour unit, 1000 DM**				*%*
Sole proprietorship (full-time farming)	102	73	71	63	-38.8
Partnerships	109	75	65	62	-43.3
Co-operatives/ Companies	12	15	23	33.2	+176.7
	*Net investment per ha land, DM**				*%*
Sole proprietorship (full-time farming)	1,323	676	543	349	-73.6
Partnerships	1,336	611	433	359	-73.1
Co-operatives/ Companies	-197	-73	80	185	+193.9
	*Net investments per labour unit, 1000 DM**				*%*
Sole proprietorship (full-time farming)	83.8	48.5	40.3	25,9	-69.1
Partnerships	89.6	53.3	34.4	29.2	-67.4
Co-operatives/ Companies	-4.9	-2.3	2.8	7.1	+244.8

* Monetary statistics are in nominal terms.

Source: Agricultural Report 1992; 1993; 1994; 1995; 1996

These investments had to be financed mainly by borrowing as it can be seen in Table 4.13. In 1991/92, debts in sole proprietorships amounts to 130,000 DM per labour unit and in partnerships to 175,000 DM per labour unit. In co-operatives and companies the debts only reached 53,800 DM per labour unit. The increase in size of sole proprietorships and partnerships is closely connected with an increase in borrowing. In 1994/95 sole proprietorships had debts of 163,000 DM and partnerships 204,400 DM per labour unit. It seems to be clear that the costs of entry may limit the size of the new established farms. Especially the transactions cost in the capital market may limit the farm size of new established farms. It might therefore be only possible for new farms to start farming at a lower size mainly using owned land, family labour and equity capital and to increase the size slowly. This discribes behaviour observed.

Costs of transformation

If former LPGs decided not to liquidate and to continue farming they had to cover the costs of transforming their enterprise and to adjust their size and organisation. They start on the right hand side of Figure 4.2. They already had land in use, buildings, machines and employees and had to made new contracts for land, labour and capital. The major problem was to adjust the size, the organisational structure and especially the labour force. The changes were organised by agricultural producer co-operatives (APC). They had to make collective decisions about the transformation process and had to serve different interest. The theory of APC suggests that such organizations try to maximise the average return to labour per labour unit (see Ward, 1958, Domar, 1966 and Vanek, 1970). Obviously, the observed behaviour is not consitent with this goal, since, as Figure 4.2 shows, the average profit and wage per labour units is maximised in partnerships. If the LPG would have immediately adjust the labour force from point D to B in Figure 4.2 more than two thirds of the labour force would have been dismissed. Due to the fact that most of the employees were members of the co-operative this restructuring is unlikely because of the collective decision making. Nevertheless, the former LPGs move slowly from the right hand side of Figure 4.2 to the left, may be due to the fact that some employees were dismissed or leaved the enterprise voluntarily.

As for society as a whole, it is rational that the individuals seek to reduce the costs of adjustment. Even if a co-operative will be inefficient in the long run, it may be rational to choose this organisational form in order to reduce the cost of adjustment. The co-operative is the legal form of West German law which is very similar the legal form of the LPG in the former GDR. A survey by Fiedler et al. (1994) shows that the most frequently mentioned advantages of the co-operative as a legal form was that it provided the easiest route to transform a

LPG into the closest legal form. In contrast, the advantages of limited companies or limited partnerships were seen in limited liability and better management opportunities. In case of limited partnerships, advantages of taxation were also mentioned. Civil partnerships see their advantages in better management, access to capital, production quotas, and government support (see Table 4.14).

Generalising these results, the major advantages and disadvantages of different legal forms may give some insights into the path of the transformation process of the former LPGs. In the beginning, restricted liability may be considered as a protection against the risks of transformation[7]. As a consequence, co-operatives and limited corporations were likely to be chosen. The co-operative has advantages in reducing the short-term costs of transformation, but disadvantages which respect to management. The co-operative will be chosen if the short term costs of transformation are sufficiently high. In the long run, disadvantages in management may lead to a transformation of co-operatives into limited corporations, partnerships or sole proprietorships. But, especially the close relationship between credit availability and liability may push co-operatives and also limited corporations to the legal form of sole proprietorships or partnerships. External forces like taxation or government support also provides incentives in this direction.

These developments may be supported by a survey by Fiedler et al. (1994, p. 51). According to their survey, 17 percent of co-operatives, 12 percent of limited companies, and 6 percent of the limited partnerships are looking for an additional change in their legal form in the next five years. From the co-operatives which want to transform again, 24 percent want to change into a limited companie, 24 percent into a unlimited partnership, 18 percent into a limited partnership and 11.7 percent into a sole proprietorship. It is notable that the legal form of a unlimited partnership is preferred by most of the farms.

The results of Fiedler et al. (1994, p.50f.) also provide insights into the reasons for transformation of the legal forms. The main incentive for a change of the legal form is the access to government support, which was pointed out by 26 percent of the farms. Another 21 percent reported they wanted to change their legal form to improve motivation of workers, and 14 percent to improve the management of the farm. In the fourth position, also 14 percent wanted to change because of advantages in taxation.

This confirms that the transformation with regard to the legal form of the farms is a response to both internal and external forces. The most important external force is the environment created by the German government in providing incentives to change the legal structures toward natural entities. Simultaneously, the results are compatible with the theoretical framework of transaction economics which predict higher transaction cost of corporate farms due to its particular decision-making and incentive structures.

Table 4.13
Debt in different legal forms in Eastern Germany, 1991/92 - 1994/95

Legal form of the enterprise	*Year* 91/92	92/93	93/94	94/95	*Change* 91/92 to 94/95
	Debt per ha land, DM[a]				%
Sole proprietorship (full-time farming)	2,061	2,097	2,200	2,189	+6.2
Partnerships	2,342	1,961	2,342	2,517	+7.6
Co-operatives/ Companies	2,069	1,790	1,645	1,757	-15.1
	Debt per labour unit, 1000 DM[a]				%
Sole proprietorship (full-time farming)	130.1	150.9	163.7	162.1	+27.6
Partnerships	157.0	171.4	185.6	204.4	+30.2
Co-operatives/ Companies	53.8	58.6	59.4	67.3	+25.1
	Debt ratio, % of total capital				%
Sole proprietorship (full-time farming)	35.3	39.0	39.8	39.1	+10.8
Partnerships	50.9	49.9	54.6	58.0	+16.0
Co-operatives/ Companies	55.3[b]	54.5[b]	47.4[b]	40.6[b]	-26.6

[a] Monetary statistics are in nominal terms.
[b] Including old debts.

Source: Agricultural Report 1992; 1993; 1994; 1995; 1996

Table 4.14
Advantages and disadvantages of different legal forms (in Eastern Germany)

Legal form of enterprise	*Motivation*	*Management*	*Liability*	*Access to capital*	*Access to state support*	*Taxation*
Co-operative	–	–	+	–	–	–
Limited company	–	+	+	–	–	–
Limited partnership	–	+	+	–	–	+
Unlimited partnership	+	+	–	+	+	+
Sole proprietorship	+	+	–	+	+	+

+ Advantages; – Disadvantages

Source: Own presentation

Conclusions

East Germany, of course, is unique among the countries in transition. First, the institutional framework which was already existing in West Germany could more or less simply be transferred to East Germany. Second, the structural adjustment in agriculture was heavily supported by financial aid from West Germany. Other countries in transition are not able to support the adjustment process is such a way. Third, with the unification East Germany became a member of the EU. Other counties are still looking forward to membership.

Despite this uniqueness, the path of the transition does not seem to be so different from other countries in transition. Particularly the Czech Republic and Hungary show similar patterns of structural adjustment. This may indicate that the transition process is mainly driven by internal forces, say the costs of entry and transformation and the potential for economies of size in agriculture.

As in most other countries in transition, the privatisation policy in East Germany was guided by the principle of restitution. Only the agricultural land which was expropriated during the land reform of 1945-49 could not be restituted because of international agreements between Russia and Germany. However, partly because of the complexities of the political interests in the western and eastern Länder took a long time to reach a consensus over the formula for the privatisation process. During the process of decision making, the former owners lost more and more influence and the farmers in East Germany eventually found a way to protect themselves against the entry of the former owners on the land market in East Germany.

Notes

1 In addition, there were about 357,000 small household plots with an average size of 0.75 ha. If we take these into account, 10.2 percent of the cultivated agricultural land was in private use.
2 For a more detailed criticism of the Gerster concept see Hagedorn (1996).
3 The concept has been developed first for the nationalised land that is not part of the nationalised estates operated as state-owned farms ('Volkseigene Güter' – VEG). However, in October 1993 it has also been transferred to the VEG's which will now be privatised in a similar way by the BVVG. The estates are to be divided into reasonable subunits, and will either be sold immediately or will be leased on a trial basis (see Agra-Europe, 1993b,c).
4 A more detailed discussion on the problems of the Bohl paper can be found in Hagedorn (1996).

5 For the characteristics of different legal forms of enterprises in Germany see Table 4.6. The question how the farming types developed cannot be included in this chapter.
6 Of course, this is also a comparision of different sizes due to the interrelationship between legal form and size of farms.
7 This argument is formally developed by Machnes and Schnytzer (1993).

References

Agra-Europe (1992a), 'Treuhandverwaltungsrat stimmt Verwertungskonzept zu', Vol. 33, No. 13, Länderberichte, pp. 4-6.

Agra-Europe (1992b), 'Richtlinie für die Verwertung volkseigener Flächen', Vol. 33, No. 26, Länderberichte, Sonderbeilage, pp. 1-8.

Agra-Europe (1992c), 'Wiedereinrichterprogramm noch in diesem Jahr?', Vol. 33, No. 28, Länderberichte, pp. 26-28.

Agra-Europe (1992d), 'Langfristige Verpachtung als Einstieg in die Flächenverwertung', Vol. 33, No. 50, Länderberichte, pp. 13-17.

Agra-Europe (1993a), 'BVVG legt Mustervertrag für langfristige Verpachtung vor', Vol. 34, No. 11, Länderberichte, pp. 10-13.

Agra-Europe (1993b), 'Bundesfinanzministerium legt Konzeption für die Verwertung der Güter vor', Vol. 34, No. 27, Länderberichte, pp. 1f.

Agra-Europe (1993c), 'BVVG erhält zusätzliche Aufgaben', Vol. 34, No. 27, Länderberichte, pp. 42f.

Agra-Europe (1993d), 'Treuhandflächen in „vorsichtigen Schritten" privatisieren', Vol. 34, No. 38, Länderberichte, pp. 12f.

Agra-Europe (1993e), 'Entwurf des Entschädigungsgesetzes stößt auf breite Ablehnung', Vol. 34, No. 38, Länderberichte, pp. 13-15.

Agra-Europe (1993f), 'Koalition erzielt Kompromiß für Entschädigungs- und Ausgleichsregelung', Vol. 34, No. 48, Länderberichte, pp. 1f.

Agra-Europe (1994), 'BVVG ha fast 13.000 Pachtverträge abgeschlosen', Vol. 34, No. 29, Länderberichte, pp. 42-45.

Agra-Europe (1994a), 'Neuer Kompromißvorschlag zum Entschädigungsgesetz', Vol. 35, No. 36, Länderberichte, pp. 22-25.

Agra-Europe (1994b), 'Gesetz über die Entschädigung nach dem Gesetz zur Regelung offener Vermögensfragen und über staatliche Ausgleichsleistungen für Enteignungen auf besatzungsrechtlicher oder besatzungshoheitlicher Grundlage (Entschädigungs- und Ausgleichs-leistungsgesetz - EALG vom 27. September 1994)', Vol. 35, No. 42, Dokumentation, pp. 1-16.

Agra-Europe (1996a), 'Novelle zum Landwirtschaftsanpassungsgesetz vorgelegt', Vol. 37, No. 28, Länderberichte, Sonderbeilage, pp. 1-10.

Agra-Europe (1996b), 'DBV-Präsidium gegen Novelle des Landwirtschafts-anpassungsgesetz', Vol. 37, No. 37, Länderberichte, Sonderbeilage, pp. 1-5.

Agra-Europe (1996c), 'Grundsätze für die einzelbetriebliche Investitione-förderung', Vol. 37, No. 28, Markt + Meinung, pp. 1-3.

Beckmann, V. (1993), 'Zur ökonomischen Theorie der Transformation von Produktivgenossenschaften', *Zeitschrift für das gesamte Genossenschafts-wesen*, 43, pp. 217-231.

Bonus, H. (1986), 'The Co-operative Association as a Business Enterprise', *Journal of Institutional and Theoretical Economics*, 142, pp. 310-339.

Bundesministerium für Ernährung, Landwirtschaft und Forsten (ed.), *Agrarbericht*, various years.

Bundesministerium für Ernährung, Landwirtschaft und Forsten (ed.) (1995a), *Agrarwirtschaft in den neuen Bundesländern*, Bonn.

Bundesministerium für Ernährung, Landwirtschaft und Forsten (ed.) (1995b), *Die Umwandlung landwirtschaftlicher Unternehmen nach dem neuen Umwandlungsrecht*, Bonn.

DER SPIEGEL (1995), 'Belogen und Betrogen', No. 24, pp. 132-142.

Domar, E.A. (1966), 'The Sovjet Collective Farm as a Producer Co-operative', *American Economic Review*, 56, pp. 734-757.

Entwurf eines Entschädigungs- und Ausgleichsgesetzes - EALG - vom 12.05.1993. Bundestags-Drucksache 12/4887.

Ernährungsdienst (1993a), 'Vermögensabgabe soll entfallen. Zukunft des Entschädigungs-gesetzes weiter unklar. Uneinigkeit in Bonn', 28 October 1993, p. 2.

Ernährungsdienst (1993b), 'Einig über Entschädigung. Schuldverschreibungen statt Vermögensabgabe', 25 November 1993, p. 1.

Fiedler, C., B. Forstner, M. Frederking, H. Kurz, H. Frhr. von Münchhausen and A. Schefski (1994), *Die Situation der landwirtschaftlichen Unternehmen in den neuen Bundesländern: Strukturdaten und ausgewählte Fragen der Produktion und des Managements*, Braunschweig.

Forstner, B (1994a), 'Gleiche Chancen für alle Betriebsformen?', *Top Agrar Spezial*, No. 11, pp. 26-29.

Forstner, B. (1994b), 'Steuerliche und förderpolitische Behandlung unterschiedlicher Rechtsformen', in Klare, K. (ed.), *Entwicklung der ländlichen Räume und der Agrarwirtschaft in den neuen Bundesländern*, Braunschweig: Landbauforschung Völkenrode, Sonderheft 152, pp. 81-90.

Forstner, B. (1995), 'Die Behandlung der Agrargenossenschaften im Bereich der Besteuerung, der Investitionsförderung und der Sozialversicherung', *Berliner Schriften zum Genossenschaftswesen.*

Gattermann, H. H., MdB, und W. Rawe, MdB (1993), Zusammenfassende Darstellung des Ausgleichszertifikatmodells zum EALG, vorgelegt am 14.10.1993 (Grundlage für eine koalitionsinterne Anhörung der

Koordinierungsgruppe Entschädigungs- und Ausgleichsleistungsgesetz zum Alternativmodell zum Entschädigungs- und Ausgleichsleistungs-gesetz vom 19.10.1993).

'Gesetz über die strukturelle Anpassung der Landwirtschaft an die soziale und ökologische Marktwirtschaft in der Deutschen Demokratischen Republik - Landwirtschafts-anpassungsgesetz - vom 29. Juni 1990', *Gesetzblatt*, Volume I, No. 42, pp. 642- 648.

'Gesetz zur Änderung des Landwirtschaftsanpassungsgesetzes und anderer Gesetze vom 3. Juli 1991', *Bundesgesetzblatt*, Volume I, 1991, pp. 1410-1429.

'Gesetz über die Entschädigung nach dem Gesetz zur Regelung offener Vermögensfragen und über staatliche Ausgleichsleistungen für Enteignungen auf besatzungsrechtlicher oder besatzungshoheitlicher Grundlage (Entschädigungs- und Ausgleichsleistungsgesetz - EALG)', *Bundesgesetzblatt*, Volume II, 1994, pp. 2624-2639.

Grund, M. (1992), 'Landverwertung. Neues Programm - neue Hoffnung', *Top-Agrar-Spezial*, 19 December 1992, pp. 8-10.

Grund, M. (1993), 'Entschädigung: Doch noch Coupons für den Landerwerb?', *Top-Agrar-Spezial*, 18 December 1993, pp. 12-14.

Hagedorn, K. (1991), 'Gedanken zur Transformation einer sozialistischen Agrarverfassung', *Agrarwirtschaft*, 40, pp.138-148.

Hagedorn, K. (1992), 'Wirtschaftliche und politische Triebkräfte der Umformung einer sozialistischen Agrarverfassung', *Jahrbuch für Neue Politische Ökonomie*, 11, pp. 191-213.

Hagedorn, K. (1993), 'Bodenmärkte und Bodenmarktpolitik in den neuen Bundesländern', in R. v. Alvensleben, C. Langbehn and E. Schinke (eds.), *Strukturanpassungen der Land- und Ernährungswirtschaft in Mittel- und Osteuropa*. Schriften der Gesellschaft für Wirtschafts- und Sozialwissenschaften des Landbaues e.V., Vol. 29. Münster-Hiltrup, Landwirtschaftsverlag, pp. 281-287.

Hagedorn, K. (1994a), 'Changing Organisation of Agriculture as a Result of the Transformation Process', Paper presented on the COST workshop on "The Impact of CEE Agricultural Reforms on Domestic and International Markets" at Leuven, June 10-11, 1994, pp. 1-26.

Hagedorn, K. (1994b), 'Die Privatisierung volkseigenen landwirtschaftlichen Bodens in den neuen Bundesländern', *Kühn-Archiv*, 88, pp. 226-244.

Hagedorn, K. (1995), 'Politisch dirigierter institutioneller Wandel: Das Beispiel der Privatisierung volkseigenen Bodens', in K. Hagedorn (ed.), *Institutioneller Wandel und Politische Ökonomie von Landwirtschaft und Agrarpolitik. Festschrift zum 65. Geburtstag von Günther Schmitt*, Frankfurt/New York: Campus Verlag, pp. 149-192.

Hagedorn, K. (1996), 'Das Institutionenproblem in der agrarökonomischen Politikforschung', in E. v. Böventer, B. Gahlen und H. Hesse (eds.), *Schriften zur angewandten Wirtschaftsforschung Vol. 72*, Tübingen: J.C.B. Mohr (Paul Siebeck).

Hagedorn, K. and B. Klages (1994), 'Konzepte zur Privatisierung volkseigenen landwirtschaftlichen Bodens und Entwürfe zum Entschädigungs- und Ausgleichsgesetz. Analyse und Alternativen', *Landbauforschung Völkenrode 44*, No. 1, pp. 44-53.

Hagedorn, K. and G. Schmitt (1985), 'Die politischen Gründe für eine wirtschaftspolitische Vorzugsbehandlung der Landwirtschaft', *Jahrbuch für Neue Politische Ökonomie*, 4, pp. 250-295.

Henrichsmeyer, W. (1990), Anpassungsprobleme und Perspektiven der DDR-Landwirtschaft, *Agrarwirtschaft*, Vol. 39, No. 8, pp. 233-234.

Kabat, L. and K. Hagedorn (1997), 'Privatisation and De-collectivisation Policies and Resulting Structural Changes of Agricultural in Slovakia', chapter 7 of this volume.

Klages, B. (1994), 'Privatisierung der Treuhandflächen', in K. Klare (ed.), *Entwicklung der ländlichen Räume und der Agrarwirtschaft in den Neuen Bundesländern. Landbauforschung Völkenrode, No. 152*, Braunschweig: Bundesforschungsanstalt für Landwirtschaft (FAL), pp. 105-120.

Klages, B. and K. Hagedorn (1993), *Konzepte zur Privatisierung volkseigenen land- und forstwirtschaftlichen Bodens und Entwürfe zum Entschädigungs- und Ausgleichsgesetz. Eine Untersuchung unter besonderer Berücksichtigung des Zertifikatmodells* (Unveröffentlichte Stellungnahme für das Bundesministerium für Ernährung, Landwirtschaft und Forsten, 84 p.). Braunschweig: Bundesforschungsanstalt für Landwirtschaft (FAL).

Klages, B. and K. Klare (1994), 'Der neue Entwurf des Entschädigungs- und Ausgleichsleistungsgesetzes - Analyse und Bewertung', *Agra Europe*, Vol. 35, No. 23, Länderberichte, Sonderbeilage, pp. 1-11.

Klages, B. and K. Klare (1995a), 'BVVG-Flächen: kaufen oder langfristig pachten?', *Top-Agrar-Spezial*, No. 1, pp. 8-11.

Klages, B. and K. Klare (1995b), 'So werden Alteigentümer jetzt entschädigt', *Top Agrar*, No. 5, pp. 46-47.

Klages, B. and K. Klage (1995c), 'Lohnt sich der Flächenzukauf zum dreifachen Einheitswert?', *Top Agrar*, No. 5, pp. 46-47.

Köhne, M. (1990): Erfolgsvoraussetzungen für LPGen. *Agrarwirtschaft*, Vol. 39, pp. 265-266.

Kruse, S. (1995), 'Entwicklung der Struktur der landwirtschaftlichen Betriebe in den neuen Ländern', *AID-Informationen für die Agrarberatung*, Vol. 3, No. 6, pp. 2-24.

Langbehn, C. (1990), Hat die LPG eine Zukunft?, *Agrarwirtschaft*, Vol. 39, pp. 197-199.

Machnes, Y. and A. Schnytzer (1993), 'Risk and the Collective Farm in Transition', in Csaki, C. and Y. Kislev (eds), *Agriculture Cooperatives in Transition*, Boulder: Westview Press.

Mehl, P. (1995), *Reformansätze und Reformwiderstände in der Agrarsozialpolitik der Bundesrepublik Deutschland. Politikinhalte und ihre Bestimmungsgründe von 1976 bis 1990*, Dissertation, Braunschweig.

Mehl, P. and K. Hagedorn (1994), 'Die Übertragung des landwirtschaftlichen Alterssischerungssystems auf die neuen Bundesländer im Gesetzentwurf der Bundesregierung zur Reform des agrarsozialen Sicherungssytems', *Landbauforschung Völkenrode*, 44, pp. 77-90.

Pollak, R. (1985), 'A Transaction Cost Approach to Families and Households', *Journal of Economic Literature*, 23, pp. 581-608.

Schenk, K.-E. (1982), '"Institutional Choice" und Ordnungstheorie', *Walter Eucken-Institut, Vorträge und Aufsätze*, Vol. 82., Tübingen.

Schmitt, G. (1990), 'Können sich landwirtschaftliche Produktionsgenossenschaften (LPG) der DDR als wettbewerbsfähig erweisen?', *Agra-Europe*, 24, Sonderbeilage, pp. 1-14.

Schmitt, G. (1991), 'Das Scheitern der Kollektivierung der Landwirtschaft in den sozialistischen Ländern: Eine transaktionskostentheoretische Erklärung', *Berichte über Landwirtschaft*, 69, pp. 38-68.

Schmitt, G. (1992), 'Der Zusammenhang zwischen Organisationsform und Betriebsgröße in der Landwirtschaft: Eine institutionenökonomische Erklärung und deren agrar-ökonomische und agrarpolitische Implikationen', *Berichte über Landwirtschaft*, 70, pp. 505 -528.

Schulz, J. (1992), 'Zu transformationstheoretischen Ansätzen der ordoliberalen Schule und der amerikanischen Institutoneökonomik', *Zeitschrift für Wirtschaftspolitik*, 41, pp. 231-256.

Statistisches Bundesamt, *Statistisches Jahrbuch für die Bundesrepublik Deutschland*, various years, Stuttgart.

Swinnen, J.F.M. (1992), 'The Development of Agricultural Policies in Eastern Europe. An Endogenous Policy Perspective', *The Leuven Institute for Central and East European Studies Working Paper*, No. 2, pp. 1-9.

Swinnen, J.F.M. (1993a), 'Policy and Institutional Reform in Central European Agriculture', *The Leuven Institute for Central and East European Studies Working Paper*, No. 19 (1993a), pp. 1-28.

Swinnen, J.F.M. (1993b), 'The Development of Agricultural Policies in Central and Eastern Europe', *Food Policy*, 18, pp. 187-191.

Swinnen, J.F.M. (1994a), 'Political Economy of Transition Price and Trade Policy Changes in Central European Agriculture', unpublished draft, Leuven: Department of Agricultural Economics.

Swinnen, J.F.M. (1994b), 'Overview of Policy and Institutional Reform in Central European Agriculture', in Swinnen, J.F.M. (ed.), *Policy and Institutional Reform in Central European Agriculture*, Aldershot: Avebury.

Top-Agrar-Spezial (1992a), 'Ab Juli wird langfristig verpachtet und verkauft. Interview mit dem Generalbevollmächtigten der Treuhand, H.-J. Rohr in Berlin', 24 June 1992, pp. 4-7.

Top-Agrar-Spezial (1992b), 'Regierung contra Opposition beim Wiedereinrichterprogramm', 19 September 1992, p. 32.

Van den Noort, P. C. (1993), 'Reform of the Farm Tenancy System: How the Netherlands Fits into the European Scene', in Silvis, H. J. (ed.), *Capital and Finance in Western and Eastern European Agriculture*, Wageningen Economic Studies No. 29, Wageningen Agricultural University.

Vanek, J. (1970), *The General Theory of Labor-Managed Market Economies*, Ithaca.

Ward, B. (1958), 'The Firm in Illyria: Market Syndicalism', *American Economic Review*, 68, pp. 566-589.

Williamson, O. E. (1975), *Markets and Hierarchies: Analysis and Antitrust Implications*, New York and London.

Williamson, O. E. (1985), *The Economic Institutions of Capitalism*, New York.

5 Privatisation and restructuring of Hungarian agriculture

Erik Mathijs and Sándor Mészáros

Introduction

This chapter gives an overview of the restructuring of Hungarian agriculture and identifies its determinants. Hungary's pre-transition farm structure was characterised by a dominance of collective farms. The 'symbiosis' between large-scale farms and household plots led to the relative success of Hungarian agriculture under Communism. The legislation designed to restructure Hungarian agriculture after the political reforms of 1989 through the restitution and privatisation of land and capital assets has been characterised by a complex set of rules and was expected to be completed in 1996. Following this legislation, state and collective farms have been transformed into a myarid of business organisations, bnut land use data suggest that large-scale farms still dominate Hungarian agriculture.

We identify several factors which determine the emergence of individual farming and the persistence of the large-scale successor organisations of the former collective and state farms. First, markets for credit, land, and inputs are either ill developed or still monopolised. The economic environment (terms of trade, prices and interest rates) is not beneficial for new entrepreneurs, nor for the sector as a whole. Second, the incentives of members of collective farms (or their successors) to withdraw are limited as a result of the relatively high average productivity of the collective farm, compared to relatively low potential productivity in individual farming mainly due to a lack of human skills and entrepreneurship. Third, the effect of the difference between the privatisation of land and non-land assets is discussed. Finally, the effect of transaction costs linked to the withdrawal from a collective farm is assessed.

The pre-transition farm structure

Before the reforms that started in 1989 three main types of economic organisations dominated Hungarian agriculture: state farms, agricultural producer co-operatives and household farms. Their most important characteristics are summarised below.

1 *State farms* were enterprises established and owned by the state and operated by hired workers. However, since 1965 state farms did not receive production targets from the state and were intended to operate with the objective of profit maximisation. In general, they were (very) large-scale production units, both in crop production and in livestock. For example, in 1989, there were 136 state farms (139 including forestry and fishing) with an average size of 6,886 hectares (total land area)

2 Hungarian *collective farms* were very close to Western-style agricultural producer co-operatives. Production assets and three-fifths of the collective farm land were in collective ownership, while another one-third of the land was privately owned by the members (see Table 5.1). Active members, who could be both asset owners and workers, received a monthly wage and a yearly profit share based mainly on their productivity. Collective farms were run by a general assembly and a management staff elected by the members. They numbered 1,245 in 1989 (1,333 including forestry and fishing) and their average area amounted to 4,180 hectares.

3 *Household plots* were small plots cultivated by members of collective farms or state farm workers. The amount of land and livestock was limited by law. The collective farms contributed to their members' private production by providing machinery services and selling feed and roughage to the farms. Together with other private farmers their number was estimated at 1,435,000 in 1989 with an average size of 0.6 hectare.

Under Communism, many large-scale collective and state farms were inefficient. Production was not planned in accordance to the specificities of the local conditions. Farming was set up in some marginal areas (e.g., on sandy soils and in a very dry climate) where agriculture is inherently unprofitable. The government policy was to maximise output to provide exports in exchange for hard currencies and fuel. However, collective and state farms in these marginal areas had to be supported heavily.

Despite these inefficiencies, Hungarian collectives performed well compared to large-scale farms in other Communist economies. This good performance led to a strong belief of policy-makers in large-scale agricultural production.

Moreover, the success of private farming under Communism was to a great extent the result of the 'symbiotic' relationship between large- and small-scale farms. Hungarian collective farms fulfilled other economic and social functions besides agricultural production. For example, inhabitants of rural villages became highly dependent on the collectives as the latter provided an extensive social security network to the rural population.

Table 5.1
Ownership distribution of land farmed by Hungarian collective farms, in percent, 1968-1989

Year	*State*	*Collective*	*Members**
1968	27.7	0.1	72.2
1975	4.4	44.7	50.9
1980	3.4	51.6	45.0
1985	4.0	56.7	39.3
1989	3.8	61.1	35.1

* and those having the same legal status: partners in matrimony, living together with a member, being the widow of a member, aged persons entitled to land rent, and persons with usufructuary right.

Source: Harcsa, 1991

Privatisation and transformation legislation

Simplifying, the legislation to restructure Hungarian agriculture consists of five parts: First, members of a collective farm who had always retained title to part of the land were permitted to withdraw their land freely from the collective. This was already achieved under the last Communist government in February 1990. Second, former land owners who had lost their land were compensated for their losses. Approximately half of the collectively owned land (which makes up more than three-fifths of all the land cultivated by collective farms, see Table 5.1) was used for this purpose. This was regulated by the Compensation Laws. Third, the Transitional Co-operative and Unified Co-operative Laws dealt with the transformation of collective farms and the privatisation of the non-land assets and the remaining land. Fourth, the privatisation of state farms is part of the overall privatisation legislation. Finally, a land law regulates, among other things, the exchange of property rights of land.

Restitution of privately owned land

Whereas most capital assets were state or collectively owned in 1989, part of the land was still owned by individual members, as in most other Central and Eastern European countries (CEECs). Collective farm members held title to one-third of the land used by collectives (see Table 5.1). However, this land was subject to land consolidation and improvement investments by the collective. The members owned a share of this land proportional to the land they brought in. Hence this land is referred to as 'proportional' land. Restoration of full property rights proved difficult. The collectives were free to design mechanisms to restitute property rights on this land. If no agreement could be reached lotteries were organised.

Compensation Laws

Three-fifths of the land cultivated by collective farms was collectively owned. Part of its privatisation was covered by the *Compensation Laws*, another part by the Co-operative Transformation Laws. The Compensation Laws were meant to provide indemnification for individuals whose property was confiscated under Communism. People eligible for compensation were farmers whose land was seized just after the Second World War and farmers who were forced to sell their land to the collective farm for a low price in the 1970s and 1980s. Eventually, people who received compensation under the three Compensation Laws included individuals who had lost property after May 1, 1939 and persons deprived of freedom or life for political reasons between 11 March 1939 and 23 October 1989. The government decided in 1992 to issue a Fourth Compensation Law to speed up the implementation of the previous ones.

Former owners (or their descendants) who claimed compensation, received *vouchers* based on the estimated value of their lost property. For property up to the value of 200,000 HUF (roughly 10 hectares), former owners were compensated 100 percent, with a degressive scale of compensation thereafter. These vouchers could be used (1) to buy physical assets and shares in newly privatised companies; (2) to buy, at auctions, land designated for compensation; (3) to buy apartments owned by the state or local authorities; (4) to claim a life-annuity from the state (for elderly people); or (5) to sell directly or through the stock exchange at a market determined daily rate.

The face value of the compensation vouchers was based on the original valuation of the lost land, which was calculated on the basis of the cadastral net income of the arable land, expressed in Gold Crown units. The average Gold Crown value of Hungarian land is 20 Gold Crowns per hectare. On the assumption that the average hectare of farm land was worth 20,000 HUF in 1991, one Gold Crown's worth of land equals 1,000 HUF of compensation.

Land for compensation originates from two sources: part of collectively owned land in collective farms and state farms set aside for privatisation. Land can be acquired by individuals through land auctions. If the purchased land is not cultivated for a minimum of five years it may be confiscated. However, the land may be sold or leased to others who will cultivate it. The land must be held for five years before it can be resold. Land can be bought using compensation vouchers only by the first bearer of vouchers.

The land set aside for compensation is allocated through auctions. Up to December 25, 1994, land worth 36.8 million in Gold Crown value (about 1.8 million hectares) was sold in exchange for compensation vouchers in more than 21,000 auctions. A second land fund for compensation worth 2.3 million in Gold Crown value was auctioned from April 1995 on. Since the number of buyers was 527,000 up to the end of 1994, the average land size acquired through the compensation process was very small: less than four hectares (Privinfo, 1995a).

Transformation of collective farms

The two laws dealing with the transformation and privatisation of collectives were accepted in Parliament in January 1992: the *Co-operative Transition Law* and the *Unified Co-operative Law*. These laws allowed the collective farms to choose between three alternative directions of transformation: (1) to transform the collective into a share holding company and sell the shares, (2) to divide the property of the collective among its members, or (3) to transform the collective farm into a free co-operative, in conformity with the principles adopted at the 1966 Vienna Congress. These laws contained also a schedule of transformation including such steps as the distribution of all non-land property among members, creating a land bank for compensation and land auctions.

All the collectively owned non-land assets were valued as they appeared in the accounts and divided as business shares. However, this evaluation was often followed by a re-evaluation. People eligible to get these business shares were existing members, people who had been members for at least five years before the Transition Law came into force, or their heirs. Business shares could also be given to employees and family of collective farm members by decision of the General Assembly of the new co-operative. As property was to be returned to those who had contributed to its accumulation, criteria to be used by the General Assembly were to be drawn up. These were: the length of membership, the level of salary and the value of any collectivised property. These business shares could then be exchanged for physical assets by consensus or by auction (OECD, 1994).

The distribution of land was treated differently. All the land of the collective farms was divided into three land funds: (1) approximately one-third of the land was still privately owned; (2) one-third was collectively owned land set aside for members and employees determined by law, i.e. 30 Gold Crowns in value of land for each member and 20 Gold Crowns for each employee; (3) the residual one-third of collectively owned land was set aside for compensation under the Compensation Laws. Any land left over had to be distributed among the members.

There were many possible ways to transform the collective farms. The two extreme ways were the transformation to the new type of co-operatives and the liquidation by bankruptcy in which case members turned into individual farmers. Between these two extremes it was possible to opt for other ways of transformation:

1 To split into two or more new co-operatives,

2 To transform into one (or more) limited liability company,

3 To transform into a new type co-operative loosing some of the members,

4 More complex combinations of the previous ways.

According to figures provided by MOSZ (National Federation of Agricultural Co-operators and Producers), which slightly deviate from official statistics, of the 1,319 collective farms in 1989, 174 were liquidated and 1,987 new organisations emerged by 1993. Of these new organisations, there were 1,387 agricultural producer co-operatives, 351 limited liability companies, 24 companies limited by shares and 31 partnerships, as well as some organisations in forestry and food processing.

Although the transformation process of former collective farms can be considered as legally completed, three major problems remain: (1) the present organisational structure is unlikely to be the final one and further transformations are expected under pressure of economic circumstances; (2) the distribution of co-operative property among active and retired members as well as to outsiders (heirs of members) has generated a conflict of interests in profit allocation; (3) as a result of the transformation process labour, land and capital were distributed according to different rules.

Transformation of state farms

The transformation and privatisation of state farms is the responsibility of the *State Property Agency* and not of the Ministry of Agriculture. People who can buy shares of state farms include existing workers and tenants, on-farm and outside owners of compensation vouchers, creditors (banks) and domestic and foreign investors. Not all of the 120 state farms left over in 1994 were to be privatised. Twenty-four farms, specialised in seed propagation, animal stock breeding and research remain under state ownership. They were transformed into enterprises controlled up to some degree by the *State Asset Holding Company*. Properties to remain in state ownership were listed in a government decree of August 1992. The other 97 state farms have been broken up into viable farm units. The state farm's commissioner had to submit a rationalisation or decentralisation plan by 30 June 1992. The transformed farms were then offered for sale at an auction or by tender.

The transformation of state farms into companies took place during 1992 and 1993. The major part of them are operated as private companies. Some of the state farms were eliminated through bankruptcy. For state farms the typical process was transformation to joint stock companies: 94 percent of the privatised state farms chose this form while only 6 percent became a limited liability company. The progress in these processes by 1994 is shown in Table 5.2. About 40 percent of the former state farms were designated for privatisation (incl. bankruptcy) in 1995. In the majority of the privatised companies the shares were bought by the managers and employees of former state farms. Land was excluded from this privatisation procedure. Land remains in public ownership held by the State Property Agency and the companies are now renting (part of) their land from the Agency.

Table 5.2
Progress in the privatisation of Hungarian state farms, 1994

	State farms		*Asset value (excl. land)*	
	number	*percent*	*million HUF*	*percent*
Remaining in state property	25	20.8	25,902	28.7
Privatised (partially or fully)	37	30.9	20,663	22.8
To be privatised	26	21.7	25,932	28.7
Liquidated	10	8.3	6,557	7.2
To be liquidated	22	18.3	11,395	12.6
Total	120	100.0	90,449	100.0

Source: Ministry of Agriculture, 1995

The land of the state farms was taken away from them before their transformation and the ownership rights were allocated to the State Property

Agency. The present allocation of these 902,088 hectares of productive land (total minus uncultivated land) is as follows:

- 40 percent is used for compensation of private persons,
- 37 percent is used by companies which remain state property,
- 27 percent was leased, mainly to companies no longer state owned, and
- 6 percent was allocated to employees of state farms.

Land law

In 1994 a new Land Law was passed by the Parliament which regulates both ownership and rental of land (Magyar Közlöny, 1995). Only domestic private persons can buy land while domestic legal persons (organisations) are excluded as well as foreign private or legal persons. At the same time, there is an upper limit for private persons for buying land which is 300 hectares land (or 6,000 Gold Crowns). However, this upper limit is not applied for those who obtained land either by compensation or by the reallocation of collectively owned land. The new Land Law also regulates the rental market by giving a ten year upper limit for the renting period and setting size constraints as well. While individuals (including foreigners) can rent up to 300 hectares, Hungarian companies or co-operatives can rent land up to 2,500 hectares.

The emerging post-1989 farm structure

Different new organisational forms

Presently six new forms of agricultural organisations exist (see Table 5.3). The following discussion of the various organisations is based on the definitions in Hungarian law, which are applied by the Central Statistical Office in Budapest (HCSO, 1994d). The evolution of the number of each of these organisations is displayed in Table 5.4.

A *joint stock company* ('Részvénytársaság' or abbreviated 'Rt') is a company with legal entity defined by Laws VI and XXV of 1988. It can be established by a corporation contract and with incorporation at the Court of Registration as well as with basic capital consisting of shares. Shareholders are responsible up to the values of their shares. This type of company was founded as a result of the transformation of former state farms, but it is allowed also for co-operatives and private persons to establish a joint stock company. In general, these companies are large-scale farms, but somewhat smaller in size than the former state farms from which they have mostly been transformed. There were 186 such companies in December 1995.

A *limited liability company* ('Korlátolt felelósségú társaság' or abbreviated 'Kft') is a company with legal entity similar to a joint stock company and defined by the same laws. Generally they are small companies employing less than 20 persons. A limited liability company may even consist of a single person. The required basic capital is 1 million HUF, an amount of money a former large-scale farm manager usually can afford. The majority of these companies were created as a result of the transformation of collective and state farms. They constitute an important element of the new structure and their number rapidly grew during the period of the reform: from 67 in 1988 to 3,636 in December 1995.

A *limited partnership* ('Betéti Társaság') is an organisation without legal entity established by a corporation contract based on Law VI of 1988 for a common economic activity and incorporated by the Court of Registration. Persons or legal entities may be members of it. The number of members has to be at least two, one with an unlimited liability (inside member) and another one whose liability is limited to his/her share in capital (outside member). There were 2,624 such partnerships in December 1995.

An *agricultural co-operative* ('Mezógazdasági szövetkezet') is a new type of co-operative defined by Law I of 1992. They differ from the old collectives as the members are owners rather than employees. Moreover, they can leave the co-operative at any time (voluntary principle), and the co-operative may have any other activity in the field of agriculture, forestry and fishery besides production. These new co-operatives are typically smaller in size than their predecessors. Former collective farms included several villages and were broken up into different economic organisations. In December 1994 there were 2,048 co-operatives.

Two kinds of individual farmers are distinguished in the statistics. A *sole proprietor* is a person whose farm could be considered as a family farm and the person as a full-time farmer in the Western terminology. Although they represent only a small part of all rural families, their number is increasing rapidly. The number of sole proprietors rose from 2,804 in 1991 over 3,784 in 1992, 11,579 in 1993 to 22,493 in 1994.

The majority of the rural households are considered *small farmers*. In the EU, they would be considered part-time farmers. The Hungarian Central Statistical Office has applied a minimal size of land and livestock to decide what is a small farm (e.g., 400 m^2 for land). There were 1,675,000 rural households in October 1994 and of them 1,201,000 were considered small farms.

Table 5.3
Types of new business organisations in Hungarian agriculture

Characteristics	*Companies*		*Co-operatives*	*Without legal entity*		
English term	Joint stock company	Limited liability company	Agricultural co-operative	Limited partnership	Sole proprietorship	Part-time farmer
National term	Részvénytársaság (Rt)	Korlátolt felelõsségû társaság (Kft)	Mezõgazdasági szövetkezet	Betéti társaság	Egyéni vállalkozás	Mezõgazdasági kistermelõ
Juridical basis for existence	Laws VI and XXV of 1988	Laws VI and XXV of 1988	Law I of 1992	Law VI of 1988	Law V of 1990	-
Registration	Yes	Yes	Yes	Yes	Yes or no	No
Minimum of members	1	1	-	2	-	-
Minimum of capital	-	1 million HUF	-	-	-	-
Individual tradability of capital	Yes	Yes (with restrictions)	No	No	Yes	Yes
Liability (of members)	Limited	Limited	Unlimited	Unlimited for at least 1 member	Unlimited	unlimited
Management	Board of management and stockholder's meeting	Chief executive and general meeting	Board of management and general assembly	Unlimited partner	Self	Self

Table 5.4

Number of organisations of agriculture, forestry and fishing in Hungary, 1989-1995

Economic organisations	*1989*	*1990*	*1991*	*1992*	*1993*	*1994*	*1995*
State farms	139	147	150	154	91	36	33
Companies with legal status							
- Limited liability companies	na	na	na	1,628	2,434	3,140	3,636
- Joint stock companies	na	na	na	45	109	171	186
Co-operatives	1,333	1,348	1,352	1,631	1,972	2,048	2,117
Limited partnerships	na	na	na	na	na	2,369	2,624
Sole proprietors	-	-	2,804	3,784	11,579	22,493	29,976
Small farms (in thousands)	1,435	1,410	1,396	1,450	1,018	1,201	na

Source: HCSO, 1995

Land use and ownership

Assessing the relative importance of the various farm types based on their share in total agricultural land (TAL) is not easy in Hungary. The only complete data that are available are based on land registration and reflects ownership situation. Land ownership data for 31 May 1994 show that Hungary has 6,122,000 ha of agricultural land of which 19.0 percent is owned by so-called 'enterprises', a category which includes companies, partnerships and state farms, 37.7 percent by co-operatives and 43.3 percent by individual farms (see Table 5.5).

Table 5.5
Land area by ownership in Hungary, 31 May 1994

	Agricultural land		*Total land*	
	ha	*%*	*ha*	*%*
Enterprises*	1,163,600	19.0	3,086,100	33.2
Co-operatives	2,306,000	37.7	2,726,400	29.3
Individual farms	2,652,400	43.3	3,490,500	37.5
Total	6,122,000	100.0	9,303,000	100.0

* companies, partnerships and state farms

Source: HCSO, 1994a

According to a census by the Central Statistical Office (1995), 1,675,000 households used in 1994 only 1,500,000 hectares of total land area (TLA) or 1,349,000 hectares of agricultural land, which reflects 22 percent of TAL. The rest is rented to large-scale farms. However, there is no indication how much land is rented to companies, partnerships or co-operatives. Hence, we can only conclude that 22.0 percent of TAL is used by individual farms, while 78.0 percent is cultivated by co-operatives, companies, partnerships, and state farms. However, recent assessments by the Central Statistical Office based on total land area, suggest that individual farmers used 22.3 percent of TLA in 1994 and 29.0 percent in 1995; enterprises 33.2 percent in 1994 and 27.9 percent in 1995; and co-operatives 29.3 percent in 1994 and 23.7 percent of 1995. It is however unclear who uses the remainder (see Table 5.6).

The census data include all farms larger than 400 m^2. Hence, all kinds of part-time farmers, hobby farmers and little gardens are also included in the statistics. If we look at the size distribution of individual farmers that use land,

21 percent of the land is in farms smaller than 1 hectare and owned by 76.6 percent of all households. Of these individual farms, 50,910 are registered as full-time farmers. They own almost 200,000 hectares total land, which is only 2.1 percent of the total land area of Hungary, or 3.7 percent of TAL.

Table 5.6
Land ownership and use according to organisation, based on *total* land area in Hungary, 1994 and 1995

	1994		*1995*	
	Area ('000 ha)	*Percent*	*Area ('000 ha)*	*Percent*
Enterprises[a]	3,086	33.2	2,593	27.9
Co-operatives	2,727	29.3	2,208	23.7
Individual owners	2,785	29.9	4,035	43.4
of them: farmers[b]	2,074	22.3	2,700	29.0
Non-agricultural use	705	7.6	467	5.0
Total	9,303	100.0	9,303	100.0

[a] companies, partnerships and state farms
[b] assessment

Source: HCSO, 1996

Table 5.7 displays the origin of total land owned by individual farmers and how this land is used. Of the total 2,169,000 hectares of total land, 32 percent was privately owned before the reform, i.e. prior to 1989; 28 percent comes from compensation; 27 percent from the restitution of proportional land and 13 percent from the distribution of collectively owned land. Whereas 883,000 hectares are leased, 178,000 hectares are rented, resulting in a final use of 1.5 million hectares of total land.

Table 5.8 exhibits some statistics of so-called 'organisations with accounting'. These are predominantly large-scale companies and co-operatives. Although the number of these large-scale organisations increased by 7.4 percent in 1994, the land area used by them decreased by 22.8 percent. This area was probably taken over by smaller farms. Hence, the average size of used land by large-scale organisations decreased from 2,496 hectares in 1993 to 1,794 hectares in 1994. In 1993, there was also a shift of the share in land use between the largest and smallest farm size categories.

Table 5.7
Hungarian land data of individual farmers in 1994

Source of land	*1000 ha*	*Percent*
Owned prior to 1989	683	32
Compensation	607	28
Proportional land in co-operatives	590	27
Collectively owned land in co-operatives	289	13
Total	2,169	100
Use of land		
Leased	883	41
Used for own production	1,286	59
Total land owned	2,169	100
Owned and used for own production	1,286	86
Rented	178	12
In household ownership	35	2
Total land used	1,499	100

Source: HCSO and Ministry of Agriculture, 1995

Table 5.8
Land use by 'organisations with accounting' in Hungary, 1993-1994

Farm size (productive land in ha)	*Number*		*Productive land (in '000 ha)*		*Share in land use (in %)*	
	1993	*1994*	*1993*	*1994*	*1993*	*1994*
Below 1000	914	1,360	303	447	5.1	9.7
1000-5000	1,264	1,106	3,183	2,472	53.2	53.5
Above 5000	217	106	2,492	1,697	41.7	36.8
Total	2,395	2,572	5,979	4,625	100.0	100.0

Source: HCSO, 1994c

Table 5.9 shows a wide variety of farm organisations ranging from smaller than 1 hectare to larger than 10,000 hectares. According to the Central Statistical Office (1994c), there were 2,572 organisations with accounting in

1994 which used 4,615,400 hectares of productive land, hence with an average size of approximately 1,800 hectares. 53 percent of them are smaller than 1,000 hectares and used 9.7 percent of total productive land used by these organisations. In 1994, there were still 29 organisations larger than 10,000 hectares, that used 25.8 percent of all productive land used by organisations with accounting.

Table 5.9
Number of 'organisations with accounting' and land use by size categories in Hungary, 1994

Size categories (in ha)	*Organisations* no.	%	*Productive land* '000 ha	%
10-and smaller	104	4.0	0.5	0.0
1-50	205	8.0	5.7	0.1
51-100	149	5.8	10.8	0.2
101-200	181	7.0	26.1	0.6
201-300	113	4.4	28.7	0.6
301-400	109	4.2	38.2	0.8
401-500	111	4.3	49.7	1.1
501-1,000	388	15.1	286.9	6.2
1,001-2,000	558	21.7	821.5	17.8
2,001-3,000	310	12.1	763.6	16.5
3,001-4,000	166	6.5	569.1	12.3
4,001-5,000	72	2.8	317.8	6.9
5,001-10,000	77	3.0	504.7	10.9
larger than 10,000	29	1.1	1,192.0	25.8
Total	2,572	100.0	4,615.4	100.0

Source: HCSO, 1994c

For the land use of individual farms, Table 5.10 presents 1994 data by size categories. Of 1.5 million farms only 20.8 thousand farms had 5 or more hectares of land. Their share in the land use of individual farms is 29.7 percent. More than three quarters of individual producers produced on land of less than 0.5 hectares, which equals the average size of the earlier household plots.

The relative importance of different farm organisations can only be assessed by looking at land use, since no data exist on the production value or value added according to organisation. Only some data on livestock by organisation for 1993/94 exist and is presented in Table 5.11. The data show that 25 percent of all cattle were owned by individual producers, 48 percent of pigs, 66 percent of sheep and 67 percent of chickens.

Table 5.10
Number of individual farms and land use by size categories, 1994

Size categories (in ha)	*Number*	*%*	*Productive land '000 ha*	*%*
0.00-0.19	806.0	51.7	94.8	10.7
0.20-0.49	404.5	26.0	121.7	13.7
0.50-0.99	180.2	11.6	124.1	14.0
1.00-1.99	96.4	6.2	128.8	14.5
2,00-4.99	51.1	3.3	153.3	17.3
5.00-9.99	13.6	0.9	91.1	10.3
10.00 and larger	7.2	0.5	172.6	19.5
Total	1,559.0	100.0	886.4	100.0

Source: HCSO, 1994c

Table 5.11
Livestock by ownership categories in 1993/94

Ownership categories	*Cattle %*	*Pigs %*	*Sheep %*	*Chickens %*
Enterprises*	25	27	8	20
Co-operatives	50	25	26	13
Individual farms	25	48	66	67
Total	100	100	100	100

* companies, partnerships and state farms

Source: HCSO, 1994a

Factors contributing to the emerging farm structure

To give an indication of the adverse economic conditions of Hungarian agriculture in the transition period, Figure 5.1 displays the evolution of both agricultural and industrial prices for the period 1985-1994. Until the reforms which started in 1990, the evolution of these prices followed a similar pattern.

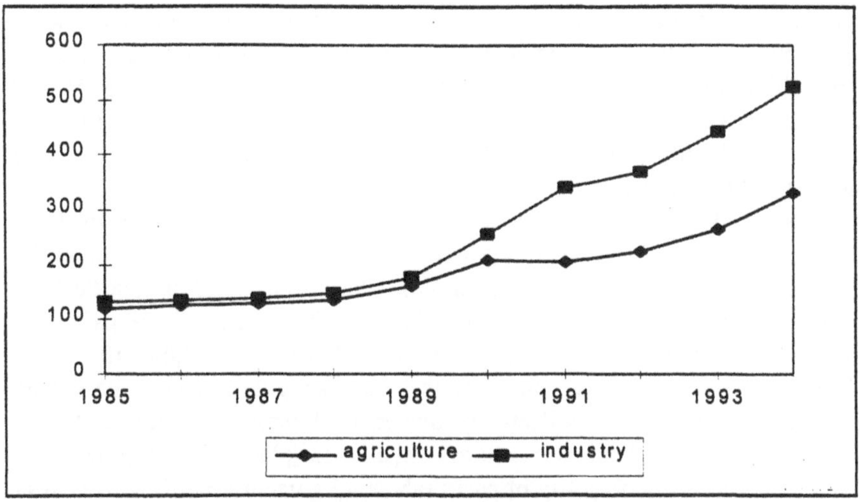

Figure 5.1 Price index for agricultural and industrial products (1980=100)

Source: HCSO, 1994a

From 1990, industrial prices increased significantly more than agricultural prices. Between 1989 and 1994, agricultural input prices tripled, whereas agricultural producer prices only doubled, resulting in a serious price squeeze and a deterioration of the terms of trade by about 50 percent. For detailed data for the terms of trade of agriculture, refer to Mészáros (1994a).

However, we want to limit our analysis to factors inherent to the agricultural sectors. In this respect, important constraints on the development of new farms are the markets for land, credit, labour, other inputs and outputs, which are either non-existent or still monopolised.

Land market

Land prices in Hungary are low: they vary between 500 and 1,000 HUF per Gold Crown, also in private transactions (Szűcs and Tanka 1994). The state wanted to stabilise the land market by forbidding the sale of land purchased at compensation auctions for five years. Many experts believe it will instead conserve the present ownership structure and postpone the necessary changes which would lead to a more rational farm size structure. In 1995, a second land fund for compensation was established with a value of 2.3 million Gold Crowns (Privinfo, 1995c).

For land prices and rents there are no statistical data, but a 1994 survey showed an average land rent of 1,533 HUF per hectare and their range varied from 600 to 4,000 HUF (Szûcs and Tanka, 1994). The data reflect the low profitability of agricultural activity. According to Toth and Varga (1995) land rent used to be calculated based on the price of wheat. The land rent equals the value of 10-20 kilograms of wheat for 1 Gold Crown, i.e. 1,800-4,500 HUF per hectare. MOSZ reports that for a hectare of land co-operatives paid 5,880 HUF in 1996 in Hajdu-Bihar (Great Plain) and 4,640 HUF in 1996 in Fejér (Transdanubia), counties where the soil value number reflecting the suitability of land for agricultural production reaches values of 50 to 60 on a scale from 10 to 77 (1982 Land assessment data), whereas only 2,560 HUF was paid for a hectare in Veszprém where soil quality only reaches a value of 20 to 40.

Policy-makers wanted to stimulate the functioning of the land market by privatising some of the collectively owned land through auctions. Private persons could buy land at auctions which took place in the process of compensation. In the majority of cases (about 75 percent according to Harcsa et al., 1994) land prices paid in auctions scarcely exceeded the minimal level of 500 HUF per Gold Crown (10,000 HUF per hectare).

Credit market

Access to credit is an important prerequisite for the develop-ment of new farms. Agricultural credit is offered mainly by five large commercial banks. The network of rural credit co-operatives is yet in a developing stage. Although special credit programmes were constructed for sole proprietors, it is still difficult for farmers to get credit. According to a survey by the Central Statistical Office, the share of farmers receiving credit decreased from 25.6 percent in the period 1985-1992 to 12.7 percent in the period 1993-1994. This share increases with farm size: whereas only 4.2 percent of farms smaller than 1 hectare were successful in getting credit, this was 47.4 percent of farms larger than 100 hectares for the period 1993-1995 (Harcsa, 1996). Another survey showed that, in 1995, 34.8 percent of full-time farmers obtained loans, but only 13.8 percent of part-time farmers (Szijjártó, 1996).

Supply of credit is limited and the requirements necessary to obtain credit are quite high: banks generally require 150-200 percent of the amount of the loan as collateral. Further, the interest rate is very high: it varied from 33.5 to 38.5 percent in the first half of 1995 in case of credit programmes but exceeded 40 percent for other types of credits (Czárlné et al., 1996). Government tries to alleviate these problems both by providing credit guarantees and interest subsidies. However, government policy was not consistent: in 1995 the deadline for loan applications was suddenly changed, such that the amount of loans for that year was drastically reduced.

Labour market

The economic restructuring in Hungary resulted in a decline in general employment: at the beginning of 1995 altogether 3.6 million people were employed, 24.2 percent less than in 1990 when the reform had started. The decrease in the number of employees was much more severe in agriculture and forestry. In agriculture the number of employees just exceeded 300 thousand in 1995, which showed a decrease of 62 percent compared to 1990. As a result, by 1995 agriculture, forestry and food industry together have only a 12.7 percentage share of national employment (see Table 5.12).

Table 5.12
Employment trends in Hungary (in 000s)

Year (January, 1)	*National economy*	*Of which*			*Total agriculture, forestry and food*	
		agricul-ture	*forestry*	*food industry*		*as % of national*
1990	4,795.2	813.3	50.0	203.0	1,066.3	22.2
1991	4,668.7	709.7	42.5	194.1	946.3	20.3
1992	4,241.8	552.6	36.3	196.8	785.7	18.5
1993	3,866.9	359.8	32.1	171.1	563.0	14.5
1994	3,700.7	326.2	18.6	148.1	492.9	13.3
1995	3,636.4	308.5	15.3	137.6	461.4	12.7
1995/1990	75.8	37.9	30.6	67.8	43.3	

Source: HCSO, 1995

As a result, unemployment turned to be relatively high in Hungary, 10.9 or 13.3 percent in 1994, depending on the method of measurement. Unemployment figures are not disaggregated for economic sectors, but it is easily understandable that in agriculture (and forestry) the rate of unemployment should exceed the average figure for the national economy and even this higher figure varies according to different regions showing very high unemployment rates in some counties (such as Borsod-Abaúj-Zemplén or Szabolcs-Szatmár).

The unemployment situation is reflected in a depressed wage level. The average gross monthly salary in agriculture, forestry and fishery was

24,641 HUF in 1994 (see Table 5.13), the lowest of all 14 sectors of the national economy (the national average was 33,289 HUF while the maximum amounted 61,712 HUF in financial activities).

Table 5.13
Evolution of monthly gross and net salaries in Hungarian agricultural organisations with accounting

Year	*Gross salary*		*Net salary*	
	HUF	*% increase to previous year*	*HUF*	*% increase to previous year*
1991	13,119	-	10,210	-
1992	15,256	16.3	11,675	14.3
1993	19,230	26.0	14,127	21.0
1994	24,641	28.1	18,261	29.3

Source: HCSO, 1995

Upstream- and downstream sectors

Most new entrepreneurs have to rent the necessary machinery to run their farm: according to a survey, only 14.5 percent of all small farmers had machinery, whereas 53.5 percent had their own means of transportation. However, farmers have to deal with rather incomplete markets for inputs, such as fertilisers, pesticides and machinery. Before reform, these markets were characterised by a high degree of concentration. The privatisation of the processing industry has been more successful than the privatisation of the supply industry. Hence, the availability of inputs at the farm level still remains a problem (Jackson and Swinnen, 1995). However, also in the market for farm produce, there is still a problem of monopolisation and unequal access to markets. Transformed co-operatives still have better relations with the food industry, simply because they could buy shares of food processing companies with the vouchers they received from individuals in exchange for land as a result of the compensation process.

Demonopolisation of the downstream sectors began in 1990 when monopolistic state boards for milk processing and meat processing were broken up into independent firms organised in every county. In the same year the process of privatisation started so that 24 percent of earlier state firms of food industry were in private ownership by 1992 (Mészáros, 1994b) while 68 percent of capital assets were privatised by the end of 1994 (see Table 5.14). Foreign investors had a 38 percentage share in the ownership of total capital assets in 1994 and they had dominating (from 68 to 100 percent) shares in the vegetable oil, sweets, tobacco and beer industry. The share of domestic investors was only 30 percent from the total capital assets of food industry in 1994.

Table 5.14
Privatisation of Hungarian food processing firms (end of 1994)

Sector	*Capital assets (in billion HUF)*	*State*	*Private*
Edible oil	7.0	100.0	0.0
Sweets	5.9	96.3	3.7
Spirits	8.0	84.7	15.3
Canning	9.4	80.1	19.9
Tobacco	8.0	79.0	21.0
Frozen foods	5.8	76.1	23.9
Beer	6.5	71.5	28.5
Milling	16.1	68.1	31.9
Meat	14.2	64.4	35.6
Wine	4.1	58.1	41.9
Poultry	5.3	57.3	42.7
Sugar	24.0	57.0	43.0
Milk	9.1	50.6	49.4
Baking	10.1	49.1	50.9
Total	133.5	68.3	31.7

Source: Ministry of Agriculture

Privatisation of food marketing also started in 1990 and is in an advanced stage now. Of 11 state firms engaged in wholesale trade of foodstuffs, eight (73 percent) were privatised by April 1995. Similarly, of 40 state firms for food retailing 34 (85 percent) were in private ownership by June, 1995 (Szabó, 1995). Food marketing was an attracting field for domestic investors because of the relatively limited amount of capital needed and the relatively high turnover rate. Foreign investors also participated in the privatisation of food marketing: their share of capital assets was 20.5 percent for wholesale trade and 32.4 percent for retail trade in 1993.

Privatisation has proceeded also in the upstream sectors but domestic production fell back sharply as a consequence both of the decline in agricultural production and of the import competition in agricultural inputs. The use of fertilisers, previously 218 kilogram plant nutrient per hectare agricultural area in 1988, decreased substantially and was only 46 kilogram in 1994. Therefore, fertiliser production was ceased in four of five previous plants by the end of 1991. Similarly, about half of the former machinery manufacturing firms had to close. New elements of agricultural input supply system are possibilities for instalment payment in machinery purchasing as well as using machinery services from companies, co-operatives or private proprietors.

Labour productivity

It is reasonable to suggest that farmers will leave the 'safe umbrella' of the collective farm or its successor only if they have a reasonable prospect of earning a higher income as an independent producer. Leaving the collective requires a relatively high potential productivity compared to a relatively low productivity on the collective farm. However, potential productivity in Hungary is rather low outside collective farms, compared to the productivity achieved by Hungarian collectives. This can be explained by (1) the high specialisation of labour which resulted in high productivity inside the collective farm, but low potential productivity in individual farming; (2) the age structure; and (3) the level of education of the rural population (Swinnen and Mathijs, chapter 10).

Hungarian Communist agriculture was characterised by a high degree of labour specialisation, which resulted in a highly unequal distribution of skills or human capital among people employed in agriculture. Workers were trained only to do very specific tasks resulting in a high productivity inside the collective farm. However, they have no skills to set up a commercial farm. Surveys show that only a very low percentage (5-10 percent) of people getting land or active in agriculture have the necessary skills to begin commercial farming. When given a piece of land they will use it for gardening, part-time farming or simply rent it back to the collective farm. On the other hand, people with more skills, such as managers and specialists, have more incentives to leave. In some cases, they were driven out of agriculture as a result of government policies directed against them and because they could earn a better salary outside agriculture, e.g. in banking.

Another factor contributing to the low potential productivity in individual farming is the age structure of the rural population. A survey shows the dramatic shift in rural population structure from 1988 to 1993 (Csite and Kovách, 1995). Whereas in 1988, 17.7 percent of the rural population was retired, this was 31.2 percent in 1993. Csite and Kovács (1995) argue that some of them preferred retiring instead of being unemployed. Education is also an important factor explaining why a substantial part of entrepreneurs come from the top layer of earlier large-scale farms.

Privatisation of non-land assets

Different rules for allocating assets. A law voted in 1989 by the last Communist government made it possible for members to leave the collective farm. However, they could only take out land and not other assets. The law therefore strengthened the situation of the collective farms and their successors. The laws voted by the centre-right, anti-Communist Antall government – the four Compensation Laws and the Transitional Co-operative and Unified Co-operative Laws – also resulted in different rules for allocating land and non-land assets, which may have led to a mismatch of physical and human capital. Members, or former members, can receive land from three sources: compensation, proportional land and land allocated under the Transitional and Unified Co-operative Laws. As a result of the latter they also receive part of the non-land assets. However, claimants who are not members or who are not even engaged in agriculture can only receive land under the Compensation Laws. Hence, they often have no choice but to lease back the land to the new co-operatives or companies since there is a prohibition of selling land received as compensation for five years.

It was not so much the objective of the Antall government to destroy the state and collective farms and establish family farms, but this was most certainly the objective of one of the coalition partners, the Independent Smallholders' Party (ISP). As the Compensation Laws were a compromise in not giving back all of the land, the ISP tried to break up the collectives by amending the Transitional and Unified Co-operative Laws (Mathijs, 1997). As a result, 40 percent of the capital assets were allocated to active members, 40 percent to retired members and 20 percent to people outside agriculture. Hence, the active members can only use 40 percent of their earlier property for future farming while about 10 billion HUF worth of property was withdrawn from agriculture by outsiders and retired members (Harcsa et al., 1994).

The allocation of assets was in many cases also a process of asymmetric information. Especially in the case of non-land assets, collective farm managers were in a good position to take advantage of their privileged situation. Their rent-seeking behaviour during the redistribution process took many forms. Some managers manipulated the process in such a way that they could get the best pieces of land and capital assets. Managers undervalued the capital assets in order to buy them at a low price and transferred them into a limited liability company in which they had a majority stake. In some cases, members who wanted to leave the collective were misinformed by the management.

Inefficient distribution of production factors. The redistribution of land was inefficient. Both a plot of land set aside for compensation and a parcel of proportional land that had to be reallocated should be of both good and bad quality. In some cases large parcels were broken up and scattered around in order to fulfil these requirements. The land left to the new co-operative and its remaining members was cut up and could not be cultivated efficiently unless they could buy (in the case of proportional land) or rent it (in the case of compensated land) back. Otherwise, cultivation is highly inefficient as large parcels of co-operative land encapsulate small plots of land owned by individuals who are not members of the co-operative.

The Hungarian compensation programme failed to allocate land in an optimal way, primarily as a result of collusion among bidders. In most auctions, claimants agreed not to bid, such that the land could be bought for the minimum price of 10,000 HUF per hectare. As the government did not try to prevent collusion by imposing an anti-cartel reserve price, it can be concluded that auctions served not so much as a mechanism to allocate land to the agents with the highest valuation for land, but rather as a mechanism to allocate land to as many claimants as possible.

On the other hand, the auctions did succeed in allocating land to the highest bidder in some cases. The highest prices were paid in the county of Veszprém, near lake Balaton, where land is expected to be used for tourism activities in the future. Also in the surroundings of the capital, in Pest county, extremely high prices were paid as the buyers anticipate getting high returns from future non-agricultural activities on the land.

Transaction costs of withdrawal or 'exit costs'

In general, if a member could leave a collective taking with him all his assets without any problems or cost, the stability of the collective farm would be seriously endangered. That is why certain obstacles are devised to make it more costly for a member to withdraw and are part of the co-operative legislation. However, the amount of obstacles can be increased, thus increasing the 'exit costs' for a member, beyond the bylaws of a co-operative. In this section, we discuss several elements which led to an increase in the exit costs a member faces when he wants to leave the collective farm or its successor.

1 People who got land under the Compensation process and who are not farm workers or heirs of farm workers did not get capital assets as a rule. Hence, it is more difficult for them to start up an individual farm and most of them rent their newly acquired land back to the collective farm.

2 Conversely, about 60 percent of the ownership of capital assets of collective farms has been transferred to people not or no longer active in agriculture in the form of business shares. They mostly receive some kind

of rent for these assets. Collective farm members who want to start up an individual farm hence have no access to these capital assets or have to compete with collective farm management for the use of these assets. In this process, they are in a disadvantageous position, since they have less information and higher search costs than management.

3 After the transformation of collective farms into producer co-operatives, members are free to leave. However, they are not allowed to withdraw any assets, even if they are the legal owner. Instead, they receive a financial compensation. Furthermore, members who want to leave have to pay their part of the collective farm's debt.

4 The restoration of property rights on formally owned or 'proportional' land was often incomplete, as it was decided in many cases not to break up existing parcels, but to share the property rights between several owners. This also happened in the Compensation process in a limited number of cases.

5 Many collective farm managers had better information on the capital assets than their members and took advantage of their privileged position. Some of them succeeded to transfer the ownership of the assets from the collective farm and its members to themselves in exchange for an artificially low price.

6 The amount of land distributed to farm workers, restituted to former owners or allocated through the compensation auctions often did not suffice to start up a viable farm unit. Hence, many potential entrepreneurs had to buy or rent land to expand their area.

Taking into consideration these elements, we can conclude that the combination of policies, and their implementation, chosen by the Hungarian government made it difficult or costly for members to leave their collective farm. Policies meant to destroy the collective farm structure even had the opposite effect: the large transfer of land and capital assets to outsiders resulted in a conservation of the collective farms, albeit in a transformed successor organisation.

Conclusions

The privatisation of land and capital assets has been characterised by a complex set of regulations resulting in different rules and inefficiencies in the transfer of property rights. This led to much freedom and leeway for those who were already managing agricultural resources and who wanted to take advantage of the reform. Policy and legislation was not very supportive for collective farm members who wished to start up an individual farm as many collective farm assets have been restituted to outsiders.

Most collective farms were transformed into producer co-operatives. Other firm types that have been established include the limited liability and the joint stock company and to a lesser extent partnerships. Individual farmers could also be registered as sole proprietors hence benefiting from a tax reduction. Only few collective farms have been liquidated or declared bankrupt.

As a result, 78 percent of total agricultural area was still used by co-operatives, companies and partnerships in 1994. Individual farms accounted for 22 percent of total agricultural area. The institution of land renting has become widespread in Hungarian agriculture, since co-operatives rent all their land either from their members or outsiders. However, the land market still limits the amount of transactions: both renting and buying are restricted. Nevertheless, 1995 data show an increase in land use by individual farms as a result of the finishing of the compensation process.

Several factors contributed to current farm structure in Hungary: the adverse conditions in the various markets, the relatively good performance of pre-reform farms versus the lack of human capital for individual farming, the differences in privatisation between land and non-land assets and the exit costs imposed on a member who wishes to withdraw from the collective farm.

References

Csite A. and Kovách I. (1995), 'Postsocialist Transformation in the Rural Societies of Central and East Europe' (in Hungarian), *Szociológiai Szemle*, Vol. 2, p. 49-72.

Czárlné I.M., Szabó J. and Tassy S. (1996), '1995 Tendencies of Agriculture and Food Industry' (in Hungarian), *Gazdaság és Statisztika*, Vol. 8, No. 1, pp. 3-21.

European Commission (1995), *Agricultural Situation and Prospects in the Central and Eastern European Countries: Summary Report*, European Commission, Directorate-General for Agriculture: Brussels.

Harcsa, I. (1991), 'Privatization and Reprivatization in Hungarian Agriculture', *Acta Oeconomica*, Vol. 43, No. 3-4, pp. 321-348.

Harcsa, I. (1996), *Peasants and Agricultural Proprietors* (in Hungarian), Central Statistical Office: Budapest.

Harcsa, I., Kovács, I. and Szelényi, I. (1994), 'The Postsocialist Transformation Crisis in Agriculture and Rural Society' (in Hungarian), *Szociológiai Szemle*, Vol. 3, pp. 15-43.

HCSO (1994a), *Statistical Yearbook of Hungary 1994*, Hungarian Central Statistical Office: Budapest.

HCSO (1994b), *Agricultural Statistical Pocket Book* (in Hungarian), Hungarian Central Statistical Office: Budapest.

HCSO (1994c), *Agricultural Small Farming 1993* (in Hungarian), Hungarian Central Statistical Office: Budapest.

HCSO (1994d), *Classification of Economic Organizations by Legal Forms* (in Hungarian), Hungarian Central Statistical Office: Budapest.

HCSO (1995), *Survey on Small Farmers 1994* (in Hungarian), Hungarian Central Statistical Office: Budapest.

Jackson, M. and Swinnen, J. (1995), *A Statistical Analysis and Survey of the Current Situation of Agriculture in the Central and Eastern European Countries*, Report to the European Commission, The Leuven Institute for Central and East European Studies: Leuven.

Magyar Közlöny (1995), 'The Land Law and Rules of Land Registration' (in Hungarian), Magyar Közlöny, Supplement.

Mathijs, E. (1997), 'Process and Politics of Agrarian Reform in Hungary', in Swinnen, J.F.M. (ed.), *Political Economy of Agrarian Reform in Central and Eastern Europe*, Ashgate: Aldeshot.

Mészáros, S. (1994a), 'Analysis of Price and Trade Regime During the Reform of Hungarian Agriculture', *Zemedelska Ekonomika*, Vol. 40, No. 3-4, pp. 219-241.

Mészáros, S. (1994b), 'The Reform Process in Hungarian Agriculture: An Overview', in Swinnen, J.F.M. (ed.), *Policy and Institutional Reform in Central European Agriculture*, Avebury: Aldershot.

Ministry of Agriculture (1995), *Situation of Privatization of Agricultural State Property on 31 December 1994* (in Hungarian), Ministry of Agriculture, Department of Privatization: Budapest.

OECD (1994), *Review of Agricultural Policies: Hungary*, OECD: Paris.

Privinfo (1995a), 'Compensation in Figures' (in Hungarian), *Privinfo*, Vol. 4, No. 1, pp. 42-44.

Privinfo (1995b), 'Privatization of State Farms' (in Hungarian), *Privinfo*, Vol. 4, No. 2, pp. 14-17.

Swinnen, J.F.M. and E. Mathijs (1997), 'Agricultural Privatisation, Land Reform and Farm Restructuring in Central and Eastern Europe: A Comparative Analysis', this volume.

Szijjártó, A., 1996, *Situation and Intentions of Agricultural Producers* (in Hungarian), Research and Information Institute for Agricultural Economics: Budapest.

Szúcs, I. and Tanka, E. (1994), *Transformation of Land Ownership and Land Use System During the Reform in Hungarian Agriculture* (in Hungarian), Research and Information Institute for Agricultural Economics: Budapest.
Tóth, E. and Varga, G. (1995), *Situation of Agricultural Co-operatives in the Period of their Transformation* (in Hungarian), Research and Information Institute for Agricultural Economics: Budapest.

6 Restructuring of farms and agricultural systems in Romania

Alexander H. Sarris and Dinu Gavrilescu

Introduction and background

The purpose of this chapter is to analyse the forces that are currently shaping and are likely to shape in the near future the structure of producing units in Romanian agriculture. Romania has undergone a drastic agricultural transformation starting in 1989 with the fall of Communism. This transformation involved a fundamental shift in the ownership of the major agricultural factor of production, land. It also involved the transformation in the ownership of livestock, and several other systemic changes. These transformations have led to new production structures that are still evolving. It is our purpose to try to understand the factors that are involved in the performance of the various different farming structures and especially the economic forces that are faced by the individual owners or operators of the various types of farms.

Before 1948, agriculture was organised around private farms, with many small ones that had been created by the land reforms of 1921 and 1945 (Mitrani, 1931, Roberts, 1951). In 1921 holdings of more than 200 hectares (ha) were expropriated and 2.8 million ha were distributed to one million smallholders. However, many large landowners remained, because of the difficulties in implementing the land reform. In 1945, the government expropriated the land owned by German citizens and collaborators, and absentee landlords, as well as private land over 50 ha. By January 1947, 1.4 million ha had been distributed (with title) to 800,000 smallholder households with fewer than five hectares of their own. Before collectivisation about 78 percent of the farms were smaller than 2 ha.

Collectivisation started in 1948. Initially it proceeded slowly, with only 10 percent of arable land expropriated by 1956. By 1962, however, 77 percent of the land was under state control. By 1989, 12,953,000 ha of agricultural land were state owned. Only 1,816,000 ha remained in private ownership, and these mostly in one million remote and fragmented mountain farms. When land was expropriated for collectivisation, inventories were made listing the names of the former owners, the area and quality of land, and a description of its location, but no maps were drawn. Many of those lists still exist, but some were destroyed and larger landowners tended to under-declare their holdings to avoid accusations of being exploiters.

Besides state control of agricultural production, the pre-1989 Communist regime regulated through state ownership and management, all the upstream and downstream activities in the agro-industrial complex. This included all aspects of marketing, fertiliser and other chemical supply, credit, irrigation and other agricultural services (such as mechanical services through separately organised 'machine-tractor stations'), and agricultural processing. On the eve of transition about half of the population lived in rural areas, and 30 percent of the working population was occupied in agriculture. However, almost all rural households had one adult that was a member of a collective farm, and a large number of non-farm workers had farming as a part time occupation on garden plots.

The violent events of December 1989 that led to the overthrow of the Ceausescu regime, were accompanied by considerable spontaneous land and agricultural asset confiscation by peasants as well as collective farm managers. The provisional government which was opposed to breaking up of the agricultural land for fear of loss of efficiency, tried to control the spontaneous events by issuing various decrees concerning land, raising the minimum legal ownership limits to 0.5 ha and making provisions for acquisition of land. It appears that these efforts tried to legalise what had already happened spontaneously. In March 1990, for instance, an announcement by prime minister Roman said that 3 million ha of arable land had been 'leased' to 'peasants' (compared to 0.8 million ha of land in peasant hands before). Subsequently the parliament passed various laws (see below) concerning agricultural land, but it seems that the immediate events in early 1990 were crucial in setting the stage for subsequent legislation (Jackson, 1997).

After 1989, there has been considerable transformation in land ownership, with the bulk of formerly co-operative land being transferred to its previous owners, but considerably less transformation has taken place in the upstream and downstream activities. These for the most part continue to be governed by management systems and rules similar to those of the Communist era.

There are several factors that impinge on agricultural transformation. These factors include firstly the legal system and the implementation of the privatisation program. However, there are other forces that are important. These include the structure of input and output markets, especially the market for credit, the structure of consumption and incomes of farm operators, and especially the availability of non-farm employment opportunities, the market conditions in the various input and output markets and particularly the prices. These and other factors will be discussed in the sequel.

In terms of theory we can distinguish two approaches to analysing the process of privatisation. The first, we could call the top-down approach, is concerned with the institutional, political and economic factors that led to the adoption of specific privatisation policies. A good example of this type of theorising applied to Central and Eastern Europe (CEE) agriculture is Swinnen (1997) and chapter 10 of this book, and another is Jackson (1997). This approach is very useful in conceptualising the forces that led to the 'initial post-Communist conditions' of agricultural property rights in CEE. However, it is less useful in analysing the forces of subsequent change. We believe that for that purpose another approach of the 'bottom-up' type is more useful, namely one that considers the exogenous environment and incentives faced by individual operators. In what follows the discussion will try to examine some elements of that second approach.

The process of land reforms since 1989

In Romania there has been a series of laws, decrees, and other regulations governing land and non-land privatisation. It is quite difficult to pinpoint the exact process of privatisation, and especially its degree of implementation. The story here is based on Jackson (1997) and Gavrilescu (1994), as well as newer information and authors' recent field work.

The first law concerning land came very early after the events of December 1989, with decree/law 42/1990 of January 29, 1990. This law provided that in agricultural areas any family, including town dwellers could obtain up to 0.25 ha of land, provided the land could be worked and the local co-operative was paid a land-use fee. The land covered by a co-operative member's house garden and yard became the member's personal property and together with any new land could be sold and inherited up to 0.6 ha. In hill and mountain regions or areas near towns, co-operative members could receive land for 'long term use' up to a limit of that worked by a family without additional workers. Members and former members of co-operatives in plains regions could receive plots of up to 0.5 ha but only for personal use. It appears that this law was designed to formalise what had already taken place spontaneously immediately after the revolution, and was aimed at the land of co-operatives.

The Land law 18/1991 was passed by parliament more than half a year after this first decree, and was published in February 1991. Its basic provisions were that land was to be restituted to former owners in sizes from 0.5 to 10 ha, depending on the amount of land previously owned (with a maximum arable land that can be owned by a physical person of 100 ha). Land in most categories could be leased and sold, but with prior rights of purchase of co-owners, neighbours and local land authorities. However, transactions involving legal persons had to receive approval of the Ministry of Agriculture and the Ministry of Environment. Changes in the use of land must receive approval by the authorities, and agricultural land must be cultivated with the threat of expropriation after two years of non-cultivation. Land distributed was not necessarily the same as land formerly brought into the collective, and was subject to a formula of 'arable equivalent'. Claimants had to present a written request indicating the amount of land claimed including a series of papers and documents within 45 days of the publication of the law. This last requirement suggests that the law made it quite difficult for many persons to file, as in the middle of the post-revolution reorganisations, the public authorities were burdened with the demand for issuing many papers.

For non-land assets the law provided that liquidation commissions could be set up to dispose of the non-land assets of collectives. This property would be distributed firstly to members' collectives (where these were formed) and then processed. If the former members opted to form a new co-operative or commercial company, then ownership rights to the non-land assets were distributed according to the relative size of the restitution claims. Those who wanted land in order to quit the collective, were to be given land that was inefficient for the collective to farm. If a collective was not formed the non-land assets were to be auctioned off, and the proceeds distributed to former members according to their contributions. If land had been formerly taken into a state controlled agricultural enterprise, then the claimant was to be given ownership shares when the enterprise was commercialised and privatised according to law 31/1990.

Law 15/August 1990 provided for the conversion of state enterprises (and this included the state farms) into forms defined in the commercial code, namely limited liability companies, joint-stock companies, and state-autonomous companies. The privatisation law 58 of August 1991, was a follow up and distributed 30 percent of the shares of commercial companies to five 'private ownership funds' that were to act like mutual investment companies. These were to issue of each Romanian citizen one certificate that could be sold to other Romanian citizens, exchanged for shares of companies being privatised, or kept for sharing in the profits of these funds. The remaining 70 percent of the shares were turned over to one large State Ownership fund which was to divest these shares at the rate of 10 percent per year through public auctions and negotiated sales. This has not as yet (mid-1996) started. Finally the National Privatisation Agency could sell up to 75 percent of certain enterprises on a selective basis, and this has proved popular and effective.

The application of the land laws started in March 1991, with the provision of rights to all persons entitled by the laws. 6,236,507 claims for recognition of ownership rights were received within the legal period, and 4,722,681 temporary vouchers (*adeverinte* or 'acknowledgements') were given, covering 35 percent of the land involved in claims. By June 1994, 99 percent of the certificates, covering 89 percent of the owners and 89 percent of the area, had been awarded temporary vouchers. In the second stage ownership certificates were to be issued, but this phase was delayed considerably. By the end of 1992 only 2 percent of owners had received titles, and by the end of 1993 about 700,000 titles covering about 17 percent of the land had been granted. By June 1994 the number of titles is reported to have been 1,095,000 covering 23.1 percent of legal owners and 25 percent of the land claims.

There were other laws such as Law 36/1991 on agricultural companies and associations, Law 46/1992 on the rights of former owners who have shares in companies). These laws and their consequences have been reviewed thoroughly by Tourne (1993) and Crespi Reghizzi (1993), and will not be discussed here.

One consequence of the land and privatisation laws is that as the primary concern of the privatisation was justice and restitution of previously confiscated land and other property, the land was not allocated to those who could work it personally. This resulted in considerable mismatch between owners and potential farmers. According to one survey 57 percent of new owners are more than 65 years old, 43 percent live in towns, 39 percent work or are retired in villages, and only 18 percent are actively-working smallholders who perform uniquely agricultural activities. However, the legislation has facilitated the organisation of various types of associations for the purpose of cultivating land. Currently the legislation in force provides for six forms of farming activity:

1 Individual farmers

2 Family partnerships, abbreviated as FA

3 *Societé civile* (namely joint-stock companies)

4 Farming partnerships (*societatile agricole*), abbreviated as AS

5 Commercial companies (*societatile comerciale*)

6 State farms

The first four forms were provided under Law 36/1991 on farming partnerships, while the fifth one is regulated by Law 31/1990. The first four legal forms of farming organisation seem according to knowledgeable observers (Crespi Reghizzi, 1993) to provide ample room for diverse types of farming practices. Whether one or the other will predominate will depend on various other circumstances, and it seems too early to forecast this. However, the important thing from an organisational perspective is that the current legal system appears fairly flexible.

Given that there is a large number of new land owners that will not be cultivators, and given that transfer of land will be restricted in the next few years as provided by the land laws, it appears that various types of land leasing and subcontracting will arise. The land lease law was passed in 1994 but it appears that this law is too rigid, and there are currently a variety of informal leasing arrangements in force, governed primarily by the need of new owners to ensure that they will not lose their land by letting someone else cultivate it. For instance, the law states that the owner's share under a sharecropping lease arrangement is 30 percent, but the actual shares are in the vicinity of 50-60 percent. Under these circumstances, and given that land ownership deeds are not yet secure, it is probably reasonable to explain the large number of associations, both AS and FA, that at least create a group responsibility over the rights of individual members. Also, despite the fact that actual sales of land are very rigidly controlled and in some cases forbidden, there are a lot of sale like arrangements reported (for instance under the guise of donations). It thus appears that the elements of a land market are starting to appear.

The emerging agricultural production structures

There were three types of agricultural production units in the years before 1989. These were the state farms, the agricultural co-operatives and the smallholders.

First we discuss the large scale capital intensive state farms. These were called State Agricultural Enterprises (IAS). They were organised on land that had belonged to the royal family before 1945, land that belonged to prosperous individuals that were declared 'enemies of the state' and land that belonged to collective farms close to urban centres. In 1985 there were 419 IASs that farmed about 2,051 thousand hectares of land (13.7 percent of total agricultural land). After 1989 these farms were simply renamed 'Commercial Companies' (CCs) with capital belonging both to the state as well as former private owners (this is why they are also referred to also as state/private companies). In terms of operations, they remained organised as before. At the end of 1992 there were approximately 550 such state/private farms occupying the same area as before, at least half of which is irrigated. The assets of these state farms in 1993 were estimated at around 500 billion lei, and they employed between 250,000 to 300,000 people.

State/private farms consist of multiple enterprises, which are often in different locations. They operate with a management structure that is hierarchical, with the director being appointed by the Ministry of Agriculture. They operate under objectives laid out by the Ministry of Agriculture, and without concern for efficiency or profit, but mostly with the goal of maximising production, as long as there is demand. There is no accounting system in the modern sense of the term, and hence the efficiency of the various units cannot be ascertained. They obtain their inputs from state owned enterprises at prices fixed by the government, and they also sell to large scale state enterprises, also at given prices. They still obtain credit at preferential terms from state banks.

Under the 1992 privatisation law 30 percent of the shares of state farms were distributed through the Private Ownership Fund. These private owners do not own land per se, but just shares in the overall company, for which they are compensated by some 'dividend'. Also the owners of land that had been taken over by the IASs were given shares in the new CCs (an additional 15 percent of the shares). Hence about 45 percent of the shares of the CCs were transferred to private individuals. However, this transfer of ownership has not altered the way these farms operate.

The employees of CCs have the best rural standard of living. Apart from salaries, they enjoy a considerable number of other benefits, such as the ability to purchase products produced by the farm at below market prices, supplemental payments for children, holiday bonuses, up to half a hectare of land for personal use, full medical coverage, subsidised visits to health spas, and for some, apartments and meals. For this reason they attracted, and still attract, the best graduates from the agricultural universities. State farms also support other social services such as retirement programs and training for children. These benefits coupled with reported benefits due to corruption and illegal seizure of farm assets after 1989, suggest that the group of people employed on state farms have a strong vested interest in the status quo. As most of them are not former owners, there is a conflict of interest between them and the current as well as former owners, that might affect the development of this type of enterprises in the future.

We now turn to the former agricultural co-operatives (CAPs). These were established during the socialist years through confiscation of land plots, which were then combined into agricultural production units with average area of about 2,000 ha. In 1985 there were about 4,000 CAPs, and they accounted for about 9.1 million ha of agricultural land (60 percent of total).

The break-up of CAPs started in February 1990, very soon after the 1989 violent changes. With Government Decree No. 42 the CAP workers were allocated 0.5 ha each, and villagers not working in CAPs were given 0.25 ha of land. The legal framework for the dissolution of CAPs was Land Law No. 18 of 1991. According to this law previous owners of land were entitled to a maximum of ten ha of land, regardless of the size of previously confiscated land. As a result of this law privatisation of land (through restitution or distribution) occurred very rapidly. While in 1989 only 12 percent of arable land was in private hands, by the end of 1991 about 80 percent was privately owned, including land that was never collectivised. A total of 5.3 million families benefited from this land distribution. Of these, 3.5 million are actively engaged in agricultural production. The law, however, stipulates fines to land owners if they do not cultivate their land, and this implies that there was an incentive for land owners that were not employed in agriculture to find ways to have their land cultivated.

As of the end of 1994, of the total land that was to be privatised, and which formerly belonged to CAPs, 99 percent had been done so via 'acknowledgements' or *adeverinta*, which are simply papers stating the amount of land an individual is entitled to without, however, the exact location or title. Possession Certificates had been issued for about 50 percent of that land, while full ownership titles had been given to about 20 percent of the land. The problems of full and legal land privatisation included the absence of cadastral surveys, confusions about the implementation of the land law, and differences in the quality of the different types of land. While delays have been experienced due to the appeal process of the law, it appears that *de facto* agricultural land privatisation has proceeded very fast.

Concerning production in the newly acquired lands, there were considerable problems. These included, first the large number of urban residents that acquired land and had no farming experience, second the small size of landholdings that were obtained and their division into several plots. This break-up, coupled with the absence of proper machinery and equipment geared to small scale production made farming by the new owners very difficult. It is under these circumstances that many new land owners chose to form farming organisations.

There are two types of 'new' private co-operative farming organisations that have emerged in the aftermath of the break-up of the CAPs. The first is the agricultural societies or co-operatives (AS). By the end of 1992 there were about 4,000 of these with a membership of 780,000 farmers, and with land totalling around 1.9 million ha (about 470 ha per AS). These ASs are legal entities (namely have juridical status and entity just like companies do), and have a management structure much like the one of former CAPs. However, directives for the ASs are given by the general meetings of the producers and not through the state. Many of the former leaders of the CAPs are now managers of the ASs, and this is probably due to the initial lack of farm management expertise by the new owners.

The second type of farm organisation is the farmer's association or partnership (FA). These are considerably looser forms of association, as they are only registered but do not have a juridical status. By the end of 1992 there were approximately 11,500 FAs with 720,000 members, operating around 1.8 million ha (about 150 ha per FA). The differences between FA and AS, apart from size and the legal status, involve the ability to hire staff, as the FA are too small for that, while AS have on average 3-5 employees apart from the members. The management of a FA is normally done by one of its members.

Table 6.1 indicates the evolution of the number and area of the above two types of private associations as well independent private farmers. The latter includes both the formerly small independent farmers, as well as the 'new' independent farmers that appeared from the breke-up of the CAPs. It appears that between 1993 and 1995 there is a shift out of formal associations (namely the ASs) into family associations and independent type of farming. This would tend to support the view that the trend is in the direction of independent private farming. This tendency will be investigated further below.

Production decisions in FA/AS are done independently by each member, who decides what is to be produced on his land, but under joint production. This means, for instance that certain tasks that entail economies of scale such as mechanisation or harvesting or input procurement can be done jointly. The members of the FA/AS work as farm workers, for which they are paid independently as labour, but inputs and credit are obtained mostly by the association. Most of the produce is also marketed by the association, but members receive part of the product as payment in kind, and also receive part of the financial surplus.

There is considerable flexibility on the part of the farmers as to how much land to put under AS/FA, or what tasks to perform jointly. For instance a landowner of 8 Ha can operate in one year as an independent farmer on 2 ha using draft animals. The remaining 6 ha can be offered to an association (FA), an agricultural company (AS) or even leased to a commercial company (CC). In the following year he can buy a tractor and work all the land alone. Or he could sell the draft animals, keep 0.5 ha for individual cultivation for family consumption, and enter an association or lease the remaining 7.5 ha. He could also choose to work as a worker for the association or not. In other words, there seems to be a continuum of degrees of co-operation that is possible for individual landowners, ranging from completely independent farming, to complete leasing of the land to an association or CC.

As indicated in Table 6.1, by early 1993 about 3.7 million ha were farmed under the various forms of association, and this constituted only 40 percent of the land of the former CAPs. Hence many new owners decided to stay out of co-operative production. While five million people benefited from land distribution of the former CAPs, only 1.5 million are members of AS/FA. Hence the majority of new landowners have not joined any association.

The final type of farm organisation is that of independent private farmers. Apart from the new farms that came out of privatisation, there are about

1 million private smallholders that were located mostly in mountain areas, and are still there. It is hard to estimate the current number of private farms, as many new landowners who are not members of AS/FA do not farm the land themselves, and either lease it to private farmers or leave it idle. The estimated upper bound of private farms is 4.5 million (1 million former smallholders plus 3.5 million new land owners), but the actual number is probably much smaller. As there is no census on which to base estimates, it is impossible to be precise on this.

The conclusion from the above brief review is that the three main forms of farm production organisation in the communist period have been succeeded by three new forms of organisation. The first is the CC, which is a direct successor, without much change in organisation and management, of the former state farms. The second is the AS/FA type of farm production, that has arisen as a natural follow-up to the former CAPs but with important management differences, as they are run privately. Finally there is the large number of former and new individual private farmers, who are no different in structure from the well known family farmers of the west. Nevertheless, as indicated above, there is a continuum of degrees of co-operation in the middle between independent farming and CCs.

The interesting issue from a structural perspective is the stability of these three types of farm organisation. While the private mode of production has persisted for many generations in the west, and also in Romania, and seems to have the potential of remaining an important form of organisation, it is interesting to speculate on whether the other two types of farm organisation are stable in the medium to long run. While the co-operative form of farm production is almost non-existent in the west, co-operation being restricted mostly to those tasks that are subject to economies of scale, such as input procurement, and occasionally output marketing, corporate farming, which is a mode of production that is well developed in the west, has a potential counterpart in Romania in the form of the current CCs.

The current state of privatisation in Romanian agriculture

In this section we give a brief empirical picture of the state of privatisation by illustrating the degree to which inputs and outputs are controlled by private farmers or farms with private property structure.

Table 6.2 exhibits the evolution of land ownership from 1989 to 1995. It can be seen that between 1989 and 1991 the ownership of land changed dramatically. While in 1989 only 13.7 percent of agricultural land was under private ownership, by 1991 it had jumped to 70 percent. The only type of agricultural land that still has a large share of public ownership is pasture and hay fields (only 50.5 percent private). It must be noted that the various forms of associations that were the successor to the CAPs are now classified as private lands. As discussed above, this does not imply that these lands are farmed by independent farmers. Hence ownership and mode of production or use are currently two distinct things. It is difficult to distinguish among this privately owned land the amount that is farmed under AS/FAs and the amount farmed by independent farmers.

Table 6.3 exhibits the ownership structure of the different types of livestock in 1989 and 1994-95. The first thing to notice is that the population of all types of animals declined considerably between 1989 and 1994. From the table it seems that this concerns mainly the public sector, but as the former co-operatives were all privatised and their animals presumably became private property, it can be said that the stock of animals in the private sector has declined as well. In 1995, however, an increase in stocks of most types of animal appears to have taken place. By 1995 the bulk of all animal units (80 percent) were privately owned.

The privatisation of other types of agricultural assets, mainly machinery and equipment, has proceeded much slower. As of the end of 1994, only 21.1 percent of the fixed assets in agriculture was under private property, compared with 15 percent in 1989. In 1989 there were approximately 151.7 thousand agricultural tractors in the country of which 116.7 thousand were utilised by the State Machinery Companies (Agromec), and 24.7 thousand by the state farms. At the end of 1995 the tractor fleet was 163.5 thousand of which 30.0 thousand was with the Agromecs, 28.8 thousand with the successors to the state farms (the CCs), 9.0 thousand with the formal farmer associations (AS), and 95.7 thousand with private individuals and FAs. This is a reasonable but slower pace of privatisation compared with that of land and livestock. Regarding the Agromec companies, out of 521 units in 1993, 90 were privatised. At present (1996) all units are in the process of privatisation by the coupon system.

Table 6.4 exhibits the volume of production of the main crop and livestock products in 1989, 1993 and 1994, according to the sector of production. The first thing that can be noticed is the marked decline in the volumes of production of almost all crop and livestock products. The exceptions are corn and in 1993 beef and pork meat. The explanation of this according to the authorities is that, apart from area changes, there has been a decline in the provision of inputs, and hence yields have declined for most categories of producers. In the livestock products category the explanation is that after the privatisation there was considerable slaughtering of animals, and hence production of meat.

Table 6.5 exhibits the areas of the main crop products in 1989, 1993 and 1994 by sector of production. The total area devoted to the production of the major crops does not appear to have declined by much in the period after 1989, and in fact it appears to have increased in 1994. There seems to be a decline in the total area cultivated to these crops by state farms by 1993 but by 1994 the area is virtually identical to that cultivated in 1989. The changes in areas cultivated vary considerably by product, with some products such as corn exhibiting expansion (mostly by the private sector), others such as sunflower exhibiting general expansion in all sectors, while others such as sugarbeet exhibit general declines in all sectors.

Turning to yields, shown in Table 6.6, the picture is interesting. While yields have declined for all the indicated crops in state farms and for most crops in the new privatised co-operatives, the case is different for the independent farmers. Yields of corn, barley and potatoes in these sectors appear to have increased, despite the alleged lack of inputs.

It is not clear what one is to make of these trends as the data suffers from serious measurement problems that are not uniform in the pre and post 1989 periods (for an analysis of that see Jackson and Swinnen, 1995). In particular, it seems that there is underreporting of production by both the CCs, as well as the new associations of producers (both AS and FA) in order to avoid taxation, and in order to divert some products from official to unofficial marketing channels. Table 6.7 exhibits the former and current crop patterns in the major crops by the three different types of farm organisation. The most noticeable change is the large share of independent farmer land that is devoted to wheat and rye and corn. This might be due to concerns for food security.

Exogenous and policy related factors conditioning farm transformation

The legal system governing the transformation of land and non-land assets was discussed in the second section. In this section we discuss some of the other factors relevant to farm transformation.

Input distribution and markets

It is difficult to develop individual private sector agricultural production without easy availability of agricultural inputs. Under the Communist regime, the distribution of inputs (seeds, fertilisers, farm machinery, mechanisation services, chemicals, irrigation, animal feeds, etc.) was regulated by large-scale state companies that catered to the state farms and the CAPs. The same was true for credit.

Table 6.1

Evolution of private farm structures in Romania (1993-95)

	Date				
	15.06.1993	*31.12.1993*	*15.06.1994*	*31.12.1994*	*31.12.1995*
	Number of farms in each class				
Agricultural societies	4,319	4,265	4,070	3,970	3,973
Family associations	14,206	13,772	14,944	13,741	15,915
Individual households	3,382,031	3,419,736	3,555,227	3,578,234	3,597,383
	Agricultural land (ha)				
Agricultural societies	1,934,758	1,909,576	1,784,002	1,770,569	1,733,000
Family associations	1,773,941	1,762,976	1,726,493	1,536,532	1,596,000
Individual households	7,508,840	7,333,290	8,119,694	7,905,372	8,052,000
	Average area per farm (ha)				
Agricultural societies	448	448	438	446	436
Family associations	125	128	116	112	100
Individual households	2.22	2.14	2.28	2.21	2.24

Source: Ministry of Agriculture and Food

Table 6.2

Evolution of agricultural land by type of ownership (End of year figures in 000 hectares)

	1989	*1990*	*1991*	*1992*	*1993*	*1994*	*1995*
Total agricultural land	14,759	14,769	14,798	14,790	14,793	14,798	14,798
- public property	4,134	2,142	2,561	2,560	2,614	2,618	2,618
- state private property	0	0	1,912	1,835	1,841	1,808	1,338
- co-operative sector	8,602	8,433	0	0	0	0	0
- private property	2,023	4,194	10,325	10,395	10,338	10,372	10,842
Total arable land	9,458	9,451	9,424	9,357	9,342	9,338	9,347
- public property	1,964	251	456	441	411	394	395
- state private property	0	0	1,518	1,449	1,445	1,440	974
- co-operative sector	6,578	6,499	0	0	0	0	0
- private property	916	2,701	7,450	7,467	7,486	7,504	7,978
Total vineyards	278	277	285	303	303	299	290
- public property	89	44	14	13	13	14	14
- state private property	0	0	74	66	68	65	64
- co-operative sector	154	151	0	0	0	0	0
- private property	35	82	197	222	222	220	212
Total orchards	318	313	311	304	295	289	275
- public property	108	73	30	26	26	25	24
- state private property	0	0	75	74	71	71	68
- co-operative sector	140	136	0	0	0	0	0

Total pastures and hayfields	4,705	4,728	4,778	4,831	4,853	4,872	4,886
- public property	1,973	1,778	2,060	2,080	2,161	2,185	2,185
- state private property	0	0	243	239	239	232	232
- co-operative sector	1,730	1,645	0	0	0	0	0

Notes: Public property is the property of state and its organisations, designated for public use. This includes mainly grasslands, and land belonging to research stations. State private property is the land of CCs. Co-operatives are the former co-ops. Private property is the land of AS/FAs and individuals.

Source: Institute of Agrarian Econiomy

Table 6.3
Evolution of livestock by type of ownership (End of the year in '000 head)

	1989	*1994*	*1995*
Total cattle	6,290.7	3,480.8	3,550.0
- public property	1,441.1	424.4	430.0
- co-operative sector	3,064.5	0.0	0.0
- private property	2,082.1	3,056.4	3,120.0
of which: cows, buffalo cows and heifers	2,468.4	1,962.7	2,000.0
- public property	319.6	158.7	160.0
- co-operative sector	1,031.9	0.0	0.0
- private property	1,116.9	1,804.0	1,840.0
Total sheep and goats	16,452.0	11,641.7	11,800.0
- public property	2,730.0	1,081.4	1,100.0
- co-operative sector	5,516.1	0.0	0.0
- private property	8,205.9	10,560.3	10,700.0
of which: ewes, young ewes and goats	9,997.4	8,591.1	8,600.0
- public property	1,287.2	650.9	660.0
- co-operative sector	2,984.9	0.0	0.0
- private property	5,725.3	7,940.2	7,940.0
Total pigs	11,671.0	7,758.1	8,000.0
- public property	6,087.0	3,343.3	3,300.0
- co-operative sector	2,250.0	0.0	0.0
- private property	3,334.0	4,414.8	4,700.0

of which: sows	1,023.4	576.4	600.0
- public property	556.5	286.0	290.0
- co-operative sector	289.7	0.0	0.0
- private property	177.2	290.4	310.0
Total poultry	113,968.3	70,156.6	75,000.0
- public property	57,736.6	21,816.5	24,000.0
- co-operative sector	14,849.3	0.0	0.0
- private property	41,382.4	46,340.1	51,000.0
of which: laying poultry	49,390.0	36,233.1	38,000.0
- public property	17,084.3	7,041.1	8,000.0
- co-operative sector	7,181.7	0.0	0.0
- private property	25,124.0	29,191.0	30,000.0
Total horses	662.7	783.8	790.0
- public property	64.2	28.1	28.0
- co-operative sector	171.9	0.0	0.0
- private property	426.6	755.7	762.0
Total in conventional units	12,067.0	7,904.0	8,175.0
- public property	3,438.0	1,650.0	1,647.0
- co-operative sector	4,187.0	0.0	0.0
- private property	4,442.0	6,253.0	6,528.0

Source: Institute of Agrarian Economy

Table 6.4

Volume of production of the main agricultural products in 1989, 1993 and 1994 (thousand tons except where noted)

	1989				*1993*				*1994*			
Product	*Total*	*State sector (CCs)*	*Co-ops*	*Indep. farmers*	*Total*	*State sector (CCs)*	*Co-ops*	*Indep. farmers*	*Total*	*State sector (CCs)*	*Co-ops*	*Indep. farmers*
Wheat	7,935.2	1,472.0	6,191.3	271.9	5,354.5	1,325.0	1,507.1	2,522.4	6,191.5	1,428.7	1,617.0	3,145.8
Corn	6,761.8	1,233.0	3,828.7	1,700.1	7,987.4	747.0	1,435.7	5,804.7	9,343.2	688.0	1,403.1	7,252.1
Sunflower	660.3	148.0	503.7	8.6	695.8	221.5	234.1	240.2	763.7	239.5	247.3	276.9
Sugarbeet	6,771.1	960.7	5,527.2	283.2	1,776.4	389.6	621.9	764.9	2,773.8	709.5	827.6	1,236.7
Potatoes	4,420.3	465.6	1,992.1	1,962.6	3,708.9	313.6	328.1	3,067.2	2,946.7	176.8	196.8	2,573.1
Vegetables	3,726.6	1,474.1	1,063.7	1,188.8	2,872.7	442.4	108.3	2,322.0	2,568.9	305.9	55.9	2,207.1
Fruits	1,580.1	572.0	233.0	775.1	2,182.6	573.0	48.0	1,561.6				
Beef meat[a]	351.2	73.0	119.8	158.4	402.7	62.6	10.8	329.3	293.0	51.7	0.4	240.9
Pork meat[a]	878.5	366.8	72.6	439.1	907.4	347.1	58.4	501.9	690.9	302.3	0.2	388.4
Sheep meat[a]	187.5	23.1	40.4	124.0	169.5	13.1	1.8	154.6	77.4	7.3	0	70.1
Poultry[a]	470.5	152.2	28.5	289.8	406.5	136.6	13.8	256.1	261.9	85.5	0	176.4
Cow milk[b]	41.1	7.5	12.8	20.8	41.1	5.2	0.7	35.2				

[a] Live weight
[b] Thousand hl

Source: National Commission of Statistics

Table 6.5
Area of major crops by sector in Romania, 1989, 1993 and 1994
(in thousand hectares)

Product	*Total*	*State farms*	*Co-ops or AS/FA*	*Indep. farmers*
		1989		
Wheat and rye	2,359.0	359.6	1,890.8	108.6
Corn	2,733.5	407.8	1,607.8	717.9
Barley	767.8	258.2	503.7	5.9
Sunflower	433.7	86.7	339.6	7.4
Sugarbeet	255.9	38.0	208.6	9.3
Potatoes	351.4	26.4	169.5	155.5
Vegetables	252.8	85.9	95.6	71.3
Total area	7,154.1	1,262.6	4,815.6	1,075.9
		1993		
Wheat and rye	2,307.5	466.8	667.6	1,173.1
Corn	3,065.7	253.0	520.6	2,292.1
Barley	468.5	241.4	121.0	106.1
Sunflower	588.3	156.6	211.8	219.9
Sugarbeet	97.2	17.1	37.0	43.1
Potatoes	249.0	18.5	19.2	211.3
Vegetables	200.3	13.6	8.6	178.1
Total area	6,976.5	1,167.0	1,585.8	4,223.7
		1994		
Wheat and rye	2,440.7	480.7	669.1	1,291.1
Corn	2,983.4	223.7	452.0	2,307.7
Barley	785.0	317.4	205.4	262.2
Sunflower	582.2	170.3	189.8	222.1
Sugarbeet	130.0	29.7	39.1	61.2
Potatoes	248.6	13.8	15.7	219.1
Vegetables	203.7	22.4	4.8	176.5
Total area	7,373.6	1,258.0	1,575.9	4,539.9

Source: National Commission for Statistics

Table 6.6
Yield of major crops by sector in Romania in 1989, 1993 and 1994
(in kg per ha)

Product	*Total*	*State farms*	*Co-ops or AS/FA*	*Indep. farmers*
		1989		
Wheat and rye	3,364	4,093	3,274	2,504
Corn	2,472	3,014	2,381	2,368
Barley	4,475	4,941	4,267	1,802
Sunflower	1,512	1,706	1,470	1,147
Sugarbeet	26,465	25,297	26,495	30,560
Potatoes	12,578	17,612	11,755	12,619
Vegetables	14,743	17,151	11,131	16,683
		1993		
Wheat and rye	2,321	2,839	2,258	2,150
Corn	2,605	2,949	2,758	2,533
Barley	2,619	2,956	2,347	2,162
Sunflower	1,180	1,414	1,105	1,086
Sugarbeet	18,276	22,784	16,815	17,743
Potatoes	14,818	16,855	17,143	14,429
Vegetables	11,750	9,663	11,794	12,130
		1994		
Wheat and rye	2,535	2,972	2,410	2,436
Corn	3,132	3,061	3,136	3,110
Barley	2,718	2,946	2,563	2,682
Sunflower	1,309	1,406	1,302	1,247
Sugarbeet	21,260	23,889	20,934	20,208
Potatoes	11,766	12,760	12,456	11,652
Vegetables	11,455	8,878	10,628	11,804

Source: National Commission for Statistics

Table 6.7
Shares of total area devoted to different crops in Romania, 1989, 1993, and 1994 (percent)

	Total	*State farms*	*Co-ops or AS/FA*	*Indep. Farmers*
		1989		
Wheat and rye	32.97	28.48	39.26	10.09
Corn	38.21	32.30	33.39	66.73
Barley	10.73	20.45	10.46	0.55
Sunflower	6.06	6.87	7.05	0.69
Sugarbeet	3.58	3.01	4.33	0.86
Potatoes	4.91	2.09	3.52	14.45
Vegetables	3.53	6.80	1.99	6.63
Total area (000 ha)	7,154.1	1,262.6	4,815.6	1,075.9
		1993		
Wheat	33.6	40.0	42.1	27.8
Corn	44.6	21.7	32.8	54.3
Barley	5.3	20.7	7.6	2.5
Sunflower	8.6	13.4	13.4	5.2
Sugarbeet	1.4	1.5	2.3	1.0
Potatoes	3.6	1.6	1.2	5.0
Vegetables	2.9	1.2	0.5	4.2
Total area (000 ha)	6,870.4	1,167.0	1,585.8	4,223.7
		1994		
Wheat	33.1	38.2	42.5	28.4
Corn	40.5	17.8	28.7	50.8
Barley	10.6	25.2	13.0	5.8
Sunflower	7.9	13.5	12.0	4.9
Sugarbeet	1.8	2.4	2.5	1.3
Potatoes	3.4	1.1	1.0	4.8
Vegetables	2.8	1.8	0.3	3.9
Total area (000 ha)	7,373.6	1,258.0	1,575.9	4,539.9

Source: Computed from data of Table 6.5

Romania has a large fertiliser manufacturing industry that supplies both the internal demand and also exports. Also there is a large agricultural machinery manufacturing capacity. There is also a pesticide industry that supplies some generic pesticides. Similarly seed availability has not been a problem, as there is a large variety of seeds that have been developed by publicly funded research. Fuels and lubricants also are not limiting factors as Romania is an oil producer.

Since 1989 there has been a significant drop in input use. Fertiliser application has dropped from an average of 1.4 million tons in the period 1980-90, to less than half of that. This is a significant drop if it is realised that even in 1989 Romania had much lower rates of fertiliser application compared to other Western European countries. Since, however, yields have declined by much less than 50 percent, it must have been the case that much of that fertiliser was not used efficiently. The number of tractors seems to have fallen by 30 percent after 1989, while the number of combines has dropped by 34 percent. The reason is largely the change in the organisation of farming through privatisation. The large input use in Romania agriculture was supported by the extensive input intensive organisation of pre-1989 agriculture, coupled with a system of prices that entailed considerable amount of implicit subsidies.

After 1989, much agricultural production is taking place on farms of a much smaller scale. The technology utilised is much less variable input intensive. This could be the result of the post-1989 organisation of input distribution. It appears that the old system has been preserved. For instance farmers in order to obtain credit and inputs, must transact with one of the large 'integrators', like Romcereal, who undertake to purchase farmers' output at state dictated prices and in turn make available fertilisers and other variable inputs, as well as credit at prices also dictated by the state. There do not appear yet to be any significant numbers of private dealers in farm inputs, and hence the farmers must follow the price system and the terms of the integrators, in order to obtain inputs. According to reports of August 1994, of 2,204 commercial companies that were to be privatised in agriculture and food industry, only 136 had been privatised, 79 had negotiations started, while another 797 were proposed for privatisation.

Under these conditions where the small-scale individual farmer must rely for inputs on large scale national monopolists, it is likely that farmers will revert to production systems that will be much less variable input intensive, and also geared to producing largely for self consumption.

The supply of inputs to the large integrators is done at prices that still entail substantial amount of subsidy. The integrators in turn seem to provide mostly the large state farms as well as the ASs and to a lesser extent the FAs. It is likely that private individual farmers must obtain inputs in some 'residual' of 'parallel' private market, where prices are much higher than in the official channels. This would also tend to discourage the use of variable inputs.

Output marketing developments

The liberalisation of food marketing was done in three stages. The first was with the application of Law Decree 75/1990 which abolished the obligatory delivery of agricultural production to the state. As open market prices were 3 to 5 times larger than the official prices, many producers tried to sell in the open markets. The second stage of liberalisation came with Government Decision 191/1991. With that the state limited itself to the centralised fixing of prices for 12 crop products and 8 animal products. While prices to producers were raised substantially, prices to consumers were not, with the difference being absorbed by the state budget. The third stage of liberalisation came in May 1993, with the restriction of products subsidised at the consumer level to only two (bread and milk)

Under the Communist regime farmers had little choice of marketing. Most products had to be sold through the state market trading companies (such as Romcereal) at prices fixed by the state. After the revolution, and despite the liberalisations, agricultural marketing still seems to be dominated by the same large market institutions. Their activities and prices are still subject to many government controls and restrictions. However there is considerable change especially by the new private farmers.

In many villages and urban areas farmers are doing their own direct marketing to consumers by renting stalls from municipalities. However, due to the absence of a wholesale marketing system there is a lot of waste of resources. For instance, a large number of the tractors are utilised for product transportation. In many products and regions private retailers compete directly with state retailers. The latter sell at official prices and thus many times inhibit the orderly development of private retail trade.

Cereals marketing is dominated by Romcereal, which has impressive amounts of infrastructure and capital, such as large storage capacity. Marketing by Romcereal is governed by state policy and that in turn is geared toward protecting the consumer. Hence prices to producers are kept low. Figure 6.1 illustrates the cereals producer to final consumer marketing chain. While over 70 percent of the cereal volume is produced by private farms and associations, the private bakeries handle only about 20 percent of the processed product.

Fruit and vegetable markets are relatively free from state controls. Private retailers always existed and continue to do so. However, larger scale organisation of fruit and vegetable marketing is difficult as the lack of transport vehicles, storage facilities and suitable wholesale markets makes it difficult.

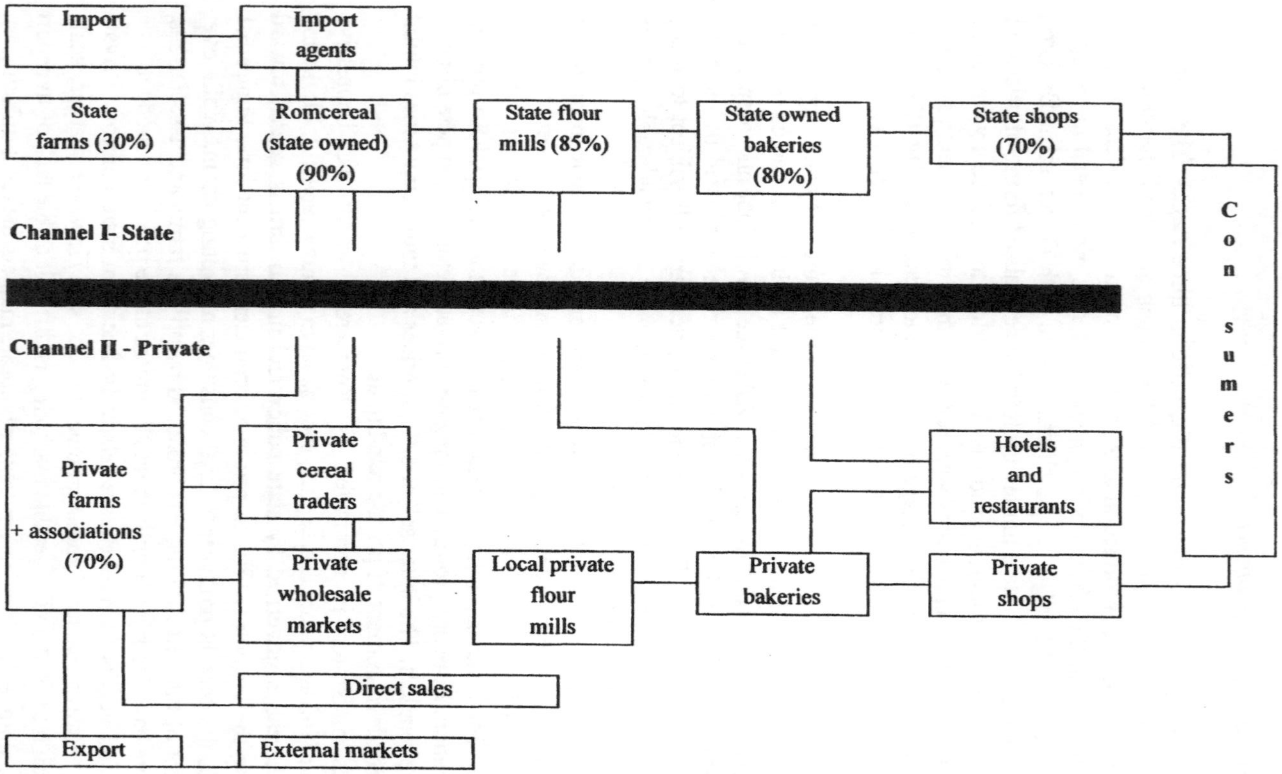

Figure 6.1 Current cereals marketing system

The state sector is the main organiser of milk distribution with the private sector operating at small scale and for the local markets. The dairies are largely state-owned, and farmers now have the option of not selling to the state dairies at the fixed prices. This is probably one of the reasons for the apparent stagnation in milk production, as the data refers mostly to the output of state dairies. The total output might be significantly understated. For livestock, private markets existed along with official ones during the Communist period, and indications are that their importance is growing.

To obtain an idea of the degree of potential market imperfections in the current open market environment, Table 6.8 presents some statistics of the dispersion of prices across various localities in 1994. The prices that form the basis for the calculations are average annual prices in each locality. The dispersion exhibited appears to be quite large for a country that is spatially well connected by roads as Romania. For instance the difference between the maximum and minimum average annual prices as a percent of the average price in all localities varies from 31.4 percent for sheep milk cheese to 122.7 percent for apples.

The conclusion from the brief review above is that the output marketing system seems to be currently a constraint to small scale private sector farm development.

Household incentives underlying new farm structures

Of utmost importance in the determination of the future farm structures are the incentives of the households making up the farm population. To investigate these we analyse a survey conducted by the Institute of Agrarian Economy (IAE) in 1994 that involved about 2,000 rural households randomly chosen in 110 villages around the country (out of a total of 3,000 villages). The questionnaire involved questions of both an economic as well as a social nature.

From an agroclimatic viewpoint Romania can be divided into three major zones. The first zone comprises the plain in the south-east of the country and the western plain. Zone II is the hilly zone around the mountains, and zone III is the mountain zone. There are seven major provinces which include agricultural land in different zones. Maramures is mostly in zone III, Banat-Crisana in zone I, Transilvania mostly in zone III, Oltenia is in zones I and II, Dobrogea mostly in zone I, Moldova is mostly in zone II, and Bucovina is in zone III. In zone III there is a predominance of individual household farms, as this is the mountain zone. In zones I and II, there is a mixture of co-operative organisations, state farms, and private farms. The FA type of farms are predominant in zones I and II.

Table 6.8

Indices of regional variations of average annual prices of some agricultural products in Romania, 1994

	Wheat grains	*Potatoes*	*Dry onion*	*Dry beans*	*Toma-toes*	*Apples*	*Pork*	*Sweet milk*	*Sheep milk cheese*	*Hen eggs*
No of localities	41	41	41	41	41	41	34	38	41	41
Average	213.3	315.0	868.9	908.1	584.1	487.9	3,445.9	483.6	3,326.3	139.5
Maximum	338.0	432.1	1,095.1	1,269.9	844.2	792.8	4,789.8	676.3	3,789.6	185.6
Minimum	152.8	202.7	588.2	558.7	384.4	194.3	2,410.8	322.6	2,745.8	112.8
(Max-Min)/Average (percent)	86.8	72.8	58.3	78.3	78.7	122.7	69.0	73.1	31.4	52.2
Standard deviation	36.0	48.2	121.7	197.5	113.2	103.0	625.8	74.5	274.0	17.4
Coef. of variation (percent)	16.9	15.3	14.0	21.8	19.4	21.1	18.2	15.4	8.2	12.4

Source: Computed from data in Romanian Statistical Yearbook, 1995

Table 6.9 presents the age structure of the surveyed household heads. The most noticeable thing is that the bulk of the household heads is over 50 years old (65 percent for the whole sample). In several provinces the age structure is even worse than that. Most households are composed of 2 persons (31 percent), while 19 percent are composed of 3 persons, and 19 percent of 4 persons. Only 20.3 percent of households are composed of 5 persons or more. In terms of the age structure of all the members of households, young persons (0-14 years old) comprise only 14.7 percent of the sample, while 22 percent of the people are more than 60 years old. Thus from a demographic perspective the bulk of rural households appear to be headed by relatively old people, and there are few young people to be found.

The educational level of rural households, however, appears to be adequate. Only 4 percent of the people in the sample is illiterate, 27 percent have graduated from primary school, 24 percent have graduated from secondary school, 30 percent have graduated from high school, and 6 percent have graduated from a university.

Labour participation rates appear to be quite high, as 68.3 percent of the total population in the sample is employed, and only 4.4 percent of the economically active population declaring that they are unemployed. Of the employed population, 54 percent work in their own household (farm or other activity), 24 percent are employees of a state CC, 17 percent have jobs in a public institution, and 5.2 percent have a job in a private commercial company (agricultural or other). Thus it appears that approximately 70-80 percent of the rural population is employed in agriculture.

Table 6.10 exhibits the current structure of land ownership of the investigated households. It can be seen that the bulk of households possess between 1 and 5 hectares of land (62 percent of the total households). However there are zonal differences. In the plain region, for instance the incidence of small plots (smaller than 0.5 ha) is very small, while small farms are more numerous in the hilly and mountain regions, and in the non-collectivised regions prior to 1989.

Table 6.9
Age structure of investigated household heads (1994 surveys)

	Total sample	*Maramures*	*Banat - Crisana*	*Transilvania*	*Oltenia*	*Muntenia*	*Dobrogea*	*Bucovina*	*Moldova*
under 30 years	3.8	-*	8.3	4.7	4.0	3.0	4.2	-*	3.0
30-40 years	9.2	17.6	11.5	10.5	8.6	5.3	10.4	12.1	9.5
40-50 years	22.2	27.5	26.0	17.9	23.2	21.8	27.1	6.1	24.8
50-60 years	24.9	21.6	19.8	22.6	35.1	25.9	31.2	24.2	18.9
60-70 years	21.9	23.5	21.9	23.7	17.2	22.6	14.6	39.4	21.3
70 years or more	18.0	9.8	12.5	20.6	11.9	21.4	12.5	18.2	22.5

* Maramures and Bucovina are historical provinces where no household head in the sample is under 30 years old.

Source: IAE 1994 Survey

Table 6.10
Structure of investigated households according to land held in possession, 1994 surveys (percent of sample)

Land property size	*< 0.5 ha*	*0.5-1ha*	*1-3 ha*	*3-5 ha*	*5-10 ha*	*>10 ha*
Total sample	8.0	12.1	40.0	22.0	14.7	3.2
Investigated zones						
Maramures	13.7	13.7	33.3	19.6	17.7	2.0
Banat-Crisana	7.3	18.8	31.2	16.7	19.8	6.2
Transilvania	9.5	7.4	41.2	24.7	13.7	3.7
Oltenia	6.6	15.2	49.7	16.6	11.2	0.7
Muntenia	9.4	10.2	41.7	22.5	13.9	2.3
Dobrogea	-	6.3	37.5	25.0	20.8	10.4
Bucovina	6.1	12.1	27.3	24.2	18.2	12.1
Moldova	6.5	14.8	37.9	25.4	14.2	1.2
Plain region	3.8	9.2	42.3	25.5	15.8	3.4
Hilly region	10.7	15.2	36.6	20.4	15.9	1.2
Mountain region	12.1	12.9	40.5	17.7	11.2	5.6
Non-co-operativised zone before 1989	16.2	13.0	39.6	14.6	10.4	6.2
Co-operativised zone before 1989	6.0	11.8	40.2	23.8	15.7	2.5

Source: IAE 1994 Survey

Table 6.11 exhibits the current degree of farm land fragmentation. The most noticeable thing is that the bulk of holdings (61.6 percent) is split among three or more parcels, the problem being rather similar in all regions and zones. This fragmentation is likely to be one of the conditioning factors in future farm organisation. The problem of farm fragmentation appears to be slightly more acute for the farmers that work the land they hold in associations. Of those farmers 66 percent had land in more than three parcels, and only 12.8 percent had only one parcel. Among individual private farmers those with more than three parcels composed 58.9 percent of the population of private farmers, while those with one parcel composed 20.2 percent of the population.

Table 6.11
Fragmentation of private land property (% of households), 1994 surveys

	Number of parcels into which property is divided			
	1	*2*	*3-4*	*> 5*
Total	17.4	17.0	33.3	28.3
Maramures	9.8	5.9	27.4	49.0
Banat-Crisana	31.2	13.5	19.8	32.3
Transilvania	18.4	15.8	24.7	40.0
Oltenia	14.0	21.2	42.4	18.5
Muntenia	18.4	16.2	40.6	20.7
Dobrogea	27.0	18.7	22.9	4.2
Bucovina	9.1	15.2	24.2	48.5
Moldova	11.2	20.7	37.3	30.2
Plain region	18.5	15.3	41.2	18.2
Hilly region	14.3	19.0	29.0	36.6
Mountain region	20.0	17.2	24.1	35.8
Non-co-operativised zone before 1989	17.2	17.7	26.0	36.0
Co-operativised zone before 1989	17.5	16.7	35.0	26.5

Source: IAE 1994 Survey

Table 6.12 exhibits the degree of ownership of mechanical equipment. It is quite obvious that the majority of new farmer-owners do not hold any mechanical equipment. Only 5.1 percent for instance own tractors. It is probably because of this lack of mechanical equipment, coupled with the old age of most owners, that a large share of household heads (38 percent) has chosen to farm the land they possess in association. The other 62 percent, nevertheless, have chosen to farm the land by themselves and with their families.

Table 6.12
Ownership of mechanical equipment (% of surveyed farmers), 1994 surveys

	Type of equipment held				
	Tractor	*Plough-tractor*	*Sower*	*Mechanical mower*	*Combine*
Total	5.1	4.6	1.5	1.4	0.6
Maramures	9.8	9.8	7.8	3.9	3.9
Banat-Crisana	4.1	4.1	2.0	2.0	1.0
Transilvania	7.3	6.8	2.1	4.7	1.0
Oltenia	1.3	1.3	-	-	-
Muntenia	6.0	4.9	0.7	0	-
Dobrogea	4.1	4.1	2.0	2.0	2.0
Bucovina	3.0	6.0	-	-	-
Moldova	4.1	2.9	1.9	-	-
Plain region	4.5	4.3	1.3	0.2	0.5
Hilly region	5.1	5.1	1.5	1.2	0.6
Mountain region	6.0	4.3	1.7	3.9	0.8
Non-co-operativised zone before 1989	4.7	3.1	1.0	2.6	0.6
Co-operativised zone before 1989	5.1	4.9	1.6	1.1	0.6

Source: IAE 1994 Survey

Table 6.13 exhibits the responses of those farmers that are associated concerning the satisfaction with their association membership. It is quite apparent that the bulk of them are quite or very satisfied with it (75 percent). It appears that only in Bucovina there is a large share of farmers (50 percent) that are dissatisfied with their association membership. The satisfaction of the farmers with the association membership is reflected in their intentions. In the total sample 81.2 percent stated that they intended to stay in the association in which they operated, while another 8.1 percent stated that they would move to another association. It is only 6.5 percent of the querried farmers who stated that they would withdraw from the association to work independently.

Table 6.13
Degree of satisfaction regarding association membership (% of those surveyed), 1994 surveys

	Very satisfied	*Quite satisfied*	*Dissatisfied*	*Extremely dissatisfied*
Total	12.8	61.8	18.6	3.7
Maramures	-	100.0	-	-
Banat-Crisana	12.5	62.5	8.3	8.3
Transilvania	13.8	72.4	13.8	-
Oltenia	11.8	60.8	23.5	2.0
Muntenia	15.1	60.0	17.1	3.3
Dobrogea	2.5	55.0	32.5	7.5
Bucovina	-	50.0	-	50.0
Moldova	14.8	64.2	17.3	2.5
Plain region	12.5	58.7	20.8	4.2
Hilly region	14.0	68.2	13.0	2.8
Mountain region	9.1	72.7	18.8	-
Non-co-operativised zone before 1989	11.1	77.8	11.1	-
Co-operativised zone before 1989	13.0	61.4	18.7	3.8

Source: IAE 1994 Survey

Of the total sample, 61 percent of those that were included in an association, and 54 percent of those that farmed independently made no investments in the last 2 years. Table 6.14 exhibits the investment behaviour of farmers in the total sample. It is evident that of those investing, the bulk invested in livestock purchases, and only a small share bought equipment or made building improvements. Table 6.15 exhibits the different investment behaviour of associated and non-associated farmers. It appears that there is no significant difference in the invesment behaviour among these two groups.

Table 6.14
Investment behaviour of farmers (% of total sample), 1994 surveys

	Bought livestock in the last 2 years	*Bought equipment in the last 2 years*	*Built annex buildings in the last 2 years*	*Established plantations in the last 2 years*
Total	17.2	3.2	4.9	2.6
Maramures	5.8	3.9	0	5.8
Banat-Crisana	10.4	3.1	5.2	4.1
Transilvania	21.0	2.6	3.2	2.1
Oltenia	11.9	0.6	12.6	0
Muntenia	13.9	2.2	5.2	3.7
Dobrogea	18.7	2.0	0	4.1
Bucovina	18.8	3.0	3.0	0
Moldova	29.6	7.7	2.4	2.3
Plain region	16.6	2.7	6.0	2.2
Hilly region	18.2	3.9	5.5	2.7
Mountain region	16.8	3.0	1.7	3.4
Non-co-operativised zone before 1989	16.6	2.6	1.5	5.2
Co-operativised zone before 1989	17.3	3.3	5.6	2.0

Source: IAE 1994 Survey

Table 6.16 exhibits the sources of financing for the investments of those people who made investments in the last two years. The overwhelming bulk of financing came from own savings (73.5 percent for those investing) with minimal amounts coming from relatives, friends, or the banks. It is not clear what the category 'other source' means, but it appears significant as it seems to provide 20 percent of the investment financing. This could be the input suppliers.

Table 6.15
Investment behaviour of associated versus non-associated farmers (% of those in each group), 1994 surveys

	Bought livestock		*Bought equipment*		*Built annex buildings*		*Established plantations*	
	Assoc.	*Non-Assoc.*	*Assoc.*	*Non-Assoc.*	*Assoc.*	*Non-Assoc.*	*Assoc.*	*Non-Assoc.*
Total	15.7	18.1	3.9	2.7	3.6	5.6	2.0	3.0
Maramures	0	6.2	0	4.1	0	0	0	6.2
Banat-Crisana	8.3	11.1	0	4.1	4.1	5.5	0	5.5
Transilvania	10.3	22.9	0	3.1	0	3.7	0	2.4
Oltenia	11.7	12.0	0	1.0	3.9	17.0	0	0
Muntenia	15.1	12.2	3.9	0	5.9	4.3	1.9	6.1
Dobrogea	22.5	0	2.5	0	0	0	5.0	0
Bucovina	0	19.3	0	3.2	0	3.2	0	0
Moldova	20.9	37.5	9.8	5.6	2.4	2.2	3.7	1.1
Plain region	16.6	16.6	3.0	2.2	4.9	7.7	2.2	2.2
Hilly region	14.0	20.3	6.5	2.7	0.9	7.6	1.8	3.1
Mountain region	9.0	17.1	0	3.1	0	1.8	0	3.6
Non-co-operativised zone before 1989	11.1	16.9	0	2.7	0	1.6	11.1	4.9
Co-operativised zone before 1989	15.8	18.6	4.0	2.7	3.7	7.2	1.8	2.2

Source: IAE 1994 Survey

Table 6.16
Investment sources (% of those investing), 1994 surveys

	Own savings	*Loan from relatives*	*Bank credit*	*Loan from other persons*	*Other sources*
Total	73.5	2.8	3.0	0.8	19.9
Maramures	75.3	0.0	6.1	0.0	18.6
Banat-Crisana	100.0	3.4	6.9	0.0	3.4
Transilvania	77.6	2.1	0.0	0.0	22.4
Oltenia	61.7	5.1	2.3	5.1	25.8
Muntenia	70.3	1.7	4.4	0.0	23.6
Dobrogea	61.5	7.4	7.4	7.4	38.5
Bucovina	100.0	0.0	0.0	0.0	16.5
Moldova	78.2	3.6	2.3	0.0	16.0
Plain region	90.5	3.3	4.4	1.8	17.8
Hilly region	65.8	4.5	2.3	0.0	27.4
Mountain region	79.1	0.0	2.1	0.0	18.9
Non-co-operativised zone before 1989	80.1	0.0	1.5	0.0	18.4
Co-operativised zone before 1989	72.5	3.2	3.2	0.7	20.3

Note: The percentages may add-up to more than 100 as some households might have more than one source of funds.

Source: IAE 1994 Survey

The bulk of those questioned did not wish to sell land (92.6 percent). However, of all farmers questioned less than 10 percent stated that they wished to buy land, the rest saying that they did not wish to buy any more. The motivation of those wishing to buy land were for the most part to ensure better income for the children. Of those not wishing to buy land about half stated that the main reason was that they were too old, while another half stated that they did not have money for buying it.

Table 6.17 exhibits the desired form of land operation by the farmers questionned. It is quite apparent that the bulk of those questionned desire to farm independently. The shares are much larger in the mountain areas that have always been characterised by independent farmers, and among areas that did not have co-operatives before 1989 (these are again the mountain areas). It is not surprising that in the regions that were characterised by co-operatives (which were largely the plains areas), there is a much larger share of new owners who desire to farm in association.

Table 6.17
Desired form of land operation (% of those interviewed), 1994 surveys

	By oneself, with family	*Juridical association*	*Family association and with neighbours*
Total	53.5	24.0	10.4
Maramures	84.3	2.0	2.0
Banat-Crisana	64.5	24.0	3.1
Transilvania	68.9	15.7	5.2
Oltenia	47.7	21.1	13.2
Muntenia	43.2	29.7	15.4
Dobrogea	16.7	58.3	12.5
Bucovina	84.8	0	6.0
Moldova	46.1	28.4	12.4
Plain region	38.7	34.9	14.2
Hilly region	52.1	24.0	10.6
Mountain region	83.6	3.0	2.6
Non-co-operativised zone before 1989	83.3	3.1	2.0
Co-operativised zone before 1989	46.4	28.9	12.3

Source: IAE 1994 Survey

Table 6.18 exhibits the desires of farmers concerning the type of marketing they want. While the bulk of farmers seem to desire marketing by themselves, a significant share desire to market products through associations and through state companies. It is rather surprising that very few farmers say that private traders is a desired form of marketing. Maybe this is because by tradition there was little of that type of marketing before, and also because there are various pre-existing perceptions concerning monopolistic behaviour of middlemen. As expected the share of farmers that desire a freer system of marketing is larger in the mountain and non-collectivised zones before 1989.

Table 6.18
Desired types of product selling organisation (% of all farmers), 1994

	By oneself	*Agricultural producers' association*	*Private trader*	*State network*
Total	26.3	8.5	2.7	11.4
Maramures	17.6	5.8	7.8	0
Banat-Crisana	27.0	14.5	3.1	11.4
Transilvania	23.1	8.0	2.1	15.2
Oltenia	27.8	8.6	2.6	11.2
Muntenia	30.0	6.4	2.2	10.1
Dobrogea	43.7	18.7	2.0	6.2
Bucovina	12.1	12.1	3.0	21.2
Moldova	22.4	6.5	2.3	12.4
Plain region	32.2	10.3	2.7	12.6
Hilly region	22.8	8.8	0.9	11.9
Mountain region	19.8	4.7	5.2	8.6
Non-co-operativised zone before 1989	18.7	6.7	3.1	4.1
Co-operativised zone before 1989	28.0	8.9	2.6	13.1

Source: IAE 1994 Survey

Table 6.19 illustrates the farmers' perceptions concerning their economic situation compared to before the revolution. The table indicates that, while the bulk of farmers estimate that their property status is better now, about 40 percent of them estimate that their overall economic situation is worse now than before the revolution, while 32 percent say that it is better. Interestingly, it is among those that were not in co-operatives before 1989 where the shares of dissatisfaction are larger. This is somewhat surprising, as one would have expected that it was these people that were already quite used to dealing in open markets. It is also interesting that the proportion of those that estimate income declines (22 percent) is much smaller than the proportion that says their overall economic situation is worse. It thus appears that there are other factors (economic and other), apart from income that condition people's satisfaction or not with the current situation. These could have to do with such factors as employment opportunities, security, and availability of goods.

Table 6.19

Perception of current conditions compared to before 1989 (Percent of those surveyed), 1994 surveys

	Economic situation			*Situation of incomes*			*Situation of private property*		
	Better	*The same*	*Worse*	*Better*	*The same*	*Worse*	*Better*	*The same*	*Worse*
Total	31.6	28.3	39.4	30.4	46.5	22.3	69.0	8.9	21.4
Maramures	23.5	23.5	51.0	21.6	56.9	17.7	85.4	2.0	13.7
Banat-Crisana	33.3	22.9	42.7	31.3	53.1	14.6	54.2	13.5	32.3
Transilvania	36.8	25.3	37.9	31,6	43.7	24.7	77.9	6.8	14.7
Oltenia	28.5	31.8	39.7	23.8	52.3	23.8	62.3	11.3	26.5
Muntenia	31.2	32.7	34.6	35.3	39.5	24.1	68.4	6.8	23.7
Dobrogea	12.5	2.1	83.3	12.5	83.3	2.1	70.8	10.4	16.7
Bucovina	30.3	51.5	18.2	36.4	39.4	24.2	84.9	0.0	15.2
Moldova	36.1	29.0	34.9	33.1	39.6	26.6	66.9	13.0	19.5
Plain region	33.6	28.6	36.7	34.0	45.1	20.1	70.7	10.4	18.2
Hilly region	30.2	23.8	45.7	27.7	50.0	21.7	65.2	9.8	24.4
Mountain region	29.7	34.1	35.8	27.2	44.4	27.6	71.1	4.7	23.3
Non-co-operativised zone before 1989	21.4	33.3	45.3	17.7	55.2	26.6	64.6	6.3	28.7
Co-operativised zone before 1989	34.0	27.1	38.1	33.4	44.5	21.3	70.1	9.5	19.7

Source: IAE 1994 Survey

A recent study of rural poverty estimated that among the peasants about 81 percent feel that their income is not enough for the strictly necessary items (Gavrilescu and Florian, 1995). This suggests a large degree of poverty and is compatible with survey results which indicate that in 1993 the average share of expenditures spent by peasant households on food is 77 percent up from 67 percent in 1990. Such large shares, if confirmed by other more detailed studies suggest levels of poverty similar to those of the poorest countries in the world. This suggests that food and income security must be very high priority objectives among rural households, and this is consistent with the high degree of co-operation found among the beneficiaries of the post-Communist period land reforms.

Conclusions

The analysis of this chapter suggests that the current economic situation in Romania seems to be quite uncertain. The massive privatisation of land and livestock that occurred in the first few years after the revolution of 1989 has not been followed quickly by the establishment of the requisite private structures in the upstream and downstream sides of the farm-food chain. Thus the current farm structure, albeit mostly in private hands, seems to have considerable tendencies to utilise production modes that were prevalent during the Communist era. This tendency is accentuated by the age structure and skill levels of those that obtained land.

On the other hand there is a large number of agricultural workers, mainly those employed in state farms, that have not received land, but have the knowledge to operate various aspects of farm production. These workers used to be the elite of the farm population, and it is not clear whether they are becoming worse off under the current conditions, and whether they will be a source of political pressure for maintaining the structure of state ownership and operation.

Despite the stated objective of the government to move toward privatised agriculture, it appears that in terms of production organisation there is a strong tendency and economic rationale for maintaining the associated forms of production structures. This seems quite reasonable as a household response in a situation of considerable uncertainty, lack of alternative employment opportunities, and a quest for food security. However, it is not clear that this mode will be stable or desirable by farm owners when the current economic uncertainties and climate are improved. In fact the initial information suggests a steady progression away from associated and into independent farming. It will be interesting to monitor whether there will be a tendency for land consolidation for operational purposes by private individuals, or whether the currently emerging farm structures will solidify and render themselves a permanent fixture of post communism Romanian agriculture.

References

Crespi Reghizzi, G. (1993), 'General Legal Framework', working paper in volume 3 of *Romania: A Strategy for the Transition in Agriculture*, Joint Romanian-International Agricultural Strategy Team.

Gavrilescu, D. (1994), 'Agricultural Reform in Romania: Between Market Priority and the Strategies for Food Security', in Swinnen, J.F.M. (ed.), *Policy and Institutional Reform in Central European Agriculture*, LICOS Studies on the Transitions in Central and Eastern Europe, vol. 1, Avebury: Aldershot.

Gavrilescu, D. and Florian, V. (1995), *Present Dimensions of Rural Poverty in Romania*, Institute of Agricultural Economics: Bucharest.

Jackson, M. (1997), 'Political Economy of Agricultural Reform in Romania', in J.F.M. Swinnen (ed.), *Political Economy of Agrarian Reform in Central and Eastern Europe*, Ashgate: Aldershot.

Jackson, M. and Swinnen, J. (1995), *A Statistical Analysis and Survey of the Current Situation of Agriculture in the Central and Eastern European Countries*, Report to the European Commission, The Leuven Institute for Central and East European Studies: Leuven.

Mitrani, D. (1931), *The Land and Peasant in Romania*, Yale University Press: New Haven.

Roberts, H. (1951), *Romania: The Politics of an Agrarian State*, Yale University Press: New Haven.

Swinnen, J.F.M. (1997), 'The Choice of Privatization and Decollectivization Policies in Central and Eastern European Agriculture: Observations and Political Economy Hypotheses', in Swinnen, J.F.M. (ed.), *Political Economy of Agrarian Reform in Central and Eastern Europe*, Ashgate: Aldershot.

Tourne, C. (1993), 'The Legal Framework for Agricultural Land Reform', working paper in volume 2 of *Romania: A Strategy for the Transition in Agriculture*, Joint Romanian-International Agricultural Strategy Team.

7 Privatisation and decollectivisation policies and resulting structural changes of agriculture in Slovakia

Ladislav Kabat and Konrad Hagedorn

Introduction

During the years since 1989 Slovak society has experienced a variety of new problems. The people's 'velvet' revolution has been replaced by a severe struggle for a new economic order. Strong tensions among political parties, a lack of esteem for democratic principles, and a general attitude of intolerance were typical for the first half of the 1990s. As a consequence, the political environment for the economic transformation process has become less favourable.

The political decision to split the Czechoslovak federation in 1993 has influenced the speed and the direction of the process of transformation and privatisation in Slovakia. This process has considerably slowed down, and the second wave which was supposed to start in October 1994 has been postponed. The initially adopted philosophy of fast privatisation by means of vouchers has been replaced by the concept of 'employees' privatisation' and 'direct sales' in which the state and its agencies are able to play a stronger role than in the originally-preferred voucher approach. Obviously, the changes in government policies toward privatisation which occurred after Slovakia separated from the Czechoslovak federation can be explained by a shift in ideological orientation towards rent-seeking structures. The privatisation process suffers from a time-consuming struggle among parties, administration, interest groups and other actors involved, who try to secure themselves a favourable economic position in the future society of Slovakia.

These changes in the strategy of privatisation and transformation can be summarised as a tendency to strengthen the role of government in the sense of the former command economy. Relatively low support is given to the development of small businesses. In contrast, the large-scale, state-owned companies (e.g., in the energy, steel and machinery industries) are passing through the transformation process very slowly, and government clearly supports their interest in preserving their large-scale and strongly centralised form. There is some evidence that government tries to delay or even to stop the restructuring process in these industries. For example, the state interfered with privately run financial intermediaries known as 'investment funds', giving a negative signal for foreign capital owners who consider investing in Slovakia.

The tendency to reinforce the influence of government and state agencies and to protect collective and state enterprises against organisational and structural changes can also be observed in the agricultural sector. In this part of the economy, government policies are clearly aiming at the objective to stabilise the large co-operative farms which prevail in Slovak agriculture. However, from an economic point of view this strategy raises a fundamental question: How long will government be able to maintain this sort of policy? Since economic efficiency of a considerable proportion of the co-operative farms decreases, and their economic losses covered by government transfers increase, this cannot be continued over a very long period of time. In other words, it seems reasonable or even inevitable that the restructuring process will have to resume in order to prevent further losses from farm production which are perpetuated by the current system.

For this reason, we first have to show that decollectivisation of Slovak agriculture is no end in itself, but is necessary in order to initiate an evolution toward a system of viable farms. This will be pointed out in the second section in which we will describe the historical background and the initial situation of the privatisation process, as it could be observed in the beginning of the 1990s. However, although restructuring is necessary, it may not be easy in certain situations. Therefore, the question arises whether privatisation of property rights and the required changes in farm organisation and structures is politically and economically feasible in Slovakia. For a discussion of this question, we try to point out some determinants of the feasibility of restructuring in the third section. In this way, we hope to collect a few elements of a positive theory to explain the process of transformation of agriculture. Then this approach is confronted with empirical observations regarding real changes in ownership, organisation and structure of Slovak agriculture in the fourth section. This includes agricultural policies which influence the incentives for various actors involved in the restructuring process. Finally, in the fifth section, we draw some conclusions with regard to both the empirical evidence for or against the theoretical approach and the present situation of Slovak agriculture.

Historical background and the initial situation of the privatisation process in the beginning of the 1990s

Historical background and development of Slovak agriculture

The problems mentioned above are especially visible in agriculture. This branch of the economy is becoming an important field of struggles between the past, present and future of Slovak society. It is not easy to understand this phenomenon without being familiar with the historical background and without considering at least some crucial periods in the history of Czechoslovak agriculture.

The first Czechoslovak state was established in 1918, after World War I. Two countries with different historical, economic and political developments merged into one state. Bohemia represented the highly developed industrial part of the Austro-Hungarian Empire with well developed agricultural structures, while Slovakia represented the industrially backward part of Hungary, a country predominantly oriented toward agriculture.

The situation of the agricultural sector in these two parts of the former Czechoslovak federation differed after World War I. As a consequence of the legislative framework for land tenure in Bohemia, based on primogeniture, in which only the oldest heir (son) was eligible for inheritance of the land, relatively large family farms survived. In contrast, in Slovakia where the Napoleonic code for inheritance applied, larger farms had disappeared, and highly fragmented, small subsistence farms became a common feature of Slovak agriculture. The typical farm in Bohemia and Moravia (i.e. the Czech part of Austro-Hungary) cultivated 20-50 ha of arable land. In Slovakia the typical size of farms would be between 2-5 ha.

During the last 45 years since World War II, and especially after the Communist Party came into power in 1948, Czechoslovak and, after separation, Slovak agriculture have experienced a complex pattern of development. After the Second World War the new, left oriented regime organised a land reform (1946). The political basis for this reform was created by partially or completely expropriating the land of many original large land owners (e.g. the Hungarian nobility, Roman-Catholic Church, German farmers in Sudeten) without compensation. This land was divided into smaller units and sold to peasants. There is no doubt that this land reform had political dimensions: The traditionally right-wing oriented peasants voted for left-wing parties and, in this way, contributed to the fact that the first election after the war held in 1946 was won by left oriented parties (in Bohemia by the Communist Party).

After the Communist Party had taken over the government in 1948, the market economy was replaced by a state-run command economy. Czechoslovak agriculture was seriously squeezed again, after the political decision was made to reorganise the farm sector in accordance with the Soviet model (i.e., *kolkhozes* and *sovkhozes*). Strong organisational measures were taken in order to speed the process of collectivisation. The period of the late 1940s and the 1950s was a dark period not only in the history of Slovak agriculture, but also in the history of the country's society as a whole. Collectivisation was strongly supported and assisted by political and ideological pressure. In this process, basic human rights were ignored. Peasants and farmers were forced to sign the contracts with co-operative farms and to give up their property rights to the land 'voluntarily'. The land and the farm assets of those farmers who did not agree to join co-operatives were, in many cases, expropriated without compensation. Moreover, many original land owners were arrested. In addition, the land belonging to the Hungarian nobility and to the Roman-Catholic Church as well as land in some less favoured areas was confiscated, and from this agricultural area new entrepreneurial entities – the state farms – were established.

By the mid-sixties collectivisation of agricultural land was generally complete. Czechoslovak agriculture entered in a new period of development. In the first period of collectivisation, co-operative farms had been created by the principle 'one village – one co-operative farm'. Later in the mid-seventies, several co-operative farms were forced to merge into large-scale units representing an average of 5,000-10,000 ha. In the beginning, production results and economic performance of co-operative farms as well as of state farms were rather poor. The main reasons were: forced membership and absence of ownership of land; strong central planning of production; lack of market incentives for producers; inefficient management; absence of individual initiative; state regulated price systems; neglect of basic economic principles (e.g., comparative advantage) and, consequently, insufficient freedom of decision-making for managers of farms.

The importance of agriculture within the overall economy and its basic structural and organisational characteristics

In 1990, Slovak agriculture absorbed about 12-13 percent the total labour force and generated about 6-8 percent of the GDP. During the last years of the old system, agriculture was heavily subsidised. The transfers amounted to approximately 15-17 percent of total national investment. These figures, however, should be considered as rough indicators rather than reliable information. They are influenced and distorted by a rigid administered system in which prices have lost their economic function as indicators of scarcity. Figures on the labour force suffer from a similar lack of reliability. Inadequate social policies forced the co-operatives and state farms to accept a special pattern of organisational and managerial structures which resulted in overstaffing and inefficient allocation of production factors.

At the beginning of the 1950s, the co-operative and state farms represented the dominant form of farm enterprises, while some private farms only survived in remote mountain areas and less favoured areas. By 1987, the large-scale production units (see Table 7.1) cultivated more than 97 percent of the agricultural land in Czechoslovakia and delivered about 80-100 percent of agricultural products, except for fruit and vegetables where nearly half of production came from smaller units (see Table 7.2).

Table 7.1
Number and average size of state and co-operative farms, Slovakia, 1975 and 1987

	1975		*1987*	
	Number	*Average size (ha)*	*Number*	*Average size (ha)*
State farms	250	5,648	235	6,162
Co-operative farms	2,736	1,548	1,658	2,597

Source: Slovak Statistical Office

Table 7.2
Share of the large-scale farms in total agricultural production, Slovakia, 1987

Product	*Cereals*	*Potatoes*	*Sugar beet*	*Fruit*	*Vegetables*	*Meat*
Percentage from large farms (%)	96.3	84.5	99.9	51.0	53.0	83.8

Source: Slovak Statistical Office

In recent decades, the agricultural sector was heavily subsidised, and wages of farm workers were also relatively favourable (as can be seen from Table 7.3). The positive development of wages in Slovak agriculture has little to do with the economic performance of its production units because they were strongly and purposefully regulated by government transfers (see also Blaas, 1995, p. 16). Sufficient food production was defined as an ideological goal, and its attainment was permanently evaluated by the Communist Party and by government.

Table 7.3
Development of average monthly salaries in several areas of the Slovak economy, 1980-1990 (in Sk*)

Branch	*1980*	*1981*	*1982*	*1983*	*1984*	*1985*	*1989*	*1990*
National economy	2,701	2,744	2,807	2,859	2,913	2,962	3,222	3,340
Industry	2,778	2,828	2,906	2,982	3,035	3,106	3,359	3,456
State farms	2,616	2,662	2,717	2,819	2,908	2,967	3,284	3,452
Food industry	2,508	2,553	2,632	2,732	2,802	2,852	3,105	3,234
Co-op farms	2,646	2,711	2,735	2,922	3,017	3,084	3,599	3,833

* For better orientation we will use the exchange rate for USD and Slovak krona 1:30 in our chapter. The real value fluctuates since 1992 till last month between 27-31 Sk for 1 USD.

Source: Slovak Statistical Office

The agro-industrial sector after 1990

Incomes in the entire agro-industrial sector have worsened since 1990 (see Kabat, 1994, 1995). The great majority of farms had negative returns (see Table 7.4). The generated losses are covered by state subsidies and some other tools. For example, an increase in the farms' debts is accepted by the banks, because they depend on the state banking system and therefore are subject to government policies. The figures in Table 7.5 can be considered as indicators of the conservative strategy by which government tries to stabilise the traditional system of co-operative farms. In 1993, financial means which were originally planned to be used for environmental programmes, research programmes and the reserve fund, were actually, to a large extent, spent for other purposes, mainly as transfers to support the existing production units, i.e. co-operative farms.

These data reveal the present economic weaknesses of Slovak agriculture and its dependency on the state budget. This means an additional motivation to explore the basic question of this chapter, i.e. how much progress has been achieved on the route towards an improvement of production structures, and what determines the feasibility of this transformation? The proclaimed 'efficiency' of the collective farms in the old system was often actually secured by a financial umbrella provided by the state. The more the process of transition to a market economy will proceed, the less government may be able to maintain these protective measures due to budget constraints and increasing competition for government assistance with other groups. If the losses and debts of many co-operative farms can no longer be compensated for by government transfers, the production units have to prove their efficiency in a more competitive economic environment.

Table 7.4
Revenue, costs, and profits of farms, Slovakia, 1992-94 (in million Sk)

Indicator	*1992*[a]	*1993*[a]	*1994*[b]
Total revenue	38,635	43,138	45,208
Total costs	47,606	49,368	50,552
Total profit	–8,940	–6,208	–5,344
Profit per hectare	–4,576	–3,174	–2,329

[a] Figures drawn from a special survey conducted by the Ministry of Agriculture in March 1994. The number of farms surveyed was 1,093 which represent 81 percent of the total agricultural land in Slovakia.
[b] Figures drawn from the Zelena sprava 1995.

Table 7.5
Government support of Slovak agriculture, 1993 (in million Sk)

Subsidy title	*Limit (ex ante)*	*Used (ex post)*	*Difference*
Support for farm restructuring and farming in less favoured areas	7,524.7	7,553.4	–28.7
Support for production units	3,495.6	4,079.3	–583.7
Environmental programmes	1,648.8	591.8	1,057.0*
Research programmes and reserve fund	1,107.1	980.5	126.6

* The positive balance shows that environmental problems are neglected but payments to production units are preferred by policy makers.

Source: Zelena sprava 1995

Having in mind what has been pointed out above, the role of agriculture in the history of the Slovak society turns out to be much more important than one would expect at first glance. In particular, it is important not to forget that the process of enforced collectivisation also brought some positive changes to rural areas, like upgrading living conditions as regards housing, roads, shops, medical and social services. In many cases, introduction of collective farms also meant for small peasants that their family incomes were increased and stabilised, they no longer had to take the risk of their small business, and their heavy everyday farm work was replaced by an easier job.

The role of agriculture was ideologically backed and defined as a priority of the national economy. This variant of agrarian fundamentalism served as a justification of adequate financial support to be distributed to this branch for several decades. Due to a lack of transparency, it was not possible for farm members to distinguish between returns originating from political subsidisation and returns based on economic efficiency during this period. From their internal perspective, many collective farms achieved relatively good production levels and good economic results.

This opinion was still popular in the Czechoslovak society after the 1989 revolution. Many people tended to interpret government plans to change the existing production structures in agriculture as a politically motivated attempt to 'punish' this branch of the national economy because of its close ideological ties with the former Communist regime. Obviously, the inefficiencies of the existing co-operative structures had not sufficiently been recognised due to the lack of information released by state authorities, and partially also because of the low transparency of the state budget.[1]

A theoretical approach for explaining the economic and political feasibility of transformation

Having in the previous section discussed the necessity to improve economic efficiency of Slovak agriculture, we will now discuss the question whether this will also be feasible. In this section, this issue will be discussed from a more general and theoretical point of view. Then we will return to the Slovak case and look for some empirical evidence. The theoretical considerations will be mainly based on approaches developed by Pryor (1992), Swinnen (1992; 1993a,b; 1994a,b) and Hagedorn (1991a,b; 1992a,b; 1993a; 1994a; 1996b,c; 1997).

Different views of the issue of transformation of agriculture

The problem of transforming collectivised or nationalised agriculture of centrally planned economies can be considered in various ways. Mainly four points of view play a role in the discussion, representing different interests of authors and actors:

1 *The normative view.* How should governments, administration, farm operators and other people involved try to change and to shape the

institutional, economic and political structure of the agricultural sector in the Central and Eastern European (CEE) countries, provided that they follow reasonable criteria of economic efficiency and of just distribution and employ adequate policies?

2 *The positive view.* How may the changes in agrarian and agricultural policy institutions as well as in farm structure and farm organisation actually occur? What kind of interests and influences will play an important role, and how can we find appropriate explanations and theories to understand these developments and determinants?

3 *The result-oriented view.* What type of ownership structure, organisation of production, policy networks and other features of agriculture will be the outcome of the transformation process, and will these results be in accordance with the objectives political and economic decision-makers want to achieve?

4 *The process-oriented view.* Which process of change in agrarian institutions, structure of agriculture, organisational forms of farming and agricultural policies will actually take place, which stages and bottlenecks can we expect. Which transitional forms of organisation may develop, and will the route of institutional and structural change be efficient and reasonable in economic and social terms?

In the first years after the old systems had collapsed, research approaches were often based on the (implicit) assumption that transformation would clearly be feasible even within a limited period of time. Thus, the result-oriented approach appeared to be adequate. However, it soon became clear that economic and political circumstances for a quick restructuring process were less favourable than expected. In particular, many of the CEE countries lack resources for arranging and financing the required changes in property rights on agricultural land and organisational forms of production. As a consequence, the question of economic and political feasibility of transformation arises and becomes one of the main issues of research with regard to these countries. This means that the results of transformation which can only be achieved in the far future are of much lower interest than the next steps of the process and the transitional forms of agricultural organisation as well as the obstacles which may counteract further progress in the process of restructuring.

Institutional economics as an approach to analyse the evolution and feasibility of transformation

What are the determinants we have to take into account when talking about feasibility of transformation? In the first place, of course, we have to distinguish between economic and political factors (see Hagedorn, 1994a, p. 3). According to the Evolutionary Theory of Constitutions (see Leipold, 1985; 1988; 1989a,b; Schmidtchen, 1989; Schüller, 1988), institutional innovation is in the first step considered to be a process of self-organisation that is only effected by economic driving forces, mainly by transaction costs. However, the approach is not limited to this point; on the contrary, it explicitly takes into account the motives and power of influential politicians and interest groups in determining the economic order, e.g., by referring to a concept of Schenk (1983).

Second, we should distinguish between a narrow and a broader set of determinants relevant to the transformation process. The first is limited to production and transaction cost considerations regarding only one institution, the farm firm as the basic organisation regulating agricultural activities. The second includes the whole institutional and political network which plays a crucial role for a farmer to be economically successful. Obviously, it is insufficient only to take into account the farm firm without emphasising the institutional and political environment.

From transaction costs economics of the farm firm toward the concept of an economic and political network

The logic of this argument has been pointed out by extending the range of determinants which may have an impact on the transformation process step by step (see Hagedorn, 1991a,b; 1992a,b; 1993a; 1994a; 1996b,c; 1997).

This procedure started from the preliminary assumption that institutional innovation is solely determined by the aim of economising transaction costs within a farm (see also Schmitt, 1990a,b). Consequently, two main questions were discussed:

1 Which *farm size* will prove superior? To what extent will decreasing production costs be accompanied by likewise decreasing transaction costs or, in contrary, be outweighed by progressively increasing transaction costs?

2 Which *organisational structure* of farming will be the most competitive, if production co-operatives, joint stock companies, limited liability

companies partnerships and family farms for example, are compared with each other?

Then an additional hypothesis was introduced. Which system of agricultural production will be dominant, depends not only on the transaction costs in the farm firms. The level of transaction costs within these economic units is interrelated with the level of transaction costs in other areas of economic activity which are in permanent interaction with farms. Accordingly, institutional solutions to reduce transaction costs of agricultural production also have to take into account product markets, factor markets, the farm family and institutions to internalise external effects with special reference to environmental problems, and many other agrarian institutions.

Nevertheless, another assumption was still valid at this stage of analysis. The evolution of agrarian institutions still appeared to be a question of minimising transaction costs by means of self-organisation within the domain of *private action.* If this was true the political economy of *collective action* would not play any role in this process. Since agricultural policies are strongly influenced by group interests (see Hagedorn and Schmitt, 1985; Hagedorn, 1991c; 1993b; 1994b; 1996a,b; Swinnen and van der Zee, 1993), this assumption had to be dropped, and political determinants of institutional change in socialist agricultural systems became a part of the game: electoral control and party competition, interpretation systems for legitimating agricultural policies, collective action by interest groups, and bureaucracy, only to mention some examples.

Determinants and feasibility of 'transforming the network'

This attempt to take into account the whole variety of institutional and political influences is in accordance with the concept developed by Swinnen (1993a,b; 1994; 1997a,b). He notes that it may be reasonable just to look at the farm unit as a transaction-cost saving entity if we are in well developed, stable farm economy which is not affected by major changes due to structural adjustment or reform policies. In a period of fundamental transformation of the farm sector, however, the whole network of relationships has to be taken into account. The evolution of farms depends on its institutional and political environment which, in turn, is influenced by the initial organisational and structural features of farms and their evolution itself. Thus, several factors have to be included into the analysis:

1 The starting situation regarding farm organisation and agricultural structures, e.g. size of the plots belonging to land owners and availability of capital,

2 Relationship to other agrarian institutions like access to inputs, marketing co-operatives and extension systems,

3 The political economy of agricultural policies, especially with regard to different interests and emerging conflicts due to privatisation and decollectivisation policies in the transition period,

4 The evolutionary characteristics and potential path dependency which may come into being during the transformation process and may lead to different routes of change in the countries in transition.

Consequently, the policy advice for the restructuring of CEE agriculture should not ignore the existing structure and/or the institutional and policy environment and should not limit the relevant variables to the initial and final equilibrium situation. Rather, it should explicitly include the *process* of change in the analysis (Swinnen, 1993b).

Thus, there are reasonable arguments which show that the isolated (and static) perspective of only the farm unit is an 'approach in a vacuum' (Swinnen, 1993b). But after we have learned that we have to consider the whole network of interrelationships between the many institutional and political elements, what does this mean for the feasibility of transformation? This means that – if we take an idealistic view – all elements and relationships within the institutional network must change in a harmonious manner when passing from collective or state ownership to private land ownership or from collectivised to decollectivised agricultural structures. Some of the components may have to change quickly, others more slowly, some may have to be reformed in early, others in later stages of the process. These depend on the kind of evolutionary interaction which will maintain and improve the workability of the system. Furthermore, this means that all actors in the various parts of the system must be willing and must be able to perform these changes in a co-ordinated or even co-operative way and do not act against the transformation process.

These conditions are not always fulfilled. Pryor (1992, p. 266) even suggests that in 'certain historical situations' collectivisation may not be reversible. For example, the Soviet Union tried to initiate decollectivisation in the late 1980s, but these measures were not successful until the beginning of the 1990s. Pryor (1992, p. 266) even maintains 'that collectivisation depends less on whether a Marxist party is in power than on a series of economic, social and political factors'.

He explains his opinion by means of an analogy: why do some technologies survive (e.g. types of keyboards), what means that people continue to use them, although other technologies have proven to be better? His answer refers to network externalities and embeddedness. These terms encompass the benefits gained by a given user when others use the same technology or standards, the interrelatedness and interdependencies of production technologies, the high short-run costs of breaking webs of interaction in comparison to benefits, and the imperfect information of the individual economic agents with regard to the intentions of others. That is, no individual would want to bear the costs of being out-of-step with others, although all would benefit if they decided together to take a particular action (Pryor, 1992, p. 268). His conclusions: societies or economies can be locked in inefficient technologies, and similarly they can also be locked in inefficient institutions.

Of course, this phenomenon of being trapped in an inadequate institutional framework, or network, may significantly increase the transaction costs of transformation. Usually, these transaction costs are 'sunk' costs (see Wegehenkel, 1981), i.e. they must be invested before the economic actors are able to use the new institutions. Furthermore, they cannot be mobilised again in case of error, i.e. when new institutions are not working and have to be replaced.

Institutions as public goods and institutional innovation as collective action

In addition, there is another serious problem: new institutions formed by the transformation process are usually public goods with the well-known characteristics of non-rivalry and non-exclusiveness. People who provide the resources to create new institutions or co-operate in some way to establish new rules cannot prevent others from using the benefits. Of course, this crucially affects the economic incentives to organise institutional innovations: if collective agreements on co-operative behaviour and cost sharing are not successfully arranged, or if there is a lack of security or communication, so that the individual farmer does not know whether or not the other members of the network will join him on the route towards institutional reforms, transformation will stagnate. In other words, political and economic actors who are considered as pioneers in the process of transformation must have insufficient information on whether those institutions they rely on will work properly. If this is not the case, they have invested sunk costs but do not obtain adequate returns.

As a consequence, there are many reasons, e.g. pointed out by Pryor (1992, p. 267), why transformation may not be successful, even if the long-term benefits of institutional change, i.e. of transforming the agricultural system, are much higher than the short-term costs. In individual calculations made by new farmers or local officials the benefits may not cover the costs, because there are losers and winners, insecurities and intransparencies, the problems of public goods and collective action, and all these together may lead to a conservation of an inefficient status quo. This conclusion may be clarified by the following explanations (see Pryor, 1992, p. 267):

1 An agricultural system usually includes many actors such as farmers, politicians, administrators, food traders, advisors and family members. Institutional change means that all these actors will have to accept the new rules and correspondingly change their behaviour, either on a voluntary basis or being urged by sanctions. Such conditions are unlikely to be always fulfilled.

2 For example, there may be some reason why a new farmer may lack certainty that other actors involved will take similar actions as he does, e.g., a local official may fear to lose his job, or a chairman of a collective farm may be afraid that his personal power will decrease.

3 Similarly, a farmer may lack certainty that others will take complementary actions, e.g., he may not be sure that alternative sources of agricultural inputs will be available to him, if the chairman of the local co-operative farm refuses to sell him the inputs he needs. Such difficulties may be typical in a phase of transformation in which competitive market structures are still to be developed.

4 Even if the farmer is certain that other participants in the network will behave in a co-operative way, he may not be sure how the benefits will be divided (by market or other mechanisms) among different actors. Thus, he may be unable to foresee and to calculate the returns he can expect, so that he cannot reasonably decide to participate in the innovation process.

Though the costs of institutional innovation may be much lower than the benefits which could be achieved by the change of the agricultural system for society as a whole, it must be taken into account that 'society as a whole' or a 'benevolent dictator' does not make the decisions. Usually, self-interested politicians decide and influential administrators take care of implementation. Both groups may be worse off in economic and political terms, if they support privatisation programmes and land reform measures:

1 Politicians may lose votes if they help to create a new group of land owners who do not vote for them or their party. As a consequence, they will be reluctant to support introduction of private property rights and of farm establishment programmes for new individual farmers.

2 Local bureaucrats may predict that less people depend on their goodwill if farming becomes a private business. Therefore, they may sabotage privatisation programmes, for example, by giving land of low quality to those members of a co-operative farm, who want to establish their own farm, by demanding high rents or by refusing to sell them agricultural inputs.

Also Swinnen (1997b) stresses the point that decollectivisation reduces the number of positions, the income, and the power of many party and state cadres in the agricultural sector. One option for government to avoid party and officials' resistance is to provide the cadres with favourable positions which enable them to become wealthy in the new institutional environment. This strategy, however, may also be dangerous, because it decreases the credibility of reforms in the view of the people concerned.

Cumulation of disposition rights in the initial situation as a transformation barrier

Having collected some elements of a theory regarding the determinants of political and economic feasibility of transformation, we now summarise and generalise the main points.

1 Institutional innovations involve sunk costs, and people are only ready to carry these costs if they feel sufficiently secure that they will obtain reasonable returns from their participation in the institutional innovation process. This sort of confidence is a public good, i.e. the more people believe in the reforms to be successful, the more this prophecy will be fulfilled.

2 In the case of the transformation of the agricultural system, institutional innovation may be performed in an incomplete or inadequate way. If the parts of the system are reorganised in an uncoordinated manner and without paying attention to a proper sequencing of transformation steps, this may seriously reduce the efficiency of the system, e.g., because old components and reformed ones are not compatible. The complete network and all relationships between its actors and organisations should be reshaped harmoniously.

3 Institutional innovations like changes of agricultural systems are collective goods. As a consequence, transformation will only be successful if a sufficient number of actors behaves in a similar way oriented towards the goal to bring about the collective good. There are two explanations for this:

first, the network must be rearranged in a consistent way, a partial or fragmented innovation may be possible but not always efficient. For example, land markets cannot be introduced, when everything else is left unchanged, e.g., the procedures of central planning. Second, cost sharing is often necessary to provide incentives for institutional innovations. For example, it will be difficult to explore a new marketing channel, if the costs have to be borne only by one or a few producers and cannot be divided up among many participants.

4 Feasibility of transformation is also influenced by the fact that there are always winners and losers. The winners will only undertake efforts to reach a consensus if they know that the state will protect the new rules of the society, particularly with regard to property rights (see Bromley, 1996). The losers will try to prevent the institutional innovations. If they are in key positions – as politicians, bureaucrats or monopolists in marketing organisations – they may be able to perpetuate the old agricultural structures with some minor modifications. The status quo will probably persist as long as the (potential) winners, i.e. the proponents of institutional innovation, cannot compensate the losers (e.g., by providing them with new positions) or cannot gain enough political power of their own, for instance, by forming an interest group or by exerting voting power.

5 The phenomenon that endangered officials try to sabotage privatisation programmes only occurs because the previous institutional arrangements have tied together economic and political influence. As a consequence, economic competence cannot be given up without being deprived of political power. Obviously, institutional arrangements and the organisational solutions defined by these institutional settings can be formed in a way that either integrates many property rights (or disposition rights) or separates these rights from each other. Integration (or cumulation) of rights can lead to inflexibility when institutional innovations are required, and separation of rights provides reform-minded actors with more flexibility in the case of institutional change.

6 Thus, a major difficulty arises from the fact that the system of central planning and of large economic units ('industrialised agriculture') has created large bundles of property rights assigned to one or only a small group of the decision-makers. For example, a leader of a collective farm simultaneously acts as an administrator who decides on the quantity and quality of land a new individual farmer is allowed to receive, and he also decides on the distribution of inputs and controls marketing of outputs.

The problem of too inflexible bundles of disposition rights not only originates from institutional decisions. Concentration of rights and competencies is often reinforced and becomes even more persistent because of the physical entities which have been established on the basis of this institutional structure. In this way, institutional and technical determinants are interrelated, e.g., large-scale equipment, big buildings, and rural infrastructure are difficult to adapt to the reformed structure of agriculture.

This causal connection between institutional and physical barriers becomes even more evident if we consider the problems originating from the different role formal and informal institutions play in the process of transformation, as has been pointed out by Swinnen (1993b, p. 234):

> the pre-reform production system is not only sustained by formal institutions and organizational arrangements, such as the imposed hierarchy and official rules regulating ownership of assets and decision-making power. The system is further sustained by an underlying network of informal institutions that form the "glue" of the formal institutional structure. Formal institutions can be suddenly changed in a process of revolution. However, informal institutions such as social taboos, customs and norms do not change so easily (...). While privatization and land reform effectively destroy and break up the previous formal arrangements, this is not the case with (1) the informal institutions and (2) the existing human capital.

The large-scale production process has shaped both non-human and human capital in collectivised agriculture. This strongly influences feasibility of transformation. In the case of physical capital, these feasibility problems will crucially depend on the availability of alternative machinery and other assets which are smaller in scale. Obviously, the main determinants are the size and variety of machines and other assets produced in the period before transition, restructuring activities in the down- and up-stream industries, and availability of credit for farmers for purchase of appropriate machinery. In addition, the particularities of the privatisation process and the features of the equipment play a role. If the machines cannot be divided up among small farmers, they may be used by specialised firms or organisations like contractors or machinery work stations which sell services to the private farmers. As Swinnen (1993b; see also Christiaensen and Swinnen, 1994) points out, small-scale or family-sized farms are able to pick up considerable scale effects without introducing a large-scale production system by appropriate institutional innovations like contractors and co-operatives for processing, purchasing inputs or trading outputs. Whether or not such institutions will be developed, depends on the privatisation of non-land assets and the restructuring of up-stream and down-stream industries.

Human capital consists of accumulated knowledge and skills and of the established informal institutions as well. In collectivised agriculture, skills and knowledge of production activities are extremely specialised. In family farms, a single person has to be familiar with the whole production process. Farm workers who are employed in collective or state farms are usually not motivated to become self-employed farmers, because they lack the capabilities and norms required in family farm operations: 'Human capital cannot be reorganized or privatized like other assets and may therefore be the most "conservative" factor in the restructuring process' (Swinnen, 1993b, p. 235).

The same is true for informal institutions as a part of human capital, which cannot be reshaped by government decisions or new laws within a short time, but will only change at a very low speed. For example, Pryor (1992, p. 281f.) explains the differences between the Russian and the Chinese attempt to decollectivise agriculture by different attitudes and behaviour with regard to entrepreneurship due to historical reasons.

Major obstacles to privatisation and decollectivisation

Pryor (1992, p. 268ff.) points out a variety of reasons why decollectivisation may fail or at least face serious difficulties. Some of these could also be important for the Slovak case:

1 *Problems of dividing the equipment.* Since farms in CEE countries predominantly used large-scale equipment, it is difficult to divide these assets among the new farms.

2 *Reliable access to inputs.* As the new farmers cannot afford to run their own network of input supply (and co-operatives for this purpose have not been founded yet), they usually have to buy the inputs from the co-operative farms which continue to be monopolist suppliers.

3 *The availability of credit.* In order to develop viable farm units the new farmers need credit for investments and for buying inputs and renting farm assets. For this reason, it is important for banks to establish local branches which can be used by the farmers.

4 *Sale of farm output.* The new farmers must be able to sell their output at profitable prices. If prices are not attractive, or if even monopolistic marketing structures are reluctant to buy the farmers' products, incentives to take the risks of farming may be insufficient.

5 *The problem of farm liabilities.* If farms are decollectivised, not only must the assets be divided but also the liabilities of the farm. In the case of highly indebted co-operative farms, many members may hesitate to become a self-employed farmer by taking their land out of the co-operative, if they then also have to take over their share of the debts.

6 *Risk of policy changes.* People who consider establishing a new farm or transforming an existing farm try to find out whether government will maintain the privatisation policy in the future or will re-collectivise the private farms some day. If the political environment does not create positive expectations, the activities of the new farmers may be paralysed. For example, if they are afraid to lose their land due to policy changes, they will not be ready to undertake major investments. Pryor (1992, p. 276) notes that every inconsistency of government policy during this transition period, every conflicting statement by different government leaders about the future of agriculture, and every broken promise to the farmers by the government increases the risks farmers face in investing their resources in any individual farming arrangement. In addition, any general political or economic difficulties facing the government that raises the possibility it could be replaced – for instance, ethnic unrest, inflation, shortages, or political dissension – would aggravate such fears.

The main precondition which is necessary to solve this problem is to introduce and to maintain strong property rights which can legally be defended by the citizens against the influences of those politicians which would like to reverse privatisation.

In addition, we have to take into account the political economy of the transformation process. One possibility to achieve a reliable basis of political support may be to create 'a large group of property owners in society' that 'will result in a long-term constituency for 'reform-minded' politicians and parties' (Swinnen, 1997a). These land owners will support those politicians and parties which will help them to obtain the highest returns to their property. However, there is also the opposite way to build a long-term constituency, and if this conservative option proves to be superior, anti-reform politicians, parties and policies will predominate. By stabilising old structures like large-scale co-operative farms a group of voters and citizens is created whose social and economic well-being depends on interventionist and protectionist agricultural policies like price support or direct payments. For example, managers as well as the workers of co-operative farms will support governments which maintain their leading positions or employment opportunities and will resist any reform policy by which they could lose their jobs. For the path the transformation process is going to take it seems to be decisive which coalition is going to dominate in the first years: reform-minded politicians and new land owners, or anti-reform politicians and co-operative farms.

Restitution and privatisation of land ownership and changes in organisation and structure of farming in Slovakia

The political and legal foundations for transforming and restructuring agriculture were laid before the decision to split of the Czechoslovak Republic was made. The legal framework for re-establishing private land ownership in Slovakia was adopted by the Federal Parliament of the Czechoslovak Republic in 1991. According to the Act of the National Council of the Slovak Republic (Corpus Juris, No. 42, 1992), the former *kolkhoz*-type co-operative farms (co-operatives) had a one-year time limit (ending on January 28, 1993) to be transformed into the private legal entities with transparent ownership relations. The majority of the socialist type co-operative farms had to pass their individual process of transformation and to adopt a new, more transparent entrepreneurial structure. In addition, privatisation of state-owned farms and church-owned land was initiated, but implementation of these decisions has been considerably delayed.

Changes in land ownership and property rights

As a consequence of the privatisation procedures property rights in land and other assets were changed fundamentally. In 1994, only 299 from 10,108 agricultural enterprises were still located in the 'public sector' and 9,809 belonged to the 'private sector' (see Table 7.6). 79.2 percent of agricultural land is cultivated by privately organised farms and only 20.8 percent is left to public enterprises. There are no major differences between the public and the private sector with regard to the types of land (arable land, vineyards, orchards, grassland), as can be seen from Table 7.6. Most of the agricultural area operated by Slovak farms consists of rented land (74.2 percent). The small share of privately owned land (1.8 percent) is due to the fact that most of the Slovak farms are co-operatives without any owned land. In addition, also the individual farmers usually own only small areas of land and rent most of their land, because land ownership is highly fragmented in Slovakia.

Nevertheless, the area of owned agricultural land seems to be underestimated in Table 7.6, because 'private' and 'public' agricultural land taken together does not equal the total area of agricultural land. This gap can be explained by difficulties in the privatisation process: Land can only be classified as belonging to the private sector if the original owner has been legally identified. In many cases, this goal has not been achieved yet. This land is used by co-operatives, state farms, in some cases also by individual farmers. Strictly speaking, an additional column would be needed for this type of land.

Whereas agricultural land is predominantly in private ownership, most of the forestry land belongs to the public sector (98.2 percent). Table 7.6 also shows that the number of livestock is similarly distributed among private and public enterprises like land is. Only horses and foals are an exception; in this case, 45 percent of the animals are kept in public enterprises.

However, these figures have to be interpreted very cautiously, because the Slovak Statistical Office defines the terms 'private sector' and 'public sector' in a special way: If more than 50 percent of the assets of an enterprise are privately (publicly) owned, it belongs to the private (public) sector. This also differs substantially from our economic conception of private or public enterprises which is oriented towards the question whether private or public actors or organisations make the entrepreneurial decisions and are also responsible for the consequences of these decisions.

Furthermore, there is an additional reason why the terms 'private sector' and 'public sector' must be interpreted with caution. All those farms are considered as being 'private' which have passed the transformation procedures in accordance with the above-mentioned Transformation Law. In particular, this means that the managers of co-operatives had to sign new lease contracts with the original owners of the land cultivated by the co-operatives. Since nearly every co-operative has finalised this transformation procedure, almost all land in Slovakia is reported in the category 'private' (1,734,968 ha). The rest of the land (465,335 ha) is referred to as 'public'. This term means that the land and other assets are owned by the state, rural communities, the Roman-Catholic Church and schools.

These developments give rise to the question whether the distribution of and changes in formal (or legal) ownership rights on land (and other assets) actually mean that also the real property rights have changed in the same way and to the same extent. 1,321,086 ha (76.1 percent) of the agricultural land which officially belongs to the 'private sector' is not cultivated by individual farmers but by co-operatives. Although there is no doubt that this land is – formally – privately owned, the question arises whether it is also operated by private farm firms. Certainly, according to the understanding of co-operatives in Western Europe, co-operatives are definitely private enterprises formed by voluntary association of individuals who pursue common economic objectives. In contrast, people and economists in CEE countries, and also in Slovakia, may consider agricultural production co-operatives as a continuation of the former collective farms based on enforced collectivisation and still strongly influenced by government and administration. In this view, co-operatives could be classified as belonging to the public sector (or to a still 'quasi-public' sector), and only individual farmers and partnerships would belong to the private sector. Accordingly, the Slovak association of private farmers and land owners maintains that the official definition of co-operatives as belonging to the private sector strongly reflects political objectives, i.e. to give the appearance that the process of transformation is proceeding faster than it actually is.

Table 7.6
Basic characteristics of Slovak agriculture in 1994

		Total	*Ownership of enterprise*	
			*private sector**	*public sector**
Number of farms	n	8,931	8,632	299
	%	100	96.6	3.4
Total area of land	ha	2,960,244	1,760,892	1,199,352
	%	100	59.5	40.5
Agricultural land	ha	2,191,303	1,734,968	456,335
	%	100	79.2	20.8
From this: Arable land	ha	1,379,021	1,100,161	278,860
Vineyards	ha	20,188	16,722	3,466
Orchards	ha	17,213	12,243	4,970
Grassland	ha	761,743	599,904	161,839
Forestry land		769,373	13,986	755,387
Rented land				
Agricultural land	ha	1,627,016	1,321,086	305,930
	%	100	81.2	18.8
From this: Arable land	ha	1,080,732	885,220	195,513
Vineyards	ha	15,867	13,710	2,158
Orchards	ha	12,868	9,759	3,108
Grassland	ha	500,253	403,224	97,028
Forestry land		134,007	4,823	129,184
Livestock (heads)				
Cattle	n	834,887	664,430	170,457
From this: Cows	n	324,388	256,526	67,862
Pigs total	n	1,693,383	1,390738	302,645
From this: Sows	n	146,882	119,904	26,978
Sheep and rams		257,888	212,503	45,385
Horses and fouls		5,082	2,818	2,264
Poultry		10,024,48 2	7,414,377	2,610,105
Machines and equipment				
Tractors		32,703	26,036	6,667
Lorries		16,175	12,759	3,416

continued next page

Table 7.6 (continued)

Harvesters		5,918	4,656	1,262
Cereal sowing machines		4,942	4,002	920
Cutting machines		2,647	2,089	558
Tractor ploughs		13,131	10,657	2,474
Workers				
Full time workers	n	158,477	122,232	36,245
at Dec. 31, 1994	%	100	77.1	22.9
From this: Women	n	55,190	43,255	11,935
Owners with family at Dec. 31, 1994	n	15,316	15,316	-
From this: women	n	5,862	5,862	-

* The terms 'private sector' and 'public sector' have to be interpreted very carefully. 'Private' means the land of all those co-operatives which have passed the transformation process as prescribed by the Transformation Law. The rest of the land is referred to as 'public'. This includes the area owned by the state, state farms, rural communities, church, schools, etc.

Source: Agricultural Cenzus 1994, Slovak Statistical Office

However, this political debate is less helpful if we want to give a reliable answer to the question whether the Slovak co-operative farms are really private or public organisations. To find this out we need to know who actually makes the entrepreneurial decisions and also implements them, and is responsible for the consequences of these decisions. Those who decide on inputs, outputs, marketing, technology, investment and profit distribution in the co-operatives are usually private actors. The same is true for the decisions on resource disposal and acquisition such as hiring and firing labour, acquisition and sale of land and capital, and, in principle, private organisations and not the state budget enjoy the profits and pick up the debts. Nevertheless, this only applies in a still incomplete manner, e.g., government partly takes over the debts and many farm managers are integrated in the traditional political and administrative network, and, in this way, their decision-making power may be partly based on the old system. Thus, we may rather speak of more or less 'mixed organisations' which find themselves in a phase of transition from public to private forms.

There are some additional reasons why we doubt that the formal changes in legal ownership of land are equally reflecting real changes in actual property rights. For example, it is often difficult for a farmer who wants to establish his farm of his own to have his land and the land of his private lessors separated from the area the local co-operative continues to cultivate. Physical access to the plots he owns or rents may be a serious problem. The land may not lie adjacent to a road. Furthermore, a farmer will hesitate to rent land from private land owners whose land belonged to a co-operative in the past if he expects that he will not get access to it or will have to cope with other obstacles maintained by the co-operative. In addition, it may be difficult to use the land he owns and rents efficiently because it is fragmented in many plots and not consolidated. A land consolidation scheme may be necessary to create a new structure of agriculture in which the development of viable farms with different organisational forms is not counteracted by the dominance of the existing co-operative farms with regard to access to and distribution of land.

A gap between formal ownership rights on land and real property rights also originates from missing or imperfect land markets. Usually the production co-operatives represent local monopsonists on the land market. This means that they can more or less determine the level of rents. They are also able to exert market power on those land owners who consider leasing their plots to individual farmers who establish or extend their private farms. It is reported that managers of co-operatives have threatened that they will not take back land once it has been leased to another farm. In Slovakia, this actually can be a serious threat because many new farmers are not willing to make medium- or long-term lease contracts. Since the Slovak government has expressed its strong preferences for 'co-operative' farming as well as its reservations towards individual farms, the new farmers are afraid that they could be treated unfairly by government and administration in the future. As a consequence, many of them do not trust in the politicians and, therefore, avoid the risks of long-term planning.

Another example which shows that formal ownership titles distributed during the transformation process may not turn out to be effective property rights is revealed by political processes concerning recent amendments to the Transformation Law (Law 42/1992). In the process of transformation of collective farms the owners' legitimate property claims had to be fulfilled. This turned out to be difficult in the case of those legitimate owners who did not become members of the private farm enterprises established during the procedure of reorganisation. The original law offered various possibilities to satisfy the owners' claims, including capitalisation of property deposits. A special amendment to the law on transformation also provided the possibility to penalise the farm enterprises for not fulfilling ownership claims. This amendment, which is usually referred to as the Sanction Law, had been adopted by the Czech Parliament in 1993. However, it has not been approved by the Slovak parliament. This difference in government policies also reflects the complicated political background of the restitution and privatisation process.

Nevertheless, according to the original version of the Transformation Law the co-operatives were obliged to release the collectivised land and also a part of the non-land assets to each eligible person, i.e. to the original owners of the land in most of the cases. However, many co-operatives were not willing to fulfil this obligation because the loss of capital would have damaged the enterprise. Due to the political influence of the production co-operatives which have close ties to politicians, the Slovak parliament passed a law which abandoned the duty to pay immediately for the assets. Instead, the co-operatives are allowed to issue so-called 'valuable papers', i.e. vouchers or certificates, which will be tradable, however, not immediately but only after seven years. In the meantime, there will probably be a very high supply of these vouchers but no 'official' demand, because trading them is not yet permitted. Consequently, the price of vouchers may be extremely low. In this situation, the current management of the production co-operatives will successfully try to buy the vouchers and to become the only owner of all assets of the co-operative in question. It is reported that in some cases three or four managers of co-operatives established a private company with limited liability which serves for the only purpose to buy or to reserve the certificates already before the deadline regarding their official tradability has elapsed.

These examples show that there are high transaction costs of using the formal ownership rights the new land owners and users of non-land assets have received. Government even tends to increase these transaction costs in order to maintain the existing structure of farming. The above-mentioned amendment to the Transformation Law can be considered as a response to a typical difficulty emerging during the transformation process – the loss of capital by the obligation to release non-land assets. In other countries, the transformed co-operatives seek private solutions to this problem; in Slovakia, they are provided with a political solution. Obviously, the political processes outlined above indicate that the change in government connected with the separation between the Czech and the Slovak Republic has considerably reduced the speed of reform and has modified its objectives.

Organisational and structural changes in the area of collectivised land

By the end of 1994, about 123,988 original owners of land submitted their claims for returning their land in order to operate it by themselves. The total area of this claimed land is 180,459 ha, which is approximately 8.5 percent of the total agricultural land in Slovakia. The size of the claimed land was predominantly (86.8 percent) less than 2 ha per farm. The majority of the original land owners, however, decided to lease their land to local co-operatives or other farm enterprises (e.g., individual farmers, companies with limited liability, share holding companies; see Table 7.7 for the legal and organisational forms of agriculture in Slovakia). This applies to 442,649

Table 7.7

Legal and organisational forms of farms in Slovakia

	Natural person (Fyzicka osoba)		
English term	Individual farmer	Tradesman Entrepreneur	Association
Slovak term	Sukromne hospodariaci rolnik	Zivnostnik	Zdruzenie
Legal basis of existence	Act on indi-vidual holders (No. 219/91 Coll.)	Act on small businesess (No. 455/91 Coll.)	Civil Code (No. 47/92 Coll.)
Registration in Company Register	No	No	No
Minimum members	1	1	1
Minimum of capital	No	No	No
Individual tradability of capital	Yes	Yes	Yes
Liability	Unlimited	Unlimited	Unlimited
Management - statutory body	Direct	Direct	Delegation

	Legal person (Pravnicka osoba)					
English term	Limited Liability Company	Public Business Company	Limited Partnership	Joint Stock Company	Co-operative	State Enterprise
Slovak term	Spolocnost s rucenim obmedzenym	Verejna obchodna spolocnost	Komanditna spolocnost	Akciova spolocnost	Druzstvo	Statny podnik
Legal basis of existence	Commercial Code (No. 513/91 Coll.)	Commercial Code (No. 513/91 Coll.)	Commercial Code (No. 513/91 Coll.)	Commercial Code (No. 513/91 Coll.)	Commercial Code (No. 513/91 Coll.)	Code on State Enterprises (No. 111/90 Coll.)
Registration in Company Register	Yes	Yes	Yes	Yes	Yes	Yes
Minimum members	1	2	2	2 if there are natural persons; 1 if it is a legal person	5 if there are natural persons 2 if there are legal persons	-
Minimum of capital	100 000,- Sk*	No	According to agreement	1 000 000,- Sk	50 000,- Sk	
Individual tradability of capital	No	No	No	No	No	No
Liability	Limited	Unlimited	Unlimited	Limited	Limited	Limited
Management - statutory body	Chief executive	Partners	Complementaries (general partners)	Board of directors elected by the general assembly	Board elected by members meeting	Appointed by state authority

Source: See the line 'legal basis of existence'

original owners of the land with a total area of 936,168 ha, which represent 44.1 percent of all agricultural land in Slovakia.

There are significant differences in compensation procedures in different regions of Slovakia. The problem of compensation was solved rather smoothly in the Western part of Slovakia. Here, the transformed co-operatives are in a better financial position than the co-operatives in the Central and Eastern territories of Slovakia. By the end of 1996 about 80 percent of the claims for land and 75 percent of the claims for financial compensation (rent) were solved. The process of restitution and transformation has surely begun to modify the structure of agriculture, but most of the land is still operated by co-operatives or state farms (see Table 7.8).

The majority of the old-fashioned co-operatives has been transformed into co-operatives of the land owners or shareholders.[2] In 1994, there were 961 co-operative farms (10.9 percent of all farms), 149 limited companies (1.7 percent), 70 share holding companies (0.8 percent), 211 state farms (2.4 percent) and 51 agricultural units completely or partly funded by the state (0.6 percent). 7,750 individual farmers were not registered in business register (86.5 percent), 9 individual farmers were registered in business register (0.1 percent). However, the distribution of the land cultivated by these categories of farms differs considerably from the distribution of the number of farms. 70.0 percent of the agricultural land in Slovakia was cultivated by co-operatives, 2.1 percent by limited companies, 1.7 percent by share holding companies, 19.4 percent by state farms and 0.9 percent by completely or partly state-funded enterprises. Only 5.5 percent of the land is operated by the two categories of individual farmers mentioned above.

This differentiation of agricultural enterprises with regard to their legal status (see also Table 7.7) which also provides some information on their organisational form shows some special features. There is a separation between individual farmers which are registered and others which are not, but most of them can be found in the second category. Partnerships among farmers which play a role in other countries are unknown as a legal form of organisation in Slovakia. The last category of Table 7.8 includes two other organisational forms: (1) agricultural units completely funded by the state, which are usually not in the production sphere but can predominantly be found in schools, the health care system, in research institutions and former state farms; (2) agricultural units partially funded by the state which in many cases belonged to the first category and include educational, health care and research institutions as well as all state farms. They have to cover 30 - 90 percent of their expenditures by their own revenues.

Table 7.8
Agricultural land according to the legal form of farm enterprises[a], Slovakia, 1994

Type of land		*All farms (private and public)*	*Individual farmers not registered in business register*	*Individual farmers registered in business register*	*Limited companies*	*Share holding companies*	*Co-operatives*	*State enterprises*	*Completely or partly state-funded enterprises[b]*
Number	n	8,931	7,750	9	149	70	961	211	51
	%	100	86.5	0.1	1.7	0.8	10.9	2.4	0.6
Total area of land	ha	2,960,244	119,270	593	63,669	36,853	1,540,507	1,180,682	18,670
	%	100	4.0	0.02	2.1	1.2	51.8	39.7	0.63
Agricultural land	ha	2,191,303	114,738	593	63,669	36,853	1,531,147	425,307	18,670
	%	100	5.2	0.03	2.9	1.7	70.0	19.4	0.9
From this:									
Arable	ha	1,378,744	79,791	517	48,522	29,681	950,986	254,350	14,897
Vineyards	ha	20,155	524		751	740	14,970	2,995	175
Orchards	ha	17,211	1,147		508	306	10,328	4,448	474
Grassland	ha	761,652	32,963		13,832	6,003	549,410	156,474	2,970
Rented agricultural	ha	1,626,695	72,665	503	61,587	24,991	1,167,193	286,405	13,351
land	%	100	4.5	0.03	3.8	1.5	71.8	17.6	0.8
Owned agricultural	ha	45,383	34,806	65	27		10,485	-	-
land	%	100	76.7	0.1	0.06		23.1		

[a] The organisational form 'partnership' is unknown as a legally defined category in Slovak agriculture.
[b] Completely or partly state-funded agricultural enterprises can be found at universities, research institutions, hospitals or even as production units.

Source: Agricultural Cenzus 1994, Slovak Statistical Office

All types of farms shown in Table 7.7 rely to a large extent on rented land (average: 80.5 percent), but there are clear differences with regard to the share of rented/owned land. The co-operatives rent 76.5 percent of the land they cultivate, the limited companies 96.7 percent, the share holding companies 67.8 percent, the state farms 67.3 percent and the completely or partly state-funded enterprises 71.5 percent. Only the individual farmers not registered in business register have a lower percentage of rented land (63.6 percent), i.e. they cultivate more owned land (33.3 percent) than the other types of agricultural enterprises do. In case of the individual farmers registered in business register, however, the percentage of rented land is considerably higher (84.8 percent). The land cultivated by co-operative farms is predominantly owned by non-members, more than by members, and non-identified ownership also plays a major role (see, for details, Table 7.9).

Table 7.9
Ownership of land cultivated by co-operative farms, Slovakia, 1993

Type of ownership	*Percent of land in collective farms*
Members of co-operatives	28.2
Non-members of co-operatives	38.8
State-owned land	6.2
Church-owned land	1.7
Village communities	5.4
Non-identified	19.6
Total	100.0

Source: Special Survey, Ministry of Agriculture, Slovak Republic, 1994

Structural changes with regard to farm size

Since the process of transformation of agriculture has begun, farm sizes have changed rather modestly in Slovakia and not as drastically as in other CEE countries. Certainly, more medium-sized and smaller co-operatives arose, many small household plots and numerous individual farms were established (see Table 7.10), and also the internal structure of the enterprises has been rearranged. From 1989 to 1993 the number of agricultural production co-operatives increased from 630 to 952 due to subdivision, while the average size decreased from 2,667 to 1,788 ha (see Wolz, 1995, p. 28). However, in 1994 (see Table 7.11) still 12.4 percent of the number of farms and 93.5 percent of the agricultural land is in farms larger than 500 ha (in the private sector this applies to 10.1 percent of the farms and 93.1 percent of the land, in the public sector to 57.3 percent of the farms and 95.7 percent of the land). The private sector contains less large-scale farms than the public sector in relative terms, but more large-scale farms in absolute terms. However, the share of farms larger than 3,000 ha is only 1.2 percent in the private sector and 22.6 percent in the public sector. The share of the agricultural area within these two categories cultivated by the farms larger than 3,000 ha amounts to 23.0 percent in the private sector and 69.2 percent in the public sector. Obviously, predominantly large units are maintained in the domain of state-run agriculture, whereas the co-operatives have undergone a process towards smaller units, although they are still large-scale farms. Small farms – including very many small 'household plots' – still play a minor role in Slovakia: although farms up to 50 ha represent 83.8 percent (up to 20 ha: 77.0 percent) of the total number of farms in the private sector, only 2.2 percent (up to 20 ha: 1.1 percent) of the agricultural area is operated by them.

Table 7.10
Distribution of agricultural enterprises according to farm size, Slovakia, 1989-1993

Farm size (ha)	*1989*	*1990*	*1991*	*1992*	*1993*
up to 500	1	7	46	55	55
501 - 1,000	30	44	159	186	192
1,001 - 1,500	91	113	201	212	226
1,501 - 2,000	101	116	164	168	169
2,001 - 2,500	105	118	115	115	115
more than 2,500	302	282	213	210	195

Source: Special Survey, Ministry of Agriculture, Slovak Republic, 1994

Table 7.12 and 7.13 show how the number of farms and the agricultural area cultivated by the farms are distributed according to classes of farm size. (These data are not available for limited companies, share holding companies and completely or partially state-funded enterprises.) 91.5 percent of the number and 98.4 percent of the land of all co-operatives belong to farms larger than 500 ha, and in the case of state farms the corresponding figures are 65.3 percent and 96.9 percent respectively. In the class '3,000 ha and more' we find only 10.0 percent of the number of co-operatives (and 24.8 percent of the land cultivated by all co-operatives), but 27.0 percent of the number of state farms (and even 70.8 percent of the land cultivated by all state farms). This substantiates the result already mentioned above that, in relative terms, there are more large enterprises among the state farms than among the co-operative farms, although the opposite is true in absolute terms. Apparently, the state farms have been less affected by the transformation process. On average, state farms cultivate 2,016 ha and co-operatives 1,593 ha. Of course, the individual farms are much smaller, i.e. their average size is 15.2 ha for the non-registered farms (considering the household plots it would be much less) and 65.9 ha for the registered farms. In the classes below 50 ha (below 20 ha) we find 95.2 percent (87.6 percent) of the number and 55.6 percent (40.0 percent) of the land of the non-registered individual farms. In the case of the registered individual farms the distribution is very uneven, and information on this category seems to be less reliable, because only 9 farms are reported to be registered.

Livestock in the transformed agricultural enterprises

If we compare the organisational types of farms with regard to the number of livestock, also some differences can be pointed out (see Table 7.14). On average, state farms and co-operatives obviously hold the largest herds of cattle and cows, which are much smaller in the case of limited and share holding companies. However, the latter show high numbers of poultry per farm, in particular the share holding companies (71,298 on average) which also have the highest average number of pigs per farm. Sheep and rams are predominantly held in co-operatives, and for horses and foals state farms seem to be the typical form of organisation. The other farm types show rather small numbers of animals, above all the individual farms with only 3 cows, 8 pigs, 6 sheep and 88 pieces of poultry on average.

The structure of agricultural employment

The distribution of the agricultural labour force according to organisational forms and farm size is shown in Table 7.15 and 7.16. These figures need no detailed comments and reveal no surprises. The farms' manpower increases in the same way as the sequence of organisational forms is arranged in Table 7.15. The individual farms have the smallest number of workers per farm, the limited and particularly the share holding companies have considerably more labourers, and the co-operatives and state farms are endowed with the largest manpower capacity. However, all types of farms – in particular the legal entities – show a broad distribution regarding their labour force, i.e. there are also co-operatives and state farms with small groups of labourers as well as limited and share holding companies which employ many workers (see Table 7.15). Table 7.16 indicates the concentration of the agricultural labour force in Slovakia on large-scale farms. 89 percent can be found in farms larger than 500 ha and 30 percent on farms over 3,000 ha. At the other end of the scale, farms smaller than 50 ha only employ 2.1 percent and farms below 20 ha account for 1.4 percent. Table 7.16 also shows some regional differences in farm size and agricultural employment.

Table 7.11
Number of farms and agricultural area according to farm size in the private and in the public sector of Slovak agriculture in 1994

Interval - Area		*Number of farms*			*Agricultural area (ha)*		
		total	*private*	*public*	*total*	*private*	*public*
total	total	8,931	8,632	299	2,191,303	1,734,968	456,335
	%	100	100	100	100	100	100
0-1 ha	total	1,679	1,679	0	730	730	0
	%	18.9	19.4	0	0	0	0
1-2 ha	total	1,104	1,104	0	1,785	1,784	1
	%	12.4	12.7	0	0.1	0.1	0
2-5 ha	total	1,899	1,896	2	6,375	6,375	0
	%	21.3	21.9	0.7	0.3	0.4	0
5-10ha	total	1,127	1,125	2	8,183	8,183	0
	%	12.6	13.0	0.7	0.4	0.5	0
10-20 ha	total	873	869	4	12,153	12,100	53
	%	9.8	10.0	1.5	0.6	0.7	0.0
20-30 ha	total	343	336	7	8,483	8,316	167
	%	3.8	3.9	2,6	0.4	0.5	0.0
30-50 ha	total	267	249	18	10,565	9,869	696
	%	3	2.9	6.8	0.5	0.6	0.2
50-100 ha	total	218	203	15	15,361	14,296	1,065
	%	2.4	2.3	5.7	0.7	0.8	0.2
100-500 ha	total	314	250	64	79,824	62,299	17,525
	%	3.50	2.9	24.1	3.6	3.6	3.8
500-1000	total	298	258	40	225,112	196,917	28,195
ha	%	3.3	3.0	15.1	10.3	11.3	6.2
1000-2000	total	455	421	34	644,731	596,846	47,885
ha	%	5.1	4.9	12.8	29.4	34.4	10.5
2000-3000	total	194	176	18	463,901	419,089	44,812
ha	%	2.2	2.0	6.8	21.2	24.4	9.8
> 3000 ha	total	160	100	60	714,099	398,198	315,901
	%	1.8	1.2	22.6	32.6	23.0	69.2

Source: Special Survey, Ministry of Agriculture, Slovak Republic 1994

Table 7.12
Number of farms according to legal forms and farm size in 1994

Interval - Area		*All farms (private and public)*[a]	*Individual farmers not registered in business register*	*Individual farmers registered in business register*	*Co-operatives*	*State enterprises*	*Other organisational forms*[b]
Total	total	8,931	7,572	9	961	211	270
	%	100	100	100	100	100	100
0-1 ha	total	1,679	1,673	1	1	-	4
	%	18.9	22.1	11.1	0.1		2.2
1-2 ha	total	1,104	1,100	1	-	-	3
	%	12.40	14.5	11.1			1.7
2-5 ha	total	1,899	1,887	1	1	3	7
	%	21.3	24.9	11.1	0.1	1.4	3.9
5-10 ha	total	1,127	1,115	1	2	-	9
	%	12.6	14.7	11.1	0.2		5.1
10-20 ha	total	873	861	-	2	3	7
	%	9.8	11.4		0.2	1.4	3.9
20-30 ha	total	343	333	-	1	4	5
	%	3.80	4.4		0.1	1.9	2.8
30-50 ha	total	267	242	3	-	8	14
	%	3.0	3.2	33.3		3.8	7.9
50-100 ha	total	218	193	-	6	8	11
	%	2.4	2.5		0.6	3.8	6.2

100-500 ha	total	314	143	2	69	47	53
	%	3.5	1.9	22.2	7.2	22.3	29.8
500-1000 ha	total	298	18	-	223	33	24
	%	3.3	0.2		23.2	15.6	13.5
1000-2000 ha	total	455	6	-	392	30	27
	%	5.1	0.1		40.8	14.2	15.2
2000-3000 ha	total	194	-	-	168	18	8
	%	2.2	-		17.5	8.5	4.5
> 3000 ha	total	160	1	-	96	57	6
	%	1.8	0.1		10.0	27.0	3.4

[a] Only farms larger than 0.01 ha are included (not the household plots).

[b] The category refers to 149 'limited companies' (63,669 ha), 70 'share holding companies' (36,853 ha), and 51 'completely or partly state-funded enterprises' (18,670 ha) for which data on distribution according to farm size are not available.

Source: Special Survey, Ministry of Agriculture, Slovak Republic, 1994

Table 7.13

Agricultural area according to legal forms and farm size, Slovakia, 1994

Interval - Area		*All farms (private, public and household plots)*	*Individual farmers not registered in business register and plots*	*Individual farmers registered in business register*	*Co-operatives*	*State enterprises*	*Other organisational forms**
Total	ha	2,191,303	114,738	593	1,531,102	425,307	119,563
	%	100	100	100	100	100	100
	average			65.9	1,593	2,016	
0-1 ha	ha	730	726		1	-	2
	%	0	0.6	0.0	0.0		0
	average	0.4	0.4				
1-2 ha	ha	1,785	1,778		-	-	6
	%	0.1	1.5	0.3			0
	average	1.6	1.6				
2-5 ha	ha	6,375	6,324			12	37
	%	0.3	5.5	0.8	0.0	0.0	0.0
	average	3.4	3.4			3.9	
5-10 ha	ha	8,183	8082			-	99
	%	0.4	7.0	1.5	0.0		0.1
	average	7.3	7.3				
10-20 ha	ha	12,153	11,982	-		37	133
	%	0.6	10.4		0.0	0.0	0.1
	average	14	14			12	
20-30 ha	ha	8,483	1237	-		91	7154
	%	25	7.2		0.0	0,0	6.0

	average		25			23	
30-50 ha	ha	10,565	9,606	113	-	295	551
	%	0.5	8.4	19.0		0.1	0.5
	average	40	40	38		37	
50-100 ha	ha	15,361	13,520	-	454	560	827
	%	0.7	11.8		0.0	0.1	0.7
	average	70	70.05		76	70	
100-500 ha	ha	79,824	26,906	-	24,472	11,849	16,596
	%	3.6	23.4	78.3	1.6	2.8	13.9
	average	254	188.15		355	252	
500-1000 ha	ha	225,112	12,953	-	170,766	22,994	18,399
	%	10.3	11.3		11.2	5.4	15.4
	average	755	719.63		766	697	
1000-2000 ha	ha	644,731		-	556,194	43,335	45,202
	%	29.4	6.7		36.3	10.2	37.8
	average	1,417			1,419	1,444	
2000-3000 ha	ha	463,901	-	-	399,795	44,813	19,293
	%	21.2			26.1	10.5	16.1
	average	2,391			2,380	2,489	
> 3000 ha	ha	714,099		-	379,344	301,323	33,432
	%	32.6	6.0		24.8	70.8	28.0
	average	4,463			3,951	5,286	

* The category refers to 149 'limited companies' (63,669 ha), 70 'share holding companies' (36,853 ha), and 51 'completely or partly state-funded enterprises' (18,670 ha) for which data on the distribution according to farm size are not available.

Source: Special Survey, Ministry of Agriculture, Slovak Republic, 1994

Table 7.14
Number of livestock according to the legal form of farm enterprises, Slovakia, December 1994

	Total	*Individual farmers not registered in business register*	*Limited companies*	*Share holding companies*	*Co-operatives*	*State farms*	*Completely or partly state-funded enterprises*
				Total animals			
Cattle total	834,887	24,566	16,843	17,008	615,302	154,501	10,085
cows	324,388	9,786	6,755	6,736	236,801	61,768	4,062
Pigs total	1,693,383	62,919	43,854	140,093	1,158,869	286,943	7,997
sows	146,882	7,732	3,627	10,014	99,583	25,755	665
Sheep and rams	257,888	45,526	7,719	1,224	161,645	42,521	2,593
Horses and foals	5,082	879	166	69	1,800	2,504	78
Poultry	10,024,482	672,175	882,051	4,990,876	2,005,651	1,604,365	-
				Animals per farm			
Cattle total	82.6	3.2	113.0	242.9	640.3	732.2	197.7
cows	32.1	1.3	45.3	96.2	246.4	292.7	79.6
Pigs total	167.5	8.3	294.3	2,001.3	1,205.9	1,360.0	156.8
sows	14.5	1.0	24.3	143.1	103.6	122.1	13.0
Sheep and rams	25.5	6.0	51.8	17.5	168.2	201.5	50.8
Horses and foals	0.5	0.1	1.1	1.0	1.9	11.9	1.5
Poultry	991.7	88.8	5,919.8	71,298.2	2,087.0	7,603.6	-

Source: Special Survey, Ministry of Agriculture, Slovak Republic, 1994

Table 7.15

Number of agricultural enterprises according to the legal form and the number of workers, Slovakia, 1994

Area	*Number of agricultural enterprises according to number of workers*											
	1-5	*6-10*	*11-24*	*25-49*	*50-74*	*75-99*	*100-199*	*200-299*	*300-399*	*400-499*	*500-999*	*1000-1500*
Total	8,647	76	156	246	233	169	374	119	55	20	12	1
Private sector	8,592	67	123	197	205	154	332	93	34	7	5	-
Public sector	55	9	33	49	28	15	42	26	21	13	7	1
Individual farmers not registered in business register	8,499	36	49	12		1						1
Individual farmers registered in business register	9		1									
Limited companies	43	16	33	23	14	7	7	1				
Share holding companies	7	1	5	12	14	7	15	5	2		2	
Co-operatives	32	14	37	152	179	140	311	87	33	7	4	
State farms	45	7	23	38	18	8	38	25	18	13	5	1
Completely or partly state-funded enterprises Total	9	2	8	9	8	6	3	1	2		1	

Source: Special Survey, Ministry of Agriculture, Slovak Republic, 1995

Table 7.16
Number of full-time workers according to farm size, Slovakia, 1994

Area[a]	*Total*	*Farm size in ha (from 0 ha - 1 ha to 3000 ha and more)*												
		0 1	*1 2*	*2 5*	*5 10*	*10 20*	*20 30*	*30 50*	*50 100*	*100 500*	*500 1000*	*1000 2000*	*2000 3000*	*3000 and more*
							Number of farms							
SR total	10,108	1,679	1,104	1,899	1,127	873	343	267	218	314	298	455	194	160
WS	5,037	1,199	563	742	494	379	180	140	120	158	128	221	65	57
MS	3,192	326	377	790	385	334	106	84	64	95	76	99	58	58
ES	1,879	154	164	367	248	160	57	43	34	61	94	135	71	45
							Number of full-time workers[b]							
SR total	165,272	344	170	1,061	439	408	263	893	856	5614	16,196	48,179	33,358	49,830
WS	78,332			999	283	274	160	732	325	3355	7,689	26,686	12,709	20,353
MS	41,723	8	54	59	73	105	63	148	203	1363	3,586	8,252	9,591	16,221
ES	45,217			3	83	29	40	13	328	896	4,921	13,241	11,058	13,256
							%							
SR total	100	0.2	0.1	0.6	0.3	0.2	0.2	0.5	0.5	3.4	9.8	29.2	20.2	30.2
WS	100			1.3	0.4	0.3	0.2	0.9	0.4	4.3	9.8	34.1	16.2	26.0
MS	100	0	0.1	0.1	0.2	0.3	0.2	0.4	0.5	3.3	8.6	19.8	23.0	38.9
ES	100	0.1		0	0.2	0.1	0.1	0.0	0.7	2.0	10.9	29.3	24.5	29.3
							Average number of full-time workers							
SR total	16	0	0	1	0	0	1	3	4	18	54	106	172	311
WS	16			1	1	1	1	5	3	21	60	121	196	357
MS	13	0	0	0	0	0	1	2	3	14	47	83	165	280
ES	24			0	0	0	1	0	10	15	52	98	156	295

[a] SR = Slovak Republic; WS = Western Slovakia; MS = Middle Slovakia; ES = Eastern Slovakia [b] The term 'full time workers' includes all members of the agricultural labour force, i.e. also managers, employees and individual farmers, who work at least 180 hours per month.

Source: Special Survey, Ministry of Agriculture, Slovak Republic, 1994

Privatisation of nationalised land: state farms and church-owned land

Privatisation of state-owned farms has been delayed or even stopped, because the process of farm-land restitution has turned out to be more complicated than it was anticipated. Table 7.17 summarises the situation until June 30, 1994, regarding privatisation of state farms. The *Slovak Land Fund* was authorised to manage state-farmed land and forestries. This agency is also responsible for preparing and completing privatisation of agricultural enterprises. The procedures of state farm privatisation are slowed down by specific problems related to restitution of ownership. Land ownership is often not clearly defined, and there are many restitutional claims which should be taken in account before the privatisation projects are accepted. Furthermore, economic conditions (e.g., macro-economic environment, financial insolvency) prevent further acceleration of the privatisation process. A shortage of capital sources and credit has lowered the interest in state farm privatisation, operation and investments. Privatisation of state farms assumes different forms:

1 Sale of property in *public auctions* (creating small businesses with remaining assets around 100 million SLK and farming area up to 1,500 ha),

2 *Direct sales* of the enterprises, and

3 *Voucher privatisation* or the combined privatisation method (applied to approximately 20 state farms).

Table 7.17
Privatisation of state-owned farms and biological service enterprises, Slovakia, 1994

	Total	*State farms*	*Services**
Total enterprises	188	140	48
Not in privatisation	6	1	5
Intended for liquidation	5	4	1
Bankrupting	1	1	-
Privatised state farms	2	2	-

* Services = Biological Services Enterprises (e.g., for breeding, seed production)

Source: Special Survey, Ministry of Agriculture, Slovak Republic, 1994

Enterprises farming church-owned land and those with many unsettled restitutional claims will put their operations into long-term lease, which should be considered a pre-privatisation stage caused by a lack of financial resources.

Individual farms

In 1992, approximately 5,000, in 1993, approximately 7,000, and in 1994, approximately 7,500 individual farms existed, predominantly small farms (for farm structure, see Table 7.7-7.15). These figures indicate an increase in the average size of the family farms, which has grown from 10.5 ha in 1992 to 12.1 ha in 1993 and 15.2 ha in 1994. Other sources say that by January 1, 1994, the number of individual farmers had increased to 16,000 (see Agrar-informationsdienst Osteuropa, 1995, p. 3). They operated about 3.5 percent of the agricultural land in the Slovak Republic. More significant is the participation of the individual farms in agricultural production reflected in its share in agricultural production by different commodities. For example, individual farmers produce 10.6 percent of the cereals, 10.3 percent of the milk, and more than 50 percent of the potatoes, fruit and vegetables. 10 percent of the cattle, 18 percent of the pigs, 43.2 percent of the goats and sheep, and 53.7 percent of all poultry can be found in individual farms. Wolz (1995) reports that at the end of 1989 there were about 1,000 private farms, by the end of May 1993 their number has grown to about 16,000. The general pattern of restitution looks as follows: most persons who claim their land are entitled to small household or subsidiary plots. Only a small share out of these 16,000 private farms can be regarded as the nucleus of viable family farms. Just 3 percent of individual farmers cultivate more than 50 hectares for the time being. The large majority took their land for home gardening as a survival strategy following the sharp increase in unemployment rates of the rural industrial sector since 1989. The majority seemed to be poorly trained and not to possess adequate entrepreneurial or farming expertise. In this way they ensured the food supply of the family with simple production methods. No cash inputs are applied. In this respect, a kind of subsistence farming can be observed which enables this group to be better off compared to those who have become unemployed without the option of reclaiming farm land (Wolz, 1995, p. 26f.).

From the description of the institutional and organisational changes we can conclude that Slovak agriculture has certainly been subject to a whole variety of transformation procedures but still is similar to its pre-reform structure, as far as its organisational structure is concerned. This is at least partly due to the fact that the system of agricultural production co-operatives still prevails and can make use of various advantages:

1 Budget constraints are relatively soft for the co-operatives because deficits are covered by government subsidies or 'soft loan' policies. Almost one tenth of the state budget (more than the expenditure for social policies) serves this purpose.

2 The banks seem to accept increasing liabilities of co-operatives, because they feel obliged to follow the corresponding government policy. Furthermore, they are afraid that all valuable assets (cattle, machines, other equipment, inputs like fertiliser, etc.) will disappear from farms overnight if they plan and announce to have them liquidated.

3 Many rules of dealing and business conventions which are regarded as a matter of course in market economies seem not to be valid yet. For example, if a co-operative lacks liquidity to pay rents it simply does not pay for the land, or the lessors have to settle with payment in kind, i.e. they are offered to be compensated with farm products.

4 Some managers of co-operatives try successfully to get rid of their debts by forming a new limited company to which they transfer all assets as well as the rented land from the indebted co-operative. After this sort of rearrangement has been completed, the 'old' co-operative may go bankrupt together with its debts.

Conclusions: feasibility problems slowing down the process of privatisation and decollectivisation

If we try to summarise the information on the restructuring process of Slovak agriculture outlined above, we can clearly conclude that fundamental institutional and structural changes have taken place. Although, the new structures and organisational forms already differ from the old ones, there are also certain problems which counteract further progress towards a system of viable farms (see also Lindemans, 1997; Michalek, 1995). This raises the question to what extent the feasibility of transformation is restricted by crucial factors in Slovakia, and whether these determinants of limited feasibility have led to a particular path of restructuring, if we look at this process from an evolutionary point of view. We will try to give a few answers to these question by means of the following interpretations which are related to the theoretical approach in the third section:

1 Whether or not the transformation process in agriculture will proceed quickly, crucially depends on the political coalition dominating in the first

years. If reform-minded politicians and new (self-employed) land owners form the strongest coalition, the restructuring process will be supported, but if anti-reform politicians and the successors of the former collective (or state) farms form the leading coalition, conservation of traditional structures may become a major goal of agricultural policies.

2 Which of these two coalitions is more likely to be successful, not only depends on political factors, but also on economic circumstances which prevail in the initial situation. If the economic and structural preconditions for agriculture to be privatised and decollectivised are favourable in the initial situation, the reform-minded coalition may be able to form faster than the anti-reform coalition, and vice versa.

3 In Slovakia, the legal pre-requisites for restitution of private land ownership and transformation procedures in collective farms were well-established. However, this may only insufficiently reflect the attitudes of the Slovak government, because the laws on privatisation and transformation originate from political decisions made by the still more reform-oriented Czechoslovak government. The above-mentioned facts that, first, the so-called Sanction Law has not been approved in Slovakia and, secondly, the claims on assets have considerably been devaluated by an amendment to the Transformation Law, suggest that the basic legal framework for restructuring agriculture would have been less consistent and reform-oriented in Slovakia if the two countries already had been separated in the starting situation of transformation.

4 In contrast, the economic conditions for privatisation and, in particular, for decollectivisation of the collective farms have been less favourable in the initial situation. This is primarily due to the high fragmentation of land ownership in Slovakia pointed out earlier. It is difficult to establish a viable farm when starting with only 2-5 ha of owned land. In the Czech Republic, where the average area of owned land is about 30-50 ha, land owners are more encouraged to use their property as a basis for establishing an agricultural enterprise.

5 In addition, the extremely high degree of land fragmentation also leads to many problems regarding identification of original owners, because there are so many of them (about three million in Slovakia), and usually each of them owns not only one, but several small plots. In Slovakia, there is a lack of government assistance for agencies involved in clarification of property rights, e.g., cadastre offices, which expect to be more strengthened by financial and technical support.

6 High fragmentation and slow clarification of land ownership may be one reason why a land market is missing or failing. Only if the market for land, in particular for leasing and renting land, is working properly, we can expect that competition for land and transfer of land between farms with different organisational forms will take place. Furthermore, it is difficult for new farmers to enlarge their holdings, when insufficient opportunities for renting land are offered due to deficiencies on the land market like regional monopsonies represented by large farms.

7 The emergence of land markets is also hampered by the fact that it is not allowed to use agricultural land for non-agricultural purposes. Slovakia has 0.46 ha of agricultural land and 0.28 ha of arable land per inhabitant. As a result of the formerly adopted political goal of self-sufficiency, this land is almost completely utilised for production purposes, despite the low and/or negative returns. Approximately 30 percent of agricultural land in Slovakia can be classified as submarginal (if used for intensive production), and the return on inputs is less than 70 percent (see Table 7.18). However, this land, in many cases, has high opportunity cost in alternative use, e.g., recreational and sporting facilities, development for housing, and its price on the land market would be much higher than the administratively set one. Selling and buying of such land could be a significant impetus for land markets to come into being. Present legislation does not support this view, and alternative use of land is still difficult to justify.

Table 7.18
Returns on inputs of agricultural land in the nine lowest price clusters*, Slovakia, 1994

Price cluster	*1*	*2*	*3*	*4*	*5*	*6*	*7*	*8*	*9*
Return in % of input	58.39	59.89	64.98	63.99	73.33	60.14	68.26	72.41	69.91

* These price clusters represent administrative prices, no observed market prices.

Source: Slovak Statistical Office, 1994

8 A lack of marketing structures and particularly of marketing channels for the new private farms represent another problem which still has to be overcome in the transformation process. Producers' marketing co-

operatives could not only serve the small farms, but also the larger production co-operatives, because these are usually also to small to establish a reasonable marketing organisation only for their own products.

9 Lack of capital and credit is probably one of the most important barriers which discourages individuals and companies who are interested in privatisation projects. State subsidies provided for restructuring and refurbishing agricultural units are usually tailored for the needs of the existing large-scale farms. Long-term credit at reasonable interest rates is not available at all.[3] The negative impact of capital shortage on the land markets needs no further explanation: Many farmers who want to buy or only to rent land need additional capital.

10 Admittedly, government could help to remove this obstacle against further progress in restructuring agriculture by creating adequate investment support programmes in which all legal forms and organisational types of agricultural enterprises could participate and were offered equal opportunities. This would enable new or re-established farms to be successful, so that they could develop and compete with the traditional collective and state farms. As a consequence, a broader variety of organisational forms and sizes of farms would probably come into being, which might be a better basis for an evolution towards a viable and flexible agricultural system.

11 However, such a reform-oriented strategy is in sharp contradiction with the political economy governing the relations between the dominating forces in agriculture and politics. As we already mentioned, the former socialist system had developed an ideology by which agriculture was attributed a fundamental role in Slovak society. This special variant of agrarian fundamentalism was popularised and became an 'informal institution' incorporated in the 'human capital' of the people, and therefore does not change very quickly. As a consequence, the demand to protect the agricultural producers is still popular, and 'agricultural producers' are assumed to be identical with the prevailing 'co-operative farms'. For example, self-sufficiency in food production and demographic stability in rural areas are official explanations to justify agricultural policies.

12 After 1989, these notions of agrarian fundamentalism apparently have been confused, to some extent, with the emotional basis of the new national movement in Slovakia. The unfavourable situation in the agricultural sector and the dissatisfaction with this situation have had a considerable political impact. Since 1993, agriculture has been declared one of the

priorities of the Slovak economy (see Ministry of Agriculture, 1993). This particular treatment can be considered as a significant success of the agricultural lobby based on ties with the left-wing parties which have found an agenda for their political activities here. For example, the 'Sanction Law' was rejected by the argument it could endanger the stability of co-operative farms. New investment programmes also predominantly aim at the existing large farm units.

13 This orientation towards conservation of traditional structures is reinforced by the traditional relations between the directors of the large-scale co-operative and state farms and the Ministry of Agriculture. Since Slovakia is a relatively small country, the ministry is responsible for implementation of policy measures, for instance, it distributes subsidies without any intermediate bureaucracy. In this way, politicians and administrators can directly and effectively exert influence on the agricultural sector. It would be more difficult to deal with a broader variety of farm types than with only a few large state or co-operative farms.

From these observations and interpretations we can conclude that an anti-reform coalition has succeeded to dominate agricultural policies in Slovakia. This coalition has started to shape the relationships between the political and economic actors in the agrarian network and supports the objective that the traditional agricultural system is supposed to survive in a modified way. However, the question arises whether this will be possible. Prospectively, there will be a discrepancy between political desires and economic necessities, and the question of the feasibility of the restructuring process will arise again in the future.

Notes

1 The term 'subsidy' had never been used as a term expressing financial transfer from taxpayers to farmers.

2 It should be mentioned that some agricultural production co-operatives also transformed into non-agricultural firms like a window factory, for example, and only continue farming as a second, less important activity. In Slovakia, many agricultural co-operatives traditionally had non-agricultural branches, e.g., in the building line.

3 In addition, the problem of the 'old debts' originating from pre-reform investments is unsolved in Slovakia.

References

Act of the National Council of the Slovak Republic. Corpus Juris, No. 42/1992.

Agrarinformationsdienst Osteuropa (1995), 'Die Slowakische Agrar- und Ernährungs-wirtschaft im Jahre 1994', pp. 2-4, January.

Agricultural Cenzus 1994, Slovak Statistical Office: Bratislava.

Blaas, G. (1995), 'Inlandsorientierung der Slovakischen Landwirtschaft', *Agrarische Rundschau*, No. 5, pp. 6-10.

Bromley, D.W. (1996), 'The Social Construction of Land', in Hagedorn, K. (ed.), *Institutioneller Wandel und Politische Ökonomie von Landwirtschaft und Agrarpolitik. Festschrift zum fünfundsechzigsten Geburtstag von Prof. Dr. Günther Schmitt*, Campus: Frankfurt am Main.

Christiaensen, L. and Swinnen, J.F.M. (1994), 'Economic, Institutional and Political Determinants of Agricultural Production Structures in Western Europe', *Working Paper*, No. 11, Department of Agricultural Economics, K.U.Leuven: Leuven.

Hagedorn, K. (1991a), 'Konzeptionelle Überlegungen zur Transformation der Landwirtschaft in den neuen Bundesländern', in Merl, S. and Schinke, E. (eds), *Agrarwirtschaft und Agrarpolitik in der ehemaligen DDR im Umbruch. Giessener Abhandlungen zur Agrar- und Wirtschaftsforschung des Europäischen Ostens, Bd. 178,* Duncker and Humblot: Berlin.

Hagedorn, K. (1991b), 'Gedanken zur Transformation einer sozialistischen Agrarverfassung', *Agrarwirtschaft*, Vol. 40, No. 5, pp. 138-148.

Hagedorn, K. (1991c), 'Public Choice and Agricultural Policy. The Case of the CAP', in Sen, A. (ed.), *Issues in Contemporary Economics. Proceedings of the Ninth World Congress of the International Economic Association*, Athens, Greece. Vol. 3: Dasgupta, P. (ed.): Policy and Development, Houndsmill, Basingstoke, Hampshire and London: Macmillan in association with the International Economic Association.

Hagedorn, K. (1992a), 'Wirtschaftliche und politische Triebkräfte der Umformung einer sozialistischen Agrarverfassung', in Boettcher, E., Herder-Dorneich, P., Schenk, K.-E. and Schmidtchen, D. (eds), *Ökonomische Systeme und ihre Dynamik. Jahrbuch für Neue Politische Ökonomie, Bd. 11*, J.C.B. Mohr (Paul Siebeck): Tübingen.

Hagedorn, K. (1992b), 'Transformation of Socialist Agricultural Systems', *Journal of International and Comparative Economics*, Vol. 1, pp. 103-124.

Hagedorn, K. (1993a), 'Transforming Socialist Agriculture: An Institutional Perspective', in Kabat, L. Fandel, P. Batkova, D. Bartova, L. and Kapustik, P. (eds), *Eastern European Agriculture - Problems, Goals and Perspectives. Proceedings of the 27th Seminar of the European Association of*

Agricultural Economists, March 18-20, 1993 in Nitra, Vysoke Tatry, Czechoslovakia.

Hagedorn, K. (1993b), 'Institutions and Agricultural Economics', *Journal of Economic Issues*, Vol. 27, No. 3, pp. 849-886.

Hagedorn, K. (1994a), 'Changing Organisation of Agriculture as a Result of the Transformation Process', paper presented to the COST workshop on The Impact of CEE Agricultural Reform on Domestic and International Markets at The Leuven Institute for Central and East European Studies, Leuven, Belgium, June 10-11, 1994.

Hagedorn, K. (1994b), 'Interest Groups', in Hodgson, G., Tool, M. and Samuels, W.J. (eds), *Handbook of Institutional and Evolutionary Economics*, Edward Elgar: Cheltenham.

Hagedorn, K. (1996a), *Das Institutionenproblem in der agrarökonomischen Politikforschung. Schriften zur Angewandten Wirtschaftsforschung, Bd. 72*, J.C.B. Mohr (Paul Siebeck): Tübingen.

Hagedorn, K. (1996b), 'Politisch dirigierter institutioneller Wandel: Das Beispiel der Privatisierung volkseigenen Bodens', in Hagedorn, K. (ed.), *Institutioneller Wandel und Politische Ökonomie von Landwirtschaft und Agrarpolitik. Festschrift zum 65. Geburtstag von Prof. Dr. Günther Schmitt*, Campus: Frankfurt am Main.

Hagedorn, K. (1996c), 'Concepts of Institutional Change for Understanding Privatisation and Restructuring of Agriculture in Central and East European Countries', in Institut für Agrarentwicklung in Mittel- und Osteuropa (ed.), *The Importance of Institutions for the Transition in Central and Eastern Europe - with Emphasis on Agriculture and Food Industry*, Proceedings of a seminar in Halle (Saale), March, 18-19, forthcoming.

Hagedorn, K. (1997), 'The Politics and Policies of Privatization of Nationalized Land in Eastern Germany', in Swinnen, J.F.M. (ed.), *Political Economy of Agrarian Reform in Central and Eastern Europe*, Ashgate: Aldershot.

Hagedorn, K. and G. Schmitt (1985), 'Die politischen Gründe für eine wirtschaftspolitische Vorzugsbehandlung der Landwirtschaft', in Boettcher, E. Herder-Dorneich, P. and Schenk, K.-E. (eds), *Jahrbuch für Neue Politische Ökonomie, Bd. 4*, J.C.B. Mohr (Paul Siebeck): Tübingen.

Kabat, L. (1994), 'Last Economic and Social Development in Slovakia', paper presented at the COST workshop on The Impact of CEE Agricultural Reform on Domestic and International Markets at The Leuven Institute for Central and East European Studies, Leuven, Belgium, June 10-11, 1994.

Kabat, L. (1995), 'Slovak Agriculture in Transition', unpublished working paper, Department of Statistics, University of Agriculture: Nitra.

Leipold, H. (1985), 'Ordnungspolitische Implikationen der Transaktionskostenökonomie', *ORDO - Jahrbuch für die Ordnung von Wirtschaft und Gesellschaft*, Vol. 36 , pp. 31-50.

Leipold, H. (1988), *Wirtschafts- und Gesellschaftssysteme im Vergleich*, Gustav Fischer: Stuttgart.

Leipold, H. (1989a), 'Das Ordnungsproblem in der ökonomischen Institutionentheorie', *ORDO - Jahrbuch für die Ordnung von Wirtschaft und Gesellschaft*, Vol. 40, pp. 129-146.

Leipold, H. (1989b), 'Neuere Ansätze zur Weiterentwicklung der Ordnungstheorie', in Boettcher, E. Herder-Dorneich, P. and Schenk, K.-E. (eds), *Jahrbuch für Neue Politische Ökonomie, Bd. 8*, J.C.B. Mohr (Paul Siebeck): Tübingen.

Lindemans, I. (1997), 'Process and Politics of Agricultural Privatization in the Czech and Slovak Republics', in Swinnen, J.F.M. (ed.), *Political Economy of Agrarian Reform in Central and Eastern Europe*, Ashgate: Aldershot.

Michalek, J. (1995), 'An Application of the Policy Analysis Matrix for an Evaluation of Agricultural Policies in the Slovak Republic' in *Proceedings of the 41st Seminar of the European Association of Agricultural Economists on Challange and Strategies for Re-establishing East-Central European Agricultures*, Gödöllö, Hungary, September 6-8, 1995.

Ministry of Agriculture of the Slovak Republic (1993), *Concept and Principles of Agrarian Policy*, Bratislava.

Ministry of Agriculture of the Slovak Republic (1994), *Special Survey*, Bratislava.

Ministry of Agriculture of the Slovak Republic (1995), *Zelena sprava ('Green Report') 1995*, Bratislava.

Pryor, F.L. (1992), *The Red and the Green. The Rise and Fall of Collectivized Agriculture in Marxist Regimes*, Princeton University Press: Princeton.

Schenk, K.-E. (1983), 'Institutional Choice und Ordnungstheorie', in Boettcher, E. Herder-Dorneich, P. and Schenk, K.-E. (eds), *Jahrbuch für Neue Politische Ökonomie, Bd. 2*, J.C.B. Mohr (Paul Siebeck): Tübingen.

Schmidtchen, D. (1989), 'Evolutionäre Ordnungstheorie oder: Die Transaktionskosten und das Unternehmertum', *ORDO - Zeitschrift für die Ordnung in Wirtschaft und Gesellschaft*, Vol. 40, pp. 161-182.

Schmitt, G. (1990a), 'Können sich die Landwirtschaftlichen Produktionsgenossenschaften (LPG) der DDR in der Europäischen Gemeinschaft als wettbewerbsfähig erweisen?', *Agra-Europe*, Vol. 31, No. 24, Länderberichte, Sonderbeilage, pp. 1-14.

Schmitt, G. (1990b), 'Why Collectivisation of Agriculture in Socialistic Countries Really Has Failed: A Transaction Cost Approach', paper presented at the Ninth Conference on USSR and East European Agriculture, September 11-14, 1990.

Schüller, A. (1988), 'Ökonomik der Eigentumsrechte in ordnungstheoretischer Sicht', in Cassel, D., Ramb, B.-T. and Thieme, H.-J. (eds), *Ordnungspolitik*, Vahlen: München.

Statistical Proceedings. Slovak Statistical Office: Data on Slovakian Agriculture. Bratislava, several years.

Swinnen, J.F.M. (1993a), 'The Development of Agricultural Policies in Central and Eastern Europe: An Endogenous Policy Theory Perspective', *Food Policy*, Vol. 18, pp. 187-191.

Swinnen, J.F.M. (1993b), 'The Politics and Economics of Land Reform and Privatization in Central and Eastern European Agriculture', in Buckwell, A. and Swinnen, J. (eds), *Policy and Institutional Change in Central European Agriculture and Its Likely Impacts on Domestic and International Markets*, The Leuven Institute for Central and East European Studies: Leuven.

Swinnen, J.F.M. (1994), 'Overview of Policy and Institutional Reform in Central European Agriculture', in Swinnen, J.F.M. (ed.), *Policy and Institutional Reform in Central European Agriculture*, Avebury: Aldershot.

Swinnen, J.F.M. (1997a), 'The Choice of Privatization and Decollectivization Policies in Central and Eastern European Agriculture: Observations and Political Economy Hypotheses', in Swinnen, J.F.M. (ed.), *Political Economy of Agrarian Reform in Central and Eastern Europe*, Ashgate: Aldershot.

Swinnen, J.F.M. (1997b), 'Does Compensation for Disruptions Stimulate Reforms? The Case of Agrarian Reform in Central and Eastern Europe', *European Review of Agricultural Economics*, Vol. 24, No. 2, pp. 249-266.

Swinnen, J.F.M. and van der Zee, F. (1993), 'The Political Economy of Agricultural Policies. A Survey', *European Review of Agricultural Economics*, Vol. 20, No. 3, pp. 261-290.

Transformation Law of the Slovak Republic (Law 42/1992), Bratislava.

Wegehenkel, L. (1981), *Gleichgewicht, Transaktionskosten und Evolution. Eine Analyse der Koordinierungseffizienz unterschiedlicher Wirtschaftssysteme*, J.C.B. Mohr (Paul Siebeck): Tübingen.

Wolz, A. (1995), 'The Transformation of Institutions and Organisations in Agricultural Production in Slovakia', *Diskussionsschriften der Forschungsstelle für Internationale Agrar- und Wirtschaftsentwicklung e.V. Heidelberg*, No. 45., Universität Heidelberg - Lehrstuhl für Internationale Wirtschafts- und Entwicklungspolitik: Heidelberg.

8 Agricultural privatisation and farm restructuring in Slovenia

Štefan Bojnec and Johan F.M. Swinnen

Introduction

This chapter studies the process of agricultural privatisation and farm restructuring in Slovenia. In the pre-Communist period the Slovenian farm structure was rather egalitarian and was dominated by small and medium sized farms. Under Communism, the process of collectivisation failed and small private family farms remained the dominant organisational structure in Slovenian agriculture. However, the Communist regime imposed institutional limitations on size and organisational changes of the peasant farms as well as on their economic activities. As in Poland and in the rest of former Yugoslavia, a bimodal farm structure of small-scale, traditional private peasants and large-scale state farms (in former Yugoslavia usually called 'social farms'[1] resulted under the Slovenian Communist regime. The state farms mostly used capital-intensive technologies and were in large-scale production units. More labour-intensive agricultural production was in small-scale family farms, which made up the major share of output and factor use in agriculture. Several institutional regulations and economic policies discriminated against the private sector, causing a more part-time orientation of private farms (Bojnec, 1994).

Under the recent agricultural privatisation and land reform programme, land used by the state (socially owned) farms and forestry enterprises was restituted to former owners. As part of this process the land is transferred to the Fund of Agricultural Land and Forests (FALF), which manages the land during transition. Furthermore, institutional restrictions on the size of private farms and on land and other factor markets have been largely removed. In this chapter we first discuss the land reform and farm restructuring under Communism. Then, we present the pre-transition farm structure and discuss the institutional basis for agricultural privatisation and land reform. The next section analyses the progress in privatisation and farm restructuring and factors affecting these processes. We emphasise that legal amendments to the land restitution law have mitigated its impact on the restitution of property rights and farm restructuring. The last section draws some conclusions and implications for future farm restructuring.

Land reform and farm restructuring under Communism

Small-scale family farming was widespread in Slovenia in the pre-Communist period. Large-scale farming was imposed in the first years after World War II for political reasons (socialisation of the countryside) and economic objectives (increasing agricultural production through capital intensive large-scale agriculture). Between 1945 and 1953 large-scale capital intensive state farms were created and the formation of collective farms with the land of small peasants was promoted as the key to agricultural development. Collectivisation and heavy taxation of private peasant farms were used to stimulate the 'socialisation' of the countryside and to extract resources for industrial development.

The 1945 Law on Agricultural Reform and Colonisation allowed nationalisation and confiscation of agricultural land, forests and farms as part of the 'expropriation of expropriators'. The main rule of the agricultural reform was that 'land belonged to its cultivator'. It was forbidden to sell or buy land, forests, and farms or to burden them with mortgages. Such activities were referred to as 'inadmissible speculation and economic sabotage'. Land, forests and farms of large landowners, banks, private companies and land and forests over 10 hectares owned by churches and other religious institutions were expropriated without compensation.[2] Most properties belonging to expelled minorities were confiscated without compensation. An exception were those minorities who were not involved in hostilities during World War II.

Private family farms could own between 20 and 35 hectares of cultivated land and 10 to 25 hectares of forests, or up to 45 hectares of total land on poor soil. Non-farm households could own up to 3 hectares of cultivated land or up to 5 hectares of forest land.[3] Beyond these upper limits land of private peasants farms was expropriated, and only symbolic compensation was paid.

The expropriated forests became state property. The agricultural land was mostly transferred to a state agricultural fund, and was later distributed either to small-holdings up to 5 hectares of cultivated land per household, or to collective farms and state farms. The distribution of nationalised land contributed to land fragmentation because land from (consolidated) large landholdings was broken up and distributed in small parcels. Nationalised tractors and large agricultural equipment were transferred to agricultural mechanisation stations or to state farms.

Private peasants strongly opposed collectivisation. In fact, agricultural collectivisation failed because of this opposition and because of the poor performance of collective farms. As a result of these economic and political problems, collectivisation was abandoned in 1953. However, this did not imply a full liberalisation of regulations on private agriculture. The 1953 Law on Agricultural Property and on Distribution of Land to Agricultural Enterprises expropriated ownership and user rights on peasants' land above 10 hectares of cultivated land per farm or above 15 hectares of land on poor soil. The expropriated land became state property and was distributed to the state farms.

The 1959 Law on Agricultural Land Use reduced limitations on trade and leasing of privately owned agricultural land. However, the state sector had preference over private farms to buy or lease agricultural land from the private sector. The land policy objective was to transfer land from private to state ownership. The leasing contract on land was fixed for at least 25 years for vineyards, orchards and hop gardens, for at least 15 years for plantations, and for at least 10 years in all other cases.

By the 1973 Law on Agricultural Land, a private farmer could own and rent 10 hectares of cultivated land in 'lowland'[4] districts or 20 hectares in hilly and mountain districts. The non-farm household could own and rent 1 hectare of agricultural land and forest land in lowlands and 3 hectares in hilly and mountain districts.

These imposed maxima on land ownership and use were included in the constitution in 1963, and induced a fragmentation of peasant farms. Peasants' farm land was further split into separate small holdings and smaller landplots as the owner died and the land was divided among the heirs. In 1973 the Law on Agricultural Inheritance specified that a farm could register as a so-called 'protected farm', with its heir, in which case it could also benefit from state support to agriculture. The subdivision of these 'protected' peasant farms was officially prohibited. Only heirs of a farm could legally inherit in kind the property of the 'protected' farm. In December 1995 this law was replaced by the Law on Inheritance of Agricultural Households, which specified that a 'protected farm' has a size between 5 and 100 hectares of field equivalents, as registered in the land cadastre. This law may prevent further ownership fragmentation of such farms, because only the heir of the farm who will continue to operate the farm can inherit in kind the property of the farm.

The pre-transition farm structure

A dual farm structure: large-scale state farms and small-scale private farms

As a result, between the mid 1950s and the beginning the 1990s, a bi-polar farm structure developed: large-scale state farms and small-scale private farms.

Private farms produced the major share of agricultural output and used most of its inputs. In 1989, private farms produced 65.7 percent of total agricultural output and used 82.6 percent of total agricultural land. Important differences exist(ed) between private and state farms in size, location, product specialisation, technology use and the government regulation and policies that affected them.

One should be careful in interpreting the concept 'private farm'. The statistical definition of 'farm' includes households with some agricultural activities, which would typically not be incorporated in Western statistics. They include farms with (a) either at least ten ares of arable land, or (b) less than ten ares of arable land but being the owner of at least (i) a cow and a calf, or a cow and a heifer; or (ii) a cow and two beef cattle; or (iii) three grown-up pigs; or (iv) four grown-up sheep and pigs; or (v) fifty grown-up poultry; or (vi) twenty beehives. In addition, many of the 'private farms' are part-time businesses (Bojnec, 1994). Private farms outnumber state farms: in 1991, 156,549 private farms were registered compared to only 189 state farms (Table 8.1 and 8.2). Private farms used almost 1.4 million hectares of total land and more than 100,000 people were recorded as being 'employed' on them. State farms used only 100,000 hectares of agricultural land and employed 14,600 people. There are also state forestry organisations with about 373,100 hectares of forest land (Table 8.3).

Almost 30 percent of private farms had less than 1 hectare of total land. The other 70 percent used on average 4.1 hectares of agricultural land and 7.8 hectares of total land. Around 70 percent of the private farms with more than 1 hectare had less than 5 hectares of agricultural land and only 8.6 percent of those private farms had more than 10 hectares of agricultural land (Kovacic, 1994). Notice that the data in Table 8.1 indicate that in 1991 more than half of total privately owned land was in private farms above 10 hectares. The reason is that in hilly and in mountain districts a significant share of forest land is owned by private farms bigger than 10 hectares. The average size of private farms in hilly and mountain districts exceeded the size of those in lowland regions because the size of land per private farm in lowland regions was much more restricted during Communism. As in Poland (OECD, 1995), the Slovene private farm structure has been rather stagnant or even worsening (that is, getting more fragmented) after World War II (see Table 8.4).

Part-time farming is very important in the private sector. In 1991, only around one-fifth of private farms were full-time farms. In addition, a significant share of employment on farms are elderly and retired farmers (Kovacic, 1994). In contrast, the average size of state farms (including farms of agricultural schools and pig and poultry farms with relatively little land) in 1989 was 262 hectares. The average size of state farms bigger than

100 hectares was 813 hectares. In addition, 21 state farms were bigger than 1000 hectares (Bojnec, 1994).

Table 8.1
Pre-transition bi-modal farm structure in Slovenia, 1989 and 1991

Farm size (ha)	*Number of farms*	*Share of farms (%)*	*Total utilised land* ('000 ha)*	*Share of land (%)*
		Private farms in 1991		
All land				
< 1 ha	44,428	28.4	21.1	2.3
1 - 5 ha	56,327	36.0	156.4	17.0
5 - 10 ha	28,112	17.9	207.5	22.6
10 - 20 ha	19,637	12.6	277.3	30.1
> 20 ha	8,045	5.1	257.6	28.0
Total	156,549	100.0	919.9	100.0
Agricultural land				
< 1 ha	58,782	37.5	26.1	5.2
1-5 ha	65,191	41.6	177.6	35.4
5-10 ha	24,367	15.6	175.3	35.0
10-20 ha	7,279	4.7	96.2	19.2
> 20 ha	930	0.6	26.0	5.2
Total	156,549	100.0	501.1	100.0
		State farms in 1989		
Agricultural land				
< 50 ha	149	63.1	0.7	1.2
50-500 ha	57	24.2	10.9	17.6
500-1000 ha	9	3.8	6.9	11.1
> 1000 ha	21	8.9	43.4	70.1
Total	236	100.0	61.9	100.0

* Utilised land includes both agricultural and forest land. Data on land area were collected on the base of census questionnaires in which land owners were asked the size of the land they owned. A significant part of the land was not included in the census data (this is also the reason for differences in the data of Tables 8.1, 8.2 and 8.9).

Source: Bojnec (1994) and Statistical Office of Slovenia

There are important regional differences in the share of private versus state farm land use. About half of Slovenia's land is covered with forests, mostly in hilly and mountainous regions. In these districts, private farming is the rule and milk, cattle breeding and forestry are the most important farm activities. State farms mostly operated on the best agricultural land. Only 37 percent of private farms was on the better lowland farm land and 63 percent of private farms was located in disadvantaged agricultural areas, mostly hilly and mountainous regions (MKG, 1994).

Table 8.2
Operation and ownership structures of the land area and employment of farms in Slovenia, 1989-1993

Operation structures	*Number of farms*			*Land area (in '000 ha)*			*Total employment*		
	1989	*1991*	*1993*	*1989*	*1991*	*1993*	*1989*	*1991*	*1993*
State farms	236	189	203	87	99	103	16,137	14,602	
State forestry organisations				541	530	525	12,263		
Family farms	...	156,549	...	1,397	1,396	1,398	...	101,407	...
Private firms	-	20	29	-	-	-	-	44	70
Total				2,025	2,025	2,025			

Source: Statistical Office of the Republic of Slovenia.

Table 8.3
Land use in 1991 Slovenia ('000 ha)

	Private sector		*State sector*	*Total*
	Utilised land[a]	*Land*[b]	*land*[b]	*land*[b]
Agricultural land	501.1	720.8	145.1	865.8
of which:				
Arable land	173.9	214.9	31.2	246.1
Orchards	12.4	33.1	3.1	36.2
Vineyards	13.9	17.8	3.9	21.6
Meadows	190.5	311.4	36.2	347.6
Pastures, reed-lands and moors	110.4	143.6	70.7	214.3
Forest land	418.8	643.0	373.1	1,016.1
Total land	919.9	1,363.8	518.2	1,882.0

[a] On the base of census questionnaires.
[b] On the base of cadastral data.

Source: SY SLO 1992 (1993), p. 206 and Results of surveys, 1994/617, p.362

Table 8.4
Number and size of private farms by land ownership in Slovenia, 1960-1991*

	1960		*1969*		*1981*		*1991*	
	Number	*%*	*Number*	*%*	*Number*	*%*	*Number*	*%*
< 1 ha	40,657	20.9	37,903	21.0	62,467	32.5	44,428	28.4
1- 5 ha	73,417	37.6	70,017	38.8	65,395	34.1	56,327	36.0
5-10 ha	39,130	20.1	36,306	20.2	32,746	17.0	28,112	17.9
> 10 ha	41,651	21.4	36,002	20.0	31,482	16.4	27,682	17.7
Total	194,855	100.0	180,234	100.0	192,090	100.0	156,549	100.0

* All land

Source: Bojnec, 1994

The performance of the state versus the private sector was also significantly determined by the choice of technology. Large-scale and capital intensive farming was the rule in the state sector, with an emphasis on vineyards, large-scale crop production and large-scale pig and poultry farms. The private sector produced more labour-intensive products, such as potatoes, milk and beef.

However, with the exception of land use, many Slovenian private farms are relatively capital intensive. During the 1970s and the 1980s, mechanisation substituted for the reduction in agricultural labour on private farms. The number of tractors on private farms rapidly increased in the 1970s (see Table 8.5). Several factors contributed to the mechanisation process in the private sector, including the relaxation of conditions for purchasing tractors since the 1970s and subsidised credit through the agricultural co-operative credit system. Many peasants also temporarily worked in Western Europe and invested their savings in the mechanisation of their farm. In 1991, practically all private farms larger than one hectare had a tractor and some of them had more than one tractor. During the 1990s private peasants were still relatively well mechanised, despite a significant cut in credit subsidies.

Fertiliser use per hectare was three to four times higher on state farms. Yields on state farms were also considerably higher. For example, wheat yields were 30 to 80 percent above the yields on private farms. The yields are high, even in comparison with Western yield levels, but as indicated above, sustained by very high variable-input use. Table 8.5 also indicates that fertiliser use has fallen by half between 1989 and 1993, both on private and on state farms. However, the official statistics suggest that the impact on wheat yields has been very limited, as wheat yields have increased.

Finally, government measures affecting input and output prices had an important impact on the 'profitability' of farming. Prices were higher and more stable for the state farms because of their better bargaining position with the vertically integrated state food processing units. Prices were often set at lower levels for private farms. Price discrimination was carried out through a system of premium payments, which was used to subsidise the state farms with financial difficulties (Bojnec, 1994).

The role of agricultural co-operatives

Agricultural co-operatives played an important role in the Slovenian agricultural economy before World War II. Also under the Communist regime they remained important, but their role was different than in many other Central and East European countries. Slovenian agricultural co-operatives were closer to the Western style of co-operatives than to the so-called collective farms. Collective farms were mostly abolished in 1953. Input supply (fertilisers, plant protection chemicals, machinery) and output procurement were controlled by agricultural co-operatives. In theory, agricultural co-operatives were (legally) owned by private peasants. However, in practice they were within the 'socialised' sector. As the rural commercial network they were a link between the state sector (e.g., the agricultural input industry, food processing industry) and the private peasants.[5]

Since the 1970s the agricultural co-operatives played an important role in marketing of agricultural inputs and purchasing of agricultural products from private peasants, and also in production co-operation based on voluntary contracts. The agricultural co-operatives were organised on a regional basis. Each co-operative had a regional monopoly in distribution of agricultural inputs, in procurement of outputs, in extension services and in allocating preferential credits to private farmers. Most private agricultural production was purchased by agricultural co-operatives and sold to the vertically integrated state-owned food processing industries (e.g., milk, cattle, pigs) and to the state-owned marketing organisations (e.g., potatoes, fruit and vegetables). Farmers sold only a small part of production directly to consumers on local free markets.

Agricultural privatisation and land reform

In February 1991, an amendment to the Law on Agricultural Land abolished the preference rights of the state sector for purchasing or renting land.[6] Under the amended law, the neighbouring farmer has first preference to buy the land, and after that other potential buyers. Furthermore, there was positive discrimination in favour of full-time farmers among private farmers. The maximum size of private farms was increased in 1990 and abolished in 1991. Land can now be freely traded among Slovene citizens. Private economic rights and property ownership are protected. Direct foreign ownership of land by foreign citizens is still prohibited, but foreigners can own land through a Slovenian registered company.

Agricultural privatisation only encompasses the state agricultural farms and the state forestry organisations in Slovenia as 'classical' socialist collective farms no longer existed in the 1980s. In 1991 Slovenia opted for restitution of state land in real boundaries to domestic former owners[7], or through compensation in the form of a comparable piece of land or government securities.[8]

Table 8.5
Mechanisation, fertiliser use and productivity in Slovenian agriculture, 1970-1995

	kW of double-axel tractors per hectare of arable land			*Kilograms of fertiliser per hectare of arable land*			*Yields of wheat in tonnes per hectare*		
	State	*Private*	*Ratio*	*State*	*Private*	*Ratio*	*State*	*Private*	*Ratio*
1970				453	150	3.0	3.06	2.24	1.4
1975	0.88	2.42	0.36	686	141	4.9	3.48	2.58	1.3
1980[a]				766	173	4.4	4.51	3.10	1.5
1985				1,054	206	5.1	5.04	3.39	1.5
1987				975	271	3.6	5.50	3.43	1.6
1989				927	214	4.3	5.15	3.58	1.4
1991	1.42	4.27	0.33	516	155	3.3	6.09	3.84	1.6
1993				439	103	4.3	5.83	3.32	1.8
1995				791[b]	255[b]	3.1[b]	5.40	3.80	1.4

[a] in 1981 for tractors
[b] in 1994

Source: Statistical Yearbook of Slovenia (SY SLO) and Bulletin of Slovenia (various issues)

Privatisation and transformation of the state farms and enterprises deal with both agricultural and forest land and with non-land assets. The process of state land restitution is covered by the Law on Denationalisation of 1991 and by the Law on the Fund of Agricultural Land and Forests of 1993. The latter covers the implementation of the land restitution and the (temporary) management of the land. Restitution of agricultural and forest lands is handled by the State Fund for Agricultural Land and Forests (FALF), which was established in 1993. Most of the land of state farms and enterprises, was expropriated by the communists from larger family farms, from large landowners and from churches, mostly between 1945 and 1953. This land, which is mostly to be returned to the former owners through the land reform process, is transferred to the FALF. The FALF is responsible for the restitution of the land and also for the management of the land during the transition period.

A significant part of the land to be restituted is marginal land and the forests are mostly in hilly, mountainous and remote areas. In many cases technical as well as legal problems arose with restitution in real boundaries. Former owners have problems providing sufficient evidence to claim their land. Additional problems arise in the case where former collective members received other land instead of their own land when the collectivisation was abandoned; or when nationalised land was sold to new owners; or when former owners have foreign citizenship. Furthermore, in many cases expropriated land had been used for industrial and urban projects. The state sector claims compensation for public investments in land, for land improvement (e.g., land consolidation), for investments in plantations (vineyards, orchards and hops). This complicates the implementation of the restitution process because for these reasons many state farms and enterprises claim property and user rights on the assets, including the land. The government's problem was how to resolve the conflicting demands of former owners and of the state farms and enterprises.

The land restitution process was expected to lead to a break-up of the state farm structure. However, the 1991 Law on Denationalisation introduced the concept of 'co-ownership on land' for former owners when land was currently used by the larger state farms, such as plantations or vineyards. It also allowed for an 'adjustment period' on land operation by the state farms. Later on, the concept of 'good landlord' was introduced in the law on the FALF to allow state farms to continue their operation on the land. In cases where restitution in kind could have a significant negative impact on the economic or technological functions of state enterprise, the former state enterprise/farm may continue to operate for a certain adjustment period: five years after denationalisation of the property rights to the original owner in form of co-ownership share, or until 7 December 1998 (see Bojnec and Swinnen, 1997).

These regulations have been used to limit the transfer of actual property rights on land and to minimise disruptions of the existing state farms. The additional regulations have followed a shift in the political majority from 'right' to 'left' over the period 1990 to 1993. While the initial government favoured restitution of land to private individuals and the break-up of the state farms, later governments that relied more on political support of ex-communists have adjusted regulations to reduce the disruptive impact on the former state farms. (See Bojnec and Swinnen (1997) for an extensive discussion of the political economy reasons behind the changes in the privatisation and restitution procedures.). However, the privatisation process remained under heavy political pressures from both sides: former owners want their land in historical boundaries and state farm employees try to prevent land restitution and want to continue to operate 'their' state land.

Social and political pressures from those employed in state farms in many cases has prevented restitution thus far. The regulations of 'co-ownership on land' and of 'good landlord' in the Law on the FALF allowed former state farms priority in the leasing or renting of farmland. Both regulations are important for land used by the larger state farms because of their investments in plantations (vineyards, orchards, and hop gardens) as the plantation assets remain the property of the investors (Bojnec and Swinnen, 1997). The tenancy conditions in the 'good landlord' regulation gave incentives and security to the state farms to continue operating for a certain 'adjustment period'. Subsequent regulatory changes by the Slovenian government and Parliament reinforce the economic position of the former state farm employees. In December 1995, the implementation of the Law on Denationalisation was amended by the Law on Temporary, Partial Moratorium on Restitution. This law introduced a three year moratorium and revision of the implementation of the restitution to large landowners and religious institutions with over 200 hectares of land, and to foreigners. Furthermore, the Law on the FALF of 1993 was amended in 1996. Under equal conditions, preference in leasing of land from the FALF is given to the present users, meaning mostly to the transformed and/or privatised former state farms and companies. This amendment applies for the duration of the amortisation period of the investments in agricultural land (e.g., for plantations) and at least for twenty years for forest land. The regulation of 'good landlord' in cultivating the land remained, and this concept is now also part of the amended Law on Privatisation of 1996 in order to protect the large scale farms and forestry enterprises due to their separate privatisation of land and non-land assets.

Non-land assets (e.g., buildings, mechanisation, plantations) owned by state farms and enterprises are subject to privatisation under the general Law on Privatisation:

1 Employees receive twenty percent of the shares in an internal distribution of shares via vouchers,

2 State Funds receive forty percent (ten percent of the shares is transferred to the Pension Fund, ten percent to the Compensation Fund, and twenty percent to the Development Fund and reserved for distribution through authorised investments funds),

3 The remaining forty percent could be privatised by one or a combination of privatisation methods which is chosen by the farm/enterprise itself and verified by the Agency for Privatisation. Most often an internal buy-out of the shares by payment at a discount up to fifty percent is selected.

All state owned enterprises in the privatisation programme, including state farms and food processing enterprises, had a deadline of 31 December 1994 to submit their privatisation proposals to the Agency for Privatisation. The process has been postponed. The privatised state farms will transform their organisation to a joint stock company or a company with limited liabilities. However, the uncertainty on the future of those farms during the transition remains because in many cases state land will be restituted in kind to former owners.

The process of transformation of the agricultural co-operatives is based on the Law on Co-operatives of 1991. The intention was to create a co-operative structure e.g. in marketing and in food-processing similar to that in Western European economies. The organisation and operation of the agricultural co-operatives are to be changed from state controlled co-operatives into co-operatives effectively controlled by their members. Membership in co-operatives is allowed for anyone who pays a share of the capital. The number of co-operatives is changing because many of them are in financial difficulties due to inherited inefficient structures and overemployment. The restructuring process of the co-operatives is through their liquidation, their division into smaller units and the establishment of new co-operatives, often based on product specialisation (e.g., for milk or wine).

Some small and medium sized private firms emerged in the 1990s. These new private firms in agricultural activity were established by private entities to supply agricultural inputs and to purchase agricultural outputs. Their major activity is agricultural marketing and not agricultural production itself. Some of them were established to compete with the marketing activities of the agricultural co-operatives.

Progress with agricultural privatisation and land reform

Table 8.6 illustrates the progress in land restitution. By October 1995, 25,500 hectares of agricultural land and 63,200 hectares of forest land were restituted.

The restituted agricultural and forest land represents about half of claims under the Law on Denationalisation or about one third of land subject to restitution. However, clear evidence about the land which should be restituted is still not known, because several claims are still not complete due to missing evidence (e.g., citizenship, nationalised land due to unpaid taxes, symbolic compensation of the former owners at the time of nationalisation, compensation from the foreign country).

Table 8.6
Restituted agricultural and forest land in Slovenia, Dec. 1992 - Oct. 1995 (in '000 hectares)

	December 1992	*December 1993*	*December 1994*	*October 1995*
Agricultural land	6.9	13.8	24.3	25.5
Forest land	18.9	40.7	58.5	63.2

Source: Porocevalec, 1995, and FALF, 1996

Most of the land used by the state enterprises is now transferred to the FALF, but leased back by the FALF to the state enterprises. This is not only the case for state farms but also for state forestry organisations which managed the state forestry. In 1990, 62.4 percent of forest land was privately owned. The restitution process, including church forest land restitution, should increase this to 80 percent (Lukacic, 1994). Forest land restitution is particularly important for private farms in hilly and mountain districts. However, in 1995 more than 30 percent of forest land was leased in state hands (see Table 8.7).

The privatisation in agriculture started slowly, because of the initial lack of accompanying necessary legislation for the restitution process. The Law on Denationalisation was passed in 1991, but the implementation of this law was slowed down, because the background legislation (the Law on Privatisation and the Law on the FALF) was passed only in 1993. Some major problems with restitution soon developed. First, conflicting information on land ownership slowed implementation, because the real situation is often different than shown in cadastral registers. Second, the concept of co-ownership on land in large parcels in most cases prevented restitution in real boundaries so far, while the former owners prefer restitution in real boundaries rather than compensation with government securities. Third, the former users (the state farms and forestry enterprises) claimed for compensation of the non-amortised value of investments on land improvements.[9] Fourth, there is a lack of experts, who are able or willing to carry-out the restitution process (Porocevalec, 1996). Fifth, several changes of legislation and court disputes slowed down the process of restitution.

Table 8.6 shows that the restitution process speeded up after the FALF started to operate in May 1993 and in 1994, while in 1995 the restitution process slowed down again. From the beginning, the restitution process was mostly oriented to the easier cases from the point of view of the former owners and the location of the land. This is the main reason the FALF initially mostly restituted land in real boundaries. Most disputable claims occurred when the land is within the state farms in large parcels (this is the problem of co-ownership right). Only 8 percent of agricultural land has been restituted under the form of co-ownership shares on agricultural land. Compensation payments with government securities was not popular among the former owners, because they preferred restitution in real boundaries or compensation with substituted land. However, the FALF has avoided compensation with substituted land, because of the lack of information on land which is likely to be restituted in real boundaries. The FALF will probably start compensation with substituted land, after the restitution process is finished. However, it is still unclear when the restitution process will be completed and how much land will remain in the FALF after the restitution and compensation process. It is likely that after completion of the Law on Denationalisation, more than 90 percent of agricultural land and about 85 percent of forest land will be owned by private owners, mostly by private family farmers.

The restitution process slowed in 1995 due to the new regulations. Moreover, it is likely to slow even further following a law passed by the Parliament in December 1995, which introduced a three year moratorium of restitution to big landowners and churches over 200 hectares, and to foreigners. It is likely that some land, which was already being on the way to be restituted, will now remain in public ownership.

Farm restructuring since 1989

What is the effect of the agricultural privatisation and land reform program and the other regulations on the farm structure? In general, one would expect farm restructuring in Slovenia to be less dramatic than in most other Central and East European countries because of the absence of collective farms and because the state farms made up a relatively small part of the overall agricultural economy (less than 20 percent of agricultural land and 35 percent of output in 1989). Still, we would expect that the reforms would lead a reduction in the average size of the state farms and to an increase in size and an improvement of the structure of the private farms. These changes are affected by the privatisation and restitution policies. Effective restitution of land property rights to former owners is likely to stimulate the break-up of the state farms. However, the factor which is most important is effective property rights, rather than ownership (Swinnen, 1996). Several measures taken by the Slovenian government and Parliament since 1993 have limited the transfer of effective property rights on land from the state farms to new (former) owners. These regulatory changes are likely to have a negative impact on state farm restructuring.

What do the data tell us?

There are no official data on average farm size and classification by farm sizes (as in Table 8.1 and 8.4) for recent years. Therefore we have to rely on other statistics to obtain information on farm restructuring and the changing role of the state sector in agriculture.

Farm numbers, employment and output

The number of state farms has declined by around 20 percent and employment on state farms has declined more than 25 percent between 1989 and 1995 (see Table 8.2).

The share of active agricultural population as a proxy for agricultural employment shows that agricultural employment, including private farm employment declined by 2.4 percent between 1991 and 1995.

The share of private farms in total agricultural production increased between 1989 and 1993 (see Table 8.8). In 1993, about three-quarters of agricultural production came from the private farms. Institutional and economic liberalisation changed the policy environment as the state farms lost their former preferential treatment.

Land use

Data on land use are collected by the Statistical Office on the base of census questionnaires (every ten years for private farms and yearly for former state farms). The last census was carried out in 1991. The census underestimated the agricultural and forest land of Slovenia compared to estimates based on cadastral data, which means that a significant part of the land was probably not included in the census data. This might explain the differences in data in Tables 8.1, 8.3 and 8.8. Recently, satellite estimations of land use give a somewhat different picture of land use in Slovenia. They show a much higher share of forest land (almost two thirds of Slovenia's land surface) compared to census data (about half of Slovenia's land). Forest land, particularly on private farms, has been underestimated in the past, and agricultural land has been overestimated.

Table 8.7
Forest land by ownership in Slovenia, 1980-1995

	1980		*1990*		*1994*		*1995*	
	'000 ha	*%*	*'000 ha*	*%*	*'000 ha*	*%*	*'000 ha*	*%*
State	391.2	37.4	402.5	37.6	338.0	30.9	350.6	31.9
Municipalities					0.4	0.1	1.1	0.1
Co-operatives					5.4	0.5	...*	...*
Rural communities					2.4	0.2	...*	...*
Churches					2.4	0.2	...*	...*
Private	654.2	62.6	668.7	62.4	745.6	68.1	735.1	67.0
Total forest land	1,045.4	100.0	1,071.2	100.0	1,094.2	100.0	1,097.9	100.0

* 11,119 hectares or 1.0% of total forest land was owned by legal entities (e.g. agricultural co-operatives, rural communities and churches).

Source: SY SLO, various issues, and Institute for Forests (1996)

Table 8.8
The relative sizes of the private and the state farm sector of Slovenia 1989-1994 (%)

	1989		*1993*		*1994*	
	Private	*State*	*Private*	*State*	*Private*	*State*
Total area	69.0	31.0	69.0	31.0	73.2	26.8
Agricultural land	82.6	17.4	83.5	16.5	89.9	10.1
Arable land	89.0	11.0	88.9	11.1	90.1	9.9[b]
Labor	...	...	87.4[a]	12.6[a]	...	...
Total agricultural	65.7	34.3	75.0	25.0	...	...
production	92.0	8.0	92.4	7.6	93.0[b]	7.0[b]
Milk production	72.3	27.7	71.9	28.1	80.5	19.5
Fertiliser consumption	68.3	31.7	69.9	30.1	63.1	36.9
Pesticides consumption						

[a] In 1991.
[b] In 1995.

Source: SY SLO 1990, pp. 216-237, SY SLO 1994, pp. 249-260, SY SLO 1995, p. 245 and Statistical Office of Slovenia

In contrast to output and employment, available land-use data indicate that until 1993 hardly any changes took place in land use of private and state farms and that even after 1993 land use changes are modest (see Tables 8.8 - 8.10). As in Poland, restructuring and privatisation of state farms has been slow (OECD, 1995). The number of state farms has increased slightly since 1991 because some of them have been broken up in separate units. The share of private farming was largely unchanged between 1989 and 1993. However, since 1994, the decline of the state sector is clearly reflected in the official data (Table 8.3).

Table 8.9 shows that the reduction in the use of arable land, orchards and vineyards in the state farms between 1989 and 1995 was still small and that the largest reduction of the state farms was found for pasture land mostly as a result of a process of land restitution.

The distribution of arable land use was also stable (Table 8.10). Cereals were the most important in arable land use as they take up about half of the arable land and the state farm share increased. The share of the area of industrial crops in arable land use was not very important. The majority of industrial crops (e.g., sugar beet, rapeseed) were produced on state farms in 1989 but by 1995 the majority of the area was in private farms. Vegetable and fodder production was mostly on private farms.

Number of animals

In general, changes have been relatively minor here as well. The number of cattle decreased in the early 1990s on both private and on state farms (Table 8.11). The reduction in the number of cattle was more significant on state than on private farms. The recession in the cattle production was caused by economic factors (i.e. an unprofitable economic position) and by a drought which negatively affected feed production. The fall was more important on state farms, which were affected by a significant cut of subsidies, a lack of subsidised credit and uncertainty about the future of cattle production on state farms. This is further reflected with the fact that in 1995 and 1996 cattle numbers fell further on state farms, but increased again substantially on private farms. In the past a significant share of pig production was on state farms. In the early 1990s the number of pigs on state farms decreased and on private farms increased, but the situation has stabilised since 1994. Sheep production, traditionally not very important in Slovenian agriculture, was mostly on private farms in hilly and mountain districts. The state farms were very important in poultry production. More than 88 percent in the poultry number was on the state farms and enterprises. However, poultry production dropped substantially in the 1990s because the loss of the ex-Yugoslav market where a significant surplus in poultry production was mostly sold. The data indicate that this drop was equally strong on private as on state farms.

Table 8.9
Disaggregated land use by private and state farms in Slovenia, 1989-1995

	1989				*1993*				*1995*			
	Private farms		*State farms and enterprises*		*Private farms*		*State farms and enterprises*		*Private farms*		*State farms and enterprises*	
	1000 ha	*%*	*1000 ha*	*%*	*1000 ha*	*%*	*1000 ha*	*%*	*1000 ha*	*%*	*1000 ha*	*%*
Arable land	215.4	87.2	31.7	12.8	216.1	88.0	29.4	12.0	206.5	88.1	27.9	11.9
Orchards	33.1	92.1	2.9	7.9	32.1	92.1	2.7	7.9	29.4	93.3	2.1	6.7
Vineyards	17.6	82.3	3.8	17.7	18.3	83.9	3.5	16.1	19.7	85.7	3.3	14.3
Meadows	307.9	90.5	32.5	9.5	310.8	89.6	36.2	10.4	314.5	91.2	30.3	8.8
Pasture land	143.8	64.6	78.9	35.4	142.2	67.4	68.8	32.6	126.4*	85.7*	21.2*	14.3*
Ponds and reeds	1.0	42.6	1.3	57.4	1.0	44.1	1.3	55.9	2.2*	64.2*	1.2*	35.8*
Forest land	645.1	63.9	364.4	36.1	644.8	63.2	375.3	36.8	745.6*	68.1*	348.6*	31.9*
Barren land	33.2	22.7	112.9	77.3	32.7	22.9	110.2	77.1	33.1*	23.6*	107.5*	76.4*
Total land	1,397.0	69.0	628.4	31.0	1,398.0	69.0	627.4	31.0	1,482.6*	73.2*	542.9*	26.8*

* In 1994.

Source: SY SLO 1990, p. 216, SY SLO 1994, p. 249, SY SLO 1995, p.245 and Statistical Office of Slovenia

Table 8.10

Disaggregated use of arable land and gardens by private and state farms in Slovenia, 1989-1995

	1989				*1993*				*1995*			
	Private farms		*State farms*		*Private farms*		*State farms*		*Private farms*		*State farms*	
	1000 ha	*%*	*1000 ha*	*%*	*1000 ha*	*%*	*1000 ha*	*%*	*1000 ha*	*%*	*1000 ha*	*%*
Cereals	102.8	84.0	19.6	16.0	101.1	84.2	18.9	15.8	90.6	82.6	19.1	17.4
Industrial crops	3.6	40.0	5.4	60.0	4.5	47.5	4.9	52.5	6.7	62.0	4.1	38.0
Vegetables	40.1	99.3	0.3	0.7	39.5	99.4	0.2	0.6	33.5	99.4	0.2	0.6
Fodder plants	65.9	91.6	6.0	8.4	67.9	93.4	4.8	6.6	73.9	94.9	4.0	5.1
Other*	3.0	89.6	0.4	10.4	3.1	85.1	0.6	14.9	1.8	78.3	0.5	21.7
Total arable land	215.4	87.2	31.7	12.8	216.1	88.0	29.4	12.0	206.5	88.1	27.9	11.9

* Other includes flowers, ornamental plants, nursery, fruit trees, vine slips, plantations of forest trees, fallow land.

Source: SY SLO 1990, p. 216, SY SLO 1994, p. 249 and Statistical Office of Slovenia

Table 8.11
Number of animals on private and state farms in Slovenia, 1989-1996

	1989				*1994*				*1996*			
	Private farms		*State farms*		*Private farms*		*State farms*		*Private farms*		*State farms*	
	*Number**	*%*	*Number**	*%*	*Number**	*%*	*Number**	*%*	*Number**	*%*	*Number**	*%*
Cattle	464.3	85.0	81.7	15.0	434.4	91.0	43.2	9.0	460.7	93.0	34.8	7.0
Pigs	314.9	54.6	261.3	45.4	350.6	59.3	240.9	40.7	348.5	58.9	243.5	41.1
Sheep	23.5	97.6	0.6	2.4	18.8	96.3	0.7	3.7	26.6	96.0	1.1	4.0
Poultry	1,449	10.9	11,820	89.1	1,206	11.4	9,386	88.6	1,063.6	11.4	8,256.6	88.6

* Numbers in thousands (on 1 January).

Source: SY SLO 1990, pp. 229-231, SY SLO 1994, pp. 255-257 and Statistical Office of Slovenia

Some conclusions on farm restructuring

The most striking observation from the official data is the absence of radical changes since 1989. While the output and employment share of the state farms has decreased, this is much less the case with land use. Only since 1994 the share of land used by state enterprises has decreased slightly, mostly for pasture land and forest land. Also in animal production the shift to smaller-scale private farms is modest. Only in cattle production is this shift strong and continuing.

Some additional information on leasing arrangements of the FALF are consistent with this picture: the best land was released to the former state enterprises in 1994 and 1995 while private farmers have been allowed to lease only part of the rest of the land (about one-third of agricultural land in the FALF). Most state forest land is operated by the transformed forestry organisations. Interestingly, forest land increased because low quality agricultural land was no longer used for agricultural production.

Within the private sector some important restructuring seems to take place with the removal of size and contract limitations. During the transition period a rural leasing market has quickly emerged spontaneously. Younger farmers rent land from elderly farmers with the rent often equal to the paid taxes on the rented land. This represents a new tendency in operational farm concentration, which is not captured by the statistics, but which is quite important in practice. The buying of land is less important, because farmers do not have enough money to buy the land due to the fact that the profitability of investments in agriculture was rather low and there was no subsidised credit for purchasing land. The more appropriate and easy way of operational farm concentration was short-term leasing of land, where younger farmers often rented land from one, two or even more elderly farmers in order to cultivate their land, which in other case would likely have remained out of production. This also means that the concentration of operational farm structure through land leasing is becoming an important new factor in villages where traditionally there was always a close link between family farm ownership and its operation.

Causes of the (lack of) farm restructuring

Several factors affect the nature and pace of restructuring. One is the privatisation process which redistributes the property rights on land and capital in agriculture. This process itself has several dimensions: the design of the process is important (i.e. what the laws and regulations specify), but also the implementation of the process. The latter is complicated by technical, legal, organisational and political problems. The extent to which the latter can be affected by the incentives and preferences of the people responsible for the implementation, depends on how the implementation itself is regulated and on the (intended or unintended) autonomy of the implementing institutions (Rabinowicz and Swinnen, 1997). Another important factor in the speed of restructuring is the availability and restructuring of skills and human capital for farming. Finally, the economic environment will importantly determine the risk and potential profitability of various strategies. Let us consider the importance of these factors for the Slovenian agricultural restructuring.

Reform design and implementation

The future of the former state farms, which are in the process of privatisation, will depend importantly on how the privatisation process is implemented, both regarding the land and the non-land assets. Land has been transferred to the FALF and was to be restituted to the former owners. If strictly implemented, this process is likely to severely disrupt production activities of the former state farms. In a world with perfect markets and without transaction costs, this need not be the case as new owners of the land would look for optimal contracting. Given the current investments on the land and their integration in a large-scale production structure, re-organised former state farms with better incentives might actually prove to be the best way to lease the land, at least in the short run. However, in a world with high transaction costs and imperfect contract and sales markets – which is likely to be a more realistic description of Slovenia's land market – major disruptions are likely to occur.

The implementation of the land restitution process is, at least, less radical than it looks on paper. In 1991 the demand for restitution was politically so strong that this privatisation approach was accepted by Parliament. However, there appeared many legal, organisational, technical and political problems in the implementation of the land restitution process. In fact, many additional regulations that have been added in laws following the initial Law on Denationalisation and these have watered-down the initial radical restitution law. For example, the introduction of the status of 'co-ownership' of land, and of 'good landlord', and of 'adjustment periods' in the regulations allows the former state farms to continue to operate for several years on the land which they now lease from the FALF (and which is still to be restituted). All these regulations are likely to reduce the break-up and restructuring of the former state farms, at least in the short to medium-run and compared to the situation that would exist if the land restitution were implemented strictly and fast. These developments and effects are further reinforced by the most recent law amendments to postpone restitution.

Furthermore, the disruptive impact of land restitution on former state farm activities and restructuring is most important for those activities for which land is a crucial input into the production process, such as wheat production, and also orchards and vineyards. These state farm activities have been protected explicitly by the 'good landlord' and 'co-ownership' amendments. The problem is somewhat less important for large-scale state pig and poultry production units. These units will be re-organised into limited liability companies and joint-stock enterprises under the privatisation laws. They rely less on land and could continue by purchasing most inputs for feeding either from domestic producers or from imports (at least if they are competitive). An important problem in Slovenia is the environmental problems caused by large-scale pig and poultry production.

Technical difficulties have also slowed the restructuring. For example, only domestic former owners are subject to restitution. However, in more than 80 percent of restituted claims the former owner was late in submitting claims. Also, many heirs of the former original owners live abroad. About 20 percent of forest lands are to be restituted to former owners who no longer live in Slovenia (Lukacic, 1994).

In addition, the regulations on the privatisation of non-land state farm assets are biased in favour of the management and employees of the state farms. The process of privatisation of the state non-land farm assets is covered by the Law on Privatisation. The workers and managers of state farms get preferential treatment in the privatisation of the state non-land farm assets they managed.[10] State farms and enterprises were supposed to submit their privatisation plans to the Agency for Privatisation by the end of 1994. The labour buy-out is the most popular method of privatisation because employees (i.e. workers and managers) receive 20 percent of the shares in an internal distribution of shares via vouchers and they have preferential access to buy an additional 40 percent of the shares at 50 percent of discount. However, the future of many of these farms is questionable because of the separated ownership of agricultural land privatisation.

Land markets

The initial land restitution fragmented previous state land among former owners. However, many former owners want to sell the restituted land (Lukacic, 1994). The restitution of land which has dispersed and fragmented ownership does not necessary imply fragmentation of use if there is either a well functioning land market where land can be sold and bought or if there are well-defined land-leasing regulations.

There is an important demand for land use especially by full-time private farmers. Slovenian farmers are relatively well mechanised and able to manage bigger farms. The availability of land was the most important obstacle for the existing private farms to an efficient use of own farm resources and to exploit the economies of scale of a larger farm. Thus far, the restitution process has not have a large impact on the average size of the existing private farms. First, restitution has been slowed down and the transformed state farms succeed in leasing 'their' former state land, a process which is ongoing. A more important effect has resulted from the removal of the maximum land area regulation and operating restrictions for private farms, and from the (formal) elimination of the preference of state farms over private farms on buying or leasing of land. Therefore, in the long run, market forces could have an important impact on farm structure and the average size of farms.

Land market and long-run leasing of land could therefore improve the economic performance of the viable farms in such deeply fragmented land and farm structure. In the near future, the concentration of land and therefore the farm structure will significantly depend on the purchasing and leasing circumstances, the availability of subsidised credits and financial conditions of the farms. Prices of land are very high at this moment (more than 20,000 DEM per hectare arable land) and vary between regions. Land closer to towns is more expensive. Within the region price depends on a quality of land. However, the worsening of terms of trade in the 1990s for farmers reduced the incentives and opportunities (due to a lack of capital) to buy land.

Many farmers therefore prefer to lease land. As discussed in the previous section, private farm leasing from the FALF is limited, but land leasing between private individuals is developing rapidly. These leasing arrangements are inducing a concentration of land use within the smaller scale private farms. This should improve the private farms' structure.

The quality of agricultural land and labour

The uneven topography of much of Slovenian land makes it difficult to establish large-scale farming. Small gardens and recreational farm activities are more appropriate on this land. The traditional attitude of the population gives great importance to land ownership for self-consumption and for recreational gardening. This can be a significant factor for the persistence of dispersed small land ownership. Of course it depends on inheritance rules and land market legislation, more specifically whether these rules provide incentives for the division of family farms. Some argue that the average size of the smaller-scale private farms will probably remain under 10 hectares of agricultural land per farm during next ten to fifteen years (Erjavec, 1994).

During the recession period in the 1990s employment in non-agricultural activities declined and employment in private agriculture was stable. This was particularly significant in rural areas where private farming co-exists with a state sector in non-agricultural activities. Private farms absorbed labour which was laid off in other sectors. This absorption of labour is likely to be only temporary. A decline in employment on private farms is expected with economic recovering, which started at the end of 1993, and with a growing number of older farmers retiring from agriculture. Many farms employed elderly farmers and with economic growth and a higher level of economic development some part-time farms will no longer have an interest in farming. A reduction of the number of private farms is therefore expected. However, part-time farming will still remain a significant factor in the agricultural and rural economy in Slovenia in the future.

Finally, an active structural policy promoting the transfer of underemployed labour from agriculture to non-farm activities and land consolidation measures may significantly improve the structure of farms. Those measures were developed in structural policies by West European countries which provided incentives for the entry of young and educated farmers to take over farming and incentives for older farmers to retire (the Slovenian agriculture labour structure of farmers is rather old). Both groups of measures depend on the provision of financial resources for agricultural restructuring and on general economic growth for non-agricultural employment.

Conclusions

The Slovenian pre-Communist structure of agriculture was rather egalitarian, whilst the pre-reform agricultural structure in 1989 was characterised by a bi-modal agricultural structure of many small peasant farms and some large state farms and forestry organisations. Collective farms were absent because the process of communist collectivisation failed. Only state agricultural farms and forestry organisations in Slovenia are directly affected by the land reform and privatisation. As the private farms already made up the major share of output, employment and land use, farm restructuring in Slovenia has not been so dramatic than in many other ex-socialist countries whose agriculture had been collectivised.

The current land reform is restituting land in historical boundaries to the former owners. The state land has been transferred into the FALF, which is responsible for restitution the agricultural and forest lands and for managing their use during a transitional period. The major part of the state land would be restituted to its former owners in historical boundaries. Former owners can also be compensated by land of comparable size and quality or by government securities.

State farms and enterprises, which prepared themselves to be privatised and transformed into private companies, strongly opposed land restitution as it would induce a disruption of their production system and because it would distribute land property rights to outsiders. A related problem was the conflict between land restitution claims and privatisation of non-land farm assets. The political opposition slowed down the implementation of the restitution process. Legal amendments were introduced to slow down restitution and to minimise the disruptions. Only part of state land was restituted to former owners while much of the best land was leased to the state enterprises by the FALF. In addition, non-land assets are mostly distributed to the present users and the state government funds.

Farm restructuring has been slow and available data show little change in land use by 1993. The state farms and enterprises mostly were able to consolidate their structure. In 1993 the FALF was established and the implementation of restitution began. However, since 1994 changes in land use are still modest, and are expected to remain so in the next few years mainly because of the relatively small share of the state sector and because of the legal amendments protecting the state farms against disruption.

The main factors inducing farm restructuring seems the removal of size restrictions for private farms and the emergence of land leasing markets. These induce a concentration of land use in the private sector.

Notes

1 The former Yugoslav notion of 'social' and 'state' sector was defined differently depending on the level of development of the socialist system and the autonomy of economic agents. During the initial stage of 'administrative' socialism (1945-1951), there was in practice a notion of 'state' (farm) sector which was defined as in other ex-socialist countries. However, the notion 'state' sector was abolished later by the introduction of 'social' property and a system of socialist self-management. 'Social' property was at the same time everyone's (all people's) property and no one's property. The user rights were allocated to 'social' farms and enterprises, which were managed by the system of worker self-management. In this article we will refer to this sector as the state sector instead of the 'social' sector, as it is often referred to.

2 An exception were churches and cloisters of historical value who could own up to 30 hectares of cultivated land and up to 30 hectares of forests. In 1946, the law was amended. The owner of the former large capitalist estate could own a house or a flat for living, 5 hectares of cultivated land and some equipment, if the estate was his only professional activity.

3 In 1965, it was limited to 2 hectares.

4 Land up to 600 meters height above sea level and with slope less than 25 steps was defined as a lowland, and the remaining as a land in hilly or mountain districts.

5 Besides a network of agricultural co-operative shops, there was a state-owned retail network supplying farmers and other consumers in rural areas with consumer goods. During the 1990s, the major part of rural retail trade has already been privatised.

6 The new Law on Agricultural Land is likely to be adopted in 1996 or in 1997.

7 A domestic citizen is a person who had citizenship of former Yugoslavia at the time of nationalisation of its property, recognised after 9 May 1945; with the exception of persons who had permanent residence on the territory of the present Slovenia at the time of nationalisation of their property and whose former Yugoslav citizenship was recognised after 15 September 1947.

8 A hotly debated issue was the restitution of property to foreigners (mostly Italian citizens). The debate was complicated by the Italian opposition to the Slovenian EU Association Agreement in the absence of a 'satisfactory' solution to the restitution dispute. In April 1996 the so-called Spanish compromise proposal on the ownership of Slovenian assets formerly belonging to EU citizens has been hotly debated in Parliament. This proposal claims for the necessary harmonisation of the Slovenian property legislation with those in operation within the EU, including a free property market in Slovenia for those EU citizens who lived in Slovenia for three years. This debate has caused polarisation within the Parliament, but with the adopted proposal it is expected that Slovenia will move closer to the EU, especially with the ratification of the Association Agreement.

9 However, an additional problem has occurred, because in several cases the investments in land had been financed with direct government support, and not directly from the former state farms and enterprises.

10 Conflict of interests arose due to the fact that the reprivatisation of land is divided from the privatisation of other farms assets.

References

Bojnec, Š. (1994), 'Agricultural Reform in Slovenia', in Swinnen, J.F.M. (ed.), *Policy and Institutional Reform in Central European Agriculture*, Avebury: Aldershot.

Bojnec, Š. and Swinnen, J. (1997), 'Privatization of Slovenian agriculture: Process and Politics', in Swinnen, J.F.M. (ed.), *Political Economy of Agrarian Reform in Central and Eastern Europe*, Ashgate: Aldershot.

Erjavec, E. (1994), 'Instrumenti zemljiške politike za izboljšanje posestne strukture v Sloveniji (Land policy instruments for improvement of farm structure in Slovenia)', IXth Conference of Slovenian Extension Services, November, pp. 47-63.

FALF (1995), 'Porocilo o delu Sklada kmetijskih zemljišc in gozdov Republike Slovenije v letu 1994 (FALF Report for 1994)', unpublished report.

FALF (1996), 'Porocilo o delu Sklada kmetijskih zemljišc in gozdov Republike Slovenije v letu 1995 (FALF Report for 1995)', unpublished report.

Institute for Forestry (1996), 'Porocilo o delu Zavoda za gozdove Slovenije v letu 1995' (The Report on work of the Institute for Forestry of Slovenia in 1995), unpublished report.

Kovacic, M. (1994), 'Velikostna struktura slovenskih kmetij (Size structure of Slovenian farms)', IXth Conference of Slovenian Extension Services, November, pp. 17-27.

Lukacic, M. (1994), 'Vloga sklada kmetijskih zemljišc in gozdov RS pri izboljšanju zemljiške strukture in vpliv denacionalizacije nanjo (Role of the FALF and impact of denationalization on land structure)', IXth Conference of Slovenian Extension Services, November, pp. 71-78.

MKG (Ministrstvo za kmetijstvo in gozdarstvo) (1994), *Porocilo o stanju kmetijstva v letu 1993 (Report on agriculture in 1993)*, Ministry for Agriculture and Forestry: Ljubljana.

OECD (1995), *Review of Agricultural Policies: Poland*, OECD: Paris.

OECD (1996), *Agricultural Policies, Markets and Trade in Transition Economies: Monitoring and Evaluation 1996*, Paris: OECD.

Official gazette of Slovenia (various issues).

Porocevalec (1995), 'Deveto porocilo o uresnicevanju zakona o denacionalizaciji' (Ninth Report on Implementation of the Law on Denationalization), Porocevalec Drzavnega Zbora, No. 27, pp. 19-72.

Rabinowicz, E. and Swinnen, J.F.M. (1997), 'Political Economy of Privatization and Decollectivization of Central and East European Agriculture: Definitions, Issues and Methodology', in Swinnen, J.F.M. (ed.), *Political Economy of Agrarian Reform in Central and Eastern Europe*, Ashgate: Aldershot.

Statistical Yearbook of Slovenia (SY SLO), various issues.

Swinnen, J.F.M. (ed.) (1994), *Policy and Institutional Reform in Central European Agriculture*, Avebury: Aldershot.

Swinnen, J.F.M. (ed.) (1997), *Political Economy of Agrarian Reform in Central and Eastern Europe*, Ashgate: Aldershot.

9 Experience with land reform and farm restructuring in the Former Soviet Union

Zvi Lerman

The legacy and origins of reform

The former Soviet Union spans 11 time zones. Although much of this space is uncultivatable tundra and taiga, its arable land endowment is 7 times that of East Central Europe, and its rural population is three times as large as that of its former socialist neighbours to the West (FAO, 1994). Land was nationalised in Russia and other parts of the Russian empire within days of the Bolshevik revolution in October 1917. The famous Decree on Land, transferring all land to the state and prohibiting private land ownership, was one of the first two decrees adopted by the new revolutionary government (the other decree put an end to Russia's involvement in World War I).

After a transition decade in which different forms of farming were allowed to coexist, agricultural production in most of the former Soviet Union was reorganised, by government fiat, in the form of collective and state farms. The traditional Soviet agriculture was characterised since the early 1930s by large-scale farms cultivating thousands of hectares of land and employing hundreds of workers (Medvedev, 1987). After World War II, this model was extended to the newly annexed parts of the USSR (the Baltic countries, Moldova, western Ukraine, western Belarus) and also imposed on all of East Central Europe. Contrary to the USSR, however, land was never fully nationalised in East Central Europe, and many members of collective farms continued to hold individual titles to land that they had been forced to 'invest' in co-operatives.

The only concession to individual farming in the former Soviet Union after the all-encompassing collectivisation drive of the early 1930s were the subsidiary household plots of less than 0.5 ha that members and employees of large-scale farms were allowed to keep and cultivate for subsistence purposes (Wädekin, 1973). The original intention of this policy was to enable the rural population to feed itself and thus free the central government to concentrate on the logistics of feeding the cities. Over the years, the subsidiary household plots cultivating less than 2 percent of agricultural land produced more than 25 percent of Soviet agricultural product, mainly livestock, potatoes, and vegetables. With this volume of production, not everything went to subsistence needs, and about one-third of the output always found its way to markets in nearby towns.

The Soviet government always recognised the importance of household plots as a source of food supplementing the relatively inefficient collective and state farms. Every Soviet leader since Stalin's time introduced measures and policies affecting the role and performance of subsidiary household plots. In 1986, Gorbachev, then Party Secretary for Agriculture, launched a programme intended to increase the contribution of household plots and extended the principles of household production to 'group responsibility' contracts within collective and state farms (Brooks, 1990). This programme, however, retained the principles of state-owned land and collectively owned assets, and furthermore imposed on the new, small production units the obligation to fulfill central state orders. As a result of these constraints, the programme did not meet with general success, and attempts to fine-tune it continued practically until the dissolution of the Soviet Union.

The beginning of the current phase of land reform in the former Soviet Union is traceable to the adoption of Principles of Legislation of the USSR and Union Republics on Land in February 1990, nearly two years before the final break-up of the Soviet Union (Nikonov, 1995). This Union-level law acknowledged the right of individuals to decide freely on the form of organisation of agricultural production and established the principle that members and employees of collective and state farms would be able to exit the large-scale farm with a plot of land. Like all 'principles of legislation' in the USSR, this law empowered the republics to adopt their own specific laws on land, which in fact constituted its main contribution to the beginning of the process of land reform. Russia was the trail-blazer in adopting a range of new laws, which included the Law on Land Reform and the Law on Peasant Farms in November 1990, the Law on Property and the Law on Enterprises in December 1990, and the Land Code in April 1991 (Lerman and Brooks, 1996). Other republics soon followed with their own land codes and peasant farming laws. The process of land reform and farm restructuring had begun still before the dissolution of the Soviet Union.

Ownership of land

Between 1917 and 1990 the state was the sole legal owner of land in the former Soviet Union, and access was allocated to producers through use rights.

Collective and state farms held state land in permanent use. Even the land in individually cultivated household plots was state-owned, and the families held it in so-called inheritable lifetime possession. After 1990, some former Soviet republics recognised private ownership of land, while other republics chose to declare that all land would remain state-owned. The legal attitude to ownership of land in the FSU countries as of 1996 is summarised in Table 9.1.

Table 9.1
Legal attitudes to land ownership and land transactions in the FSU countries

	Private ownership	*Transactions*
Russia	Yes: all land (No in 10 ethnic republics)	Buy and sell, mortgage (moratorium lifted by 1994 presidential decree)
Ukraine	Yes: all land	Moratorium on buy and sell; no mortgage
Moldova	Yes: all land	Moratorium until 2001
Belarus	Yes: household plots (June 93)	Buy and sell allowed Use rights non-transferable
Georgia	Yes: all land	Buy and sell allowed only in Feb. 1996 law
Armenia	Yes: all land	Buy and sell, mortgage (law of 1991)
Azerbaidjan	No	
Kazakhstan	Yes: only household plots (decree)	Buy and sell, mortgage Use rights: permanent and tradable
Turkmenistan	Yes: potentially all farm land (constitution)	Prohibited to sell, exchange, give as a gift; leasing allowed Use rights non-transferable
Kyrgyzstan	No	Use rights secure to 99 years and tradable
Uzbekistan	No: lifetime inheritable possession	Prohibited to sell, exchange, give as a gift, mortgage
Tadjikistan	No	

In November 1990, more than a year before the dissolution of the Soviet Union, Russia legalised ownership of land by individual citizens, in addition to state ownership. The 1993 Constitution of the Russian Federation recognises that 'land and other natural resources may be in private, state, and other forms of ownership'. Yet even within the Russian Federation, ten constituent republics do not recognise private ownership of land within their territories (Tatarstan, Bashkiriya, Dagestan, Komi, Mari El', Kabardino-Balkariya, North Osetiya, Tuva, Yakutiya-Sakha, and Koryakiya). These republics base their position on article 72 of the Constitution, which affirms that questions regarding ownership, use, and disposition of land, mineral deposits, water and other natural resources will be decided jointly by the Russian Federation and the constituent republics.

Following Russia, the major Slavic republics (Ukraine, Moldova, Belarus) legalised private ownership of land in 1991-1992. Belarus, however, changed its land code in June 1993 and now allows only household plots of up to 1 ha to be in private ownership: additional land must be leased from the state. In the Trans-Caucasus, Armenia was the first to recognise private ownership of land in January 1991. The neighbouring Georgia, however, delayed recognition of private ownership of land until February 1996, although a large proportion of arable and perennial land had been in the individual sector since 1992.

Among the Central Asian republics, Turkmenistan is the only state where private ownership of land is recognised by the post-Soviet constitution (adopted in May 1992). Kazakhstan allowed private ownership of household plots (but not commercial farmland) by presidential decree in December 1995. Kyrgyzstan, Tadjikistan, and Uzbekistan all retain state monopoly on land ownership, and farmers (individual or collective) must obtain use rights from the state. Although located on the other side of the Caspian Sea, the predominantly Muslim Azerbaidjan follows the same policy.

Yet even in countries that recognise private ownership of land, the current laws severely circumscribe owners' rights in the important areas of land transfer, land use, and size of holdings. Sale of privately owned farmland is often prohibited outright or is subject to a long moratorium (Table 9.1). Other restrictive provisions on transactions in land allow alienation only to residents in the same village, the local farm enterprise, or through intermediation of the state. Even the right to lease privately owned land to other farmers is still not universally accepted. Restrictions on land use typically include the obligation to farm the land, and to observe sound conservation practices. Failure to farm the land for a specified period of time (usually a year or two) can result in expropriation. Size of private holdings is restricted to 50-100 ha in Russia, Ukraine, and Moldova. In Belarus and Kazakhstan, private ownership is limited to small household plots of up to 1 ha (although additional farmland may be leased for commercial farming from the state).

Restrictions on the rights of owners are intended to prevent absentee ownership and underutilisation of land resources. They are also intended to prevent new owners from becoming victims (or perhaps beneficiaries) of speculative transactions in land. The restrictions on property rights reflect a fundamental ambiguity regarding private property in land, even where private ownership is legally recognised. The restrictions on ownership rights are among the most contested issues of agricultural reform. In Russia, the 10-year moratorium on buying and selling of privately owned land imposed by the 1991-1992 land reform legislation was eliminated by the presidential decree of October 1993. Yet not all political factions accept the president's policy, and the fierce debate around the question of buying and selling of land will probably continue until the adoption of a new land code at some unspecified time in the future, which will be the permanent law replacing the temporary presidential decrees that govern Russia's land policy today. In Georgia, with more than a third of agricultural land in individual tenure since 1992, selling and buying of land was allowed only by the new land law passed in February 1996, yet each transaction in land still requires administrative permission. In the Kyrgyz Republic, where all farmland is state-owned, user rights were made secure for 49 years and fully marketable in February 1994. A November 1995 presidential decree extended the secure lease term to 99 years. In Kazakhstan, the presidential decree of April 1994 similarly ensured full marketability of individual and collective user rights in land, although private ownership of land is limited to household plots. In Ukraine, on the other hand, buying and selling of land is still prohibited, even when it is registered as privately owned.

The ambiguity and confusion concerning property rights in land in former Soviet republics emerges with almost paradoxical clarity from a comparison between Turkmenistan and Uzbekistan. Turkmenistan's May 1992 constitution recognises private land ownership, whereas Uzbekistan's December 1992 constitution continues the Soviet tradition of the state as the sole legal owner of land. Yet in both these countries the land code prohibits the peasant to sell, exchange, or give land as a gift. This position is even retained in Turkmenistan's new 1996 draft land code. Use rights are thus totally non-transferable, regardless of whether land is legally classified as privately owned or state-owned. In Kazakhstan and Kyrgyzstan, on the other hand, use rights are freely transferable and marketable, even though these countries do not recognise full-fledged private ownership of land. The meaning of private land ownership in the former Soviet Union is thus very different from the accepted meaning in the West, while on the other hand private ownership does not appear to be a prerequisite for tradability of land rights.

The various restrictions on ownership rights, while often understandable in the historical and political context, are obviously inconsistent with the principles of market economy. There is a danger that they may become a serious obstacle to the shaping of an efficient economic structure based on private ownership and, in the final analysis, harm the competitiveness of agriculture in the region. Restrictions on buying and selling of land and on maximum size of holdings are obviously inconsistent with the need to allow the emergence of farms of optimal size. The objective of preventing non-farming uses of agricultural land and environmental concerns are best addressed through market mechanisms, such as differential taxation and ecological abuse penalties, and not through administrative threats of expropriation.

Distribution of land

Agrarian reform in the former Soviet Union, as in other former socialist countries, focuses on two interrelated issues: redistribution of land and restructuring of large-scale farms. These processes unfold regardless of the specific decisions on land ownership. The need for redistribution of land arises because the small-scale individual sector traditionally cultivated only 2 percent of agricultural land in the USSR, while 98 percent was managed by large-scale collective and state farms. This dual farming structure emerged as a result of highly restrictive government policies, and not through market mechanisms. The transition to a market-oriented economy thus necessitates reallocation of land and assets between the two subsectors of agriculture. Reallocation of assets necessarily leads to restructuring of large-scale farms, which must adjust their operations to a smaller resource base and to the disappearance of the centrally controlled administrative command system.

Of the two main options of restitution to former owners and distribution to the farming population, most former Soviet republics have chosen the strategy of 'land to the tiller'. Land is made available for distribution to workers of collective and state farms, and also to others who express their intention to farm. The obvious explanation of the different eligibility policy compared to East Central Europe is that nearly a century of Communism has washed away all traces of former land ownership, and very little tradition of private land ownership remains in most of the former Soviet Union. Voices for restitution or some form of compensation to former owners were raised in Western Ukraine, Moldova, and Western Belarus, in areas where collectivisation was imposed only after World War II. In the end though, only the Baltic states adopted the policy of restitution and compensation of the former landowners.

The distribution mechanism is fundamentally different for members of collective farms and for outsiders who wish to start farming. Collective farm members participate in a process of share distribution: the collective land is divided in the form of paper shares or certificates among the eligible members. Each eligible individual receives a share basically for the same amount of land, adjusted for quality. Distribution of land shares to members does not imply allocation of physical land plots corresponding to the shares: despite the allocation of land shares to the members, the land remains in joint cultivation pending further restructuring decisions by the 'share-owners'. A share-owner who wishes to establish an independent farming operation (individually or with a group of co-workers) in principle is entitled to receive from the collective farm a physical plot of land corresponding to the land share.

Outsiders who are not entitled to land shares can receive land for private farming from a special state reserve established for this purpose. In dense agricultural countries, such as Russia, Ukraine, or Moldova, the reserve for these private farmers was created by setting aside some of the land managed by collective and state farms (typically 15-20 percent of the total land). In the desert countries of Central Asia, a prospective private farmer may have to negotiate for a plot of irrigated land with the local collective farm (Uzbekistan) or accept some dry and rough 'virgin land' that will have to be developed from scratch (Turkmenistan).

According to a typical law of peasant farms, every individual with some minimum agricultural qualifications is entitled to apply for land to establish a private farm. Entitlement to land shares in collectives is somewhat more complicated. Most of the distribution programmes adopted in the FSU have given access to land shares to all rural inhabitants working directly in primary agricultural production on state and collective farms, as well as to those employed in the social services of the particular collectives. Rural industrial, supply, and public sector employees have been excluded from the sharing out of state and collective farm lands.

Armenia and Georgia stand out among the FSU countries in two respects. First, the entitlement to land extended to the entire rural population, and even to urban residents who had recently left the village. Second, and perhaps most importantly, physical plots of land, and not paper certificates of entitlement, were distributed to all beneficiaries. These decisions were mainly motivated by specific local conditions in countries where civil unrest and actual war contributed to rapid disintegration of central and local government and made it impossible to proceed with a more systematic and orderly allocation programme. Moldova is another former Soviet republic where recent legislation (February 1995) responded to popular pressure from various segments of the rural population and attempted to provide for a land quota to non-farming rural residents. This would naturally result in a downward revision of the previously calculated individual quota of land, but the implementation of the new procedure was blocked by the constitutional court.

Distribution of land shares in itself does not change the balance between the collective and individual sector of agriculture. The individual sector can increase only through actual transfer of land from collective and state farms to new landholders. Programmes for the creation of private farms outside the collectivist framework are one of the forces that shift land from collective to individual sector. Augmentation of household plots by internal distribution of land within the collectives is another such force. Thus, in most former Soviet republics the size of the average household plot nearly doubled since 1991. The changing role of the collective and individual sectors in Russia's agriculture is shown in Table 9.2, which illustrates the declining share of agricultural land in large-scale farm enterprises and the increase of the share of land in the individual sector. Although the data in this table are for Russia, the trends are typical of all former Soviet republics. While prior to 1990 the individual sector in the FSU cultivated 2-4 percent of agricultural land, its share today is over 10 percent of agricultural land in many of the republics (Table 9.3). The contribution of the individual sector to agricultural product has increased with the increase in land holdings, rising from the pre-1990 level of 25-30 percent to 40-50 percent, and in some of the republics (e.g., Georgia and Armenia) its share is much higher (CIS, 1994). However, despite the impressive increase in the size and contribution of the individual sector in the FSU, it is still much smaller than the individual sector in East Central Europe, where in most countries over 40 percent of agricultural land is cultivated individually (Figure 9.1).

Table 9.2
Land holdings by type of user in Russia: 1991-1995
(% of agricultural land)

Sector and enterprise type	*Jan. 1991*	*Jan. 1994*	*Jan. 1995*
State sector	58.2	17.6	16.5
Collective/co-operative sector	40.0	70.5	71.1
Incl. shareholding enterprises	0.3	52.5	54.2
Individual sector	1.8	11.9	12.5
Private farms	0.0	5.0	5.3
Associations of private farms	0.0	2.2	2.4
Household plots	1.4	2.8	2.9
Other	0.4	1.9	1.9

Source: Brooks and Lerman, 1996

Table 9.3
Land holdings by user in Former Soviet Republics: 1995
(% of agricultural land)

	Collective and state sector	*Individual sector*
Russia	87	13
Ukraine	80	15
Armenia	18	82
Georgia*	78	22
Turkmenistan	92	8
Uzbekistan	86	14
Kyrgyzstan	59	24

* In Georgia, households own 22% of agricultural land and 44% of arable land.

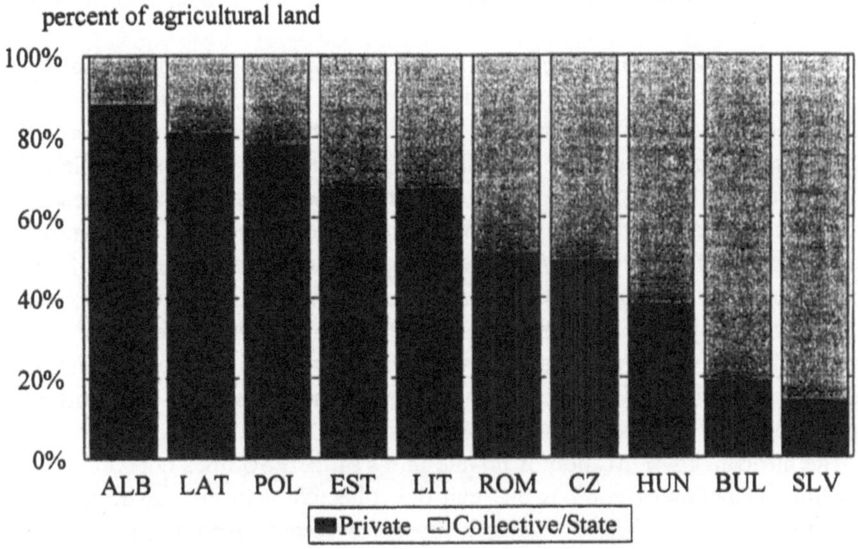

Figure 9.1 Collective and private sector in ECE countries

The size of the individual sector is not directly related to the amount of privately owned land. In Uzbekistan and Kyrgyzstan, there is no private land and the individual sector has use rights in state-owned land. But even in countries where private ownership of land is recognised, only part of the land in the individual sector is privately owned. Thus, in Armenia, 60 percent of agricultural land (including pastures) remains state-owned, but most of it is leased to individuals. As a result, the 82 percent of land in individual use in

Armenia (see Table 9.3) is made up of 30 percent privately owned land and 52 percent land leased from the state. Similarly in Russia and Ukraine, only one-third to one-half of the land in private farms is privately owned: the rest is still state-owned, although used by private individuals (Table 9.4).

Table 9.4
Land tenure in private farms (1993-1994)

	Russia	*Ukraine*
Privately owned	49%	30%
Lifetime possession	27%	52%
Usership	7%	17%
Leased	17%	2%

Source: World Bank surveys (Brooks and Lerman, 1995; Lerman et al., 1995)

So-called peasant, or private, farms are an important component of the individual sector in all former Soviet republics. While household plots are cultivated mostly on a part-time basis by 'share-owners' in collective enterprises, peasant farms are typically independent family farms established outside the collectivist framework. The emergence of independent peasant farms has been perhaps the most tangible outcome of the agrarian transformation in the FSU. While the changes in traditional large-scale farms have been slow and generally superficial, and private land ownership is still hotly debated by politicians and the public in many countries, the number of independent private farmers is rapidly growing (Table 9.5). In some countries (including Russia and Ukraine) the aggregate land holdings in private farms are approaching the total land in household plots.

The mechanism of creation of private farms in the FSU does not follow the general conception of families that exit the collective with their share of land and assets. Private farmers typically apply to receive land from specially created state reserves, and in Russia and Ukraine well over one-third of the new farmers are former urban residents who have decided to take up farming (Brooks and Lerman, 1994; Lerman et al., 1994). The average private farm in most former Soviet republics is much larger than in East Central Europe, where private farms are typically enlarged subsidiary household plots that have cut loose from the collectivist framework. Thus, private farms in East Central Europe are from 1 ha to 3 ha in size (European Commission, 1995), whereas peasant farms in the FSU typically manage from 10 to 40 ha of land (see Table 9.5). Among the FSU countries, it is only in Armenia and Georgia that land formerly cultivated by collective and state farms was distributed in small-sized plots of about 1 ha to most of the rural population, and the border line between household plots and private farms is fuzzy.

Table 9.5
Private farms in the Former Soviet Union

	Number of private farms			*Average size, ha (1994)*	*Share of agricultural land in private farms (1994)*
	1992	*1993*	*1994*		
Armenia	165,200	238,300	298,100	1.3	30.0%
Russia	49,000	182,800	270,000	42	5.0%
Kyrgyzstan	4,100	8,600	18,300	83	10.0%
Kazakhstan	3,300	9,300	16,300	410	3.0%
Ukraine	2,100	14,700	27,700	20	1.5%
Moldova	300	500	3,100	2.2	0.5%
Belarus	750	2,400	2,700	21	0.5%
Uzbekistan	1,900	5,900	7,500	10	0.5%

Source: CIS, 1994

Despite virtually equal aggregate land holdings, peasant farms still contribute much less to gross agricultural product than the household plots, mainly because of a different product mix. The land-poor household plots traditionally specialise in livestock production, relying on the collective enterprise for feed, whereas the peasant farms with their relatively generous endowments of land emphasise crops (Brooks and Lerman, 1995). As a result, the share of household plots in total agricultural production has always been disproportionately high compared to their land holdings, whereas the contribution of peasant farms to agricultural product is roughly equal to their share of land, i.e., 4-5 percent.

The rate at which private farms are spreading in the region lags far behind the original expectations (both local and international) and behind the opportunities provided by the legal framework. Politicians and experts offer different explanations of this fact, based on theoretical consideration and speculation. Recent World Bank surveys in the region actually elicited the opinion of the rural population on the subject (Brooks and Lerman, 1994; Lerman et al., 1994). The surveys indicate that the main reason for the reluctance to become a private farmer is that appropriate supportive conditions for private farming still do not exist. Thus, peasants hesitate to start an independent farm mainly because of insufficient capital, difficulty of obtaining machinery and farm inputs, and legal uncertainty (Lerman et al., 1995). The cautious behaviour is an expression of sober and realistic judgment of the situation by the peasantry. Similar findings are reported for countries in East Central Europe (Euroconsult, 1994).

It is sometimes argued that the present generation of peasants in the FSU has only a small part or none of the skills that are needed for successful operation of a private family farm. This is not necessarily true, because of the pervasiveness of household plots in rural areas, where all family members actually gained thorough farming experience since childhood. But even if this argument were true, it is not the main restraining factor: the main obstacle is that private farming in the region to day is too risky and promises a relatively modest income compared to the risks involved. At this stage, peasants prefer to keep their land and assets in the relative safety of the familiar collective organisation (old or newly reconstituted). The continuing development of functioning market services, including product marketing, input supply, and credit channels, will probably create conditions that encourage further transformation of the farming structure.

Restructuring of large-scale farms

Before land shares can be distributed to individuals, the state must give up ownership of land and transfer it to the collective. The mechanism to effect the transfer of land from the state to the collective, which is essentially part of the process of land reform, is linked in practice with the process of farm restructuring. This mechanism was basically determined by a series of Russian presidential decrees between December 1991 and March 1992. The same principles were adopted also by other republics that allow private ownership of land.

The 'land privatisation' mechanism involves re-registration of the former collective or state farm in a new organisational form, such as a limited liability partnership, an agricultural co-operative, a joint-stock company, an association, or even a collective enterprise. By this act of registration, the new organisation takes over the ownership of former state land, and can proceed to issue land shares to its members. The same mechanism generally applies to both collective and state farms. Contrary to East Central Europe, the reform legislation in the FSU does not provide different principles for state farms. These are simply transformed into collective farms following the decision of the general assembly of the workers, and after that the same rules and mechanisms apply to all large-scale enterprises.

Once the former collective or state farm re-registers and assumes ownership of its land, the next stage is internal restructuring of production organisation, management, and operations, hopefully in line with market-oriented principles. Transfer of land from state to collective ownership is a first step in a process that ultimately should lead to transfer of all property rights to individuals. The legal mechanism for transfer of property rights to individuals prescribes distribution of paper certificates of entitlement to a share of land and a share of other farm assets to each member in a collective. Distribution of land and asset shares, combined with the legal right of exit from the existing collective, establishes a fundamental mechanism for meaningful restructuring of the farm sector in the future. It is an interim mechanism, which is ultimately expected to lead to physical distribution of land plots and farm assets in kind.

External reorganisation (formal registration as a new organisational form) and distribution of shares to members must be followed by internal reorganisation involving reallocation of resources (land, assets, labour) to relatively autonomous market-oriented subunits. Actual observations of farm restructuring in the FSU indicate that the process is evolving along several alternative tracks, which are schematically presented in Figure 9.2.

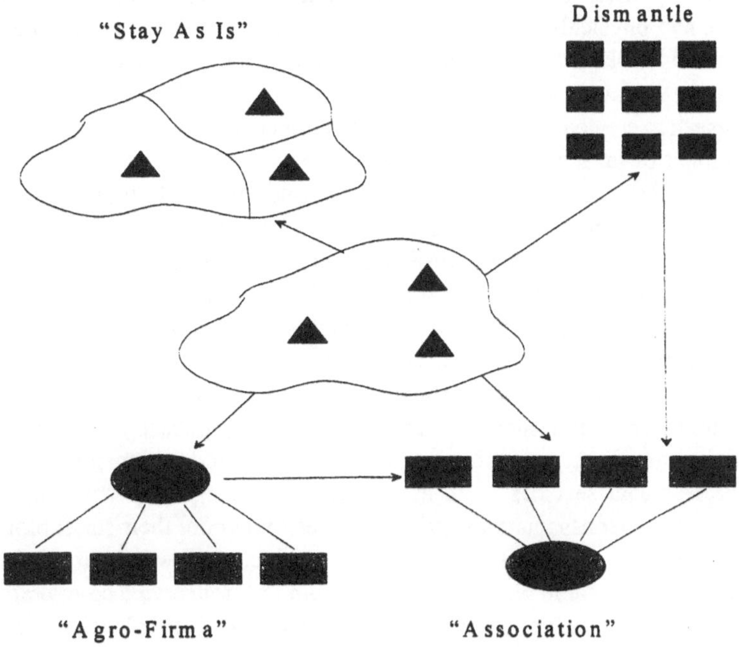

Figure 9.2 Alternative farm-restructuring modes

The most conservative and easiest procedure is to retain the prior structure under a new name. Land and assets are distributed to individuals in the form of paper shares, but this is not followed by physical allocation of individual entitlements. The recipients keep their shares of land and assets in the old collective, continuing the tradition of collective production. The collective may change its name to a joint-stock society, or a similar organisational structure, but it continues functioning as before. Another variation on this theme is the separation of a multi-village farm enterprise into several village-based enterprises, each continuing to manage the land and assets of its members collectively. In principle, even in a 'stay as is' situation, the shareholders can be expected to adopt a profit orientation in order to maximise dividends on their shares, yet distribution of dividends is still without much practical significance, and the shares are not tradable. Without appropriate incentives to shareholders through tangible returns on their investment, no true sense of ownership is created, and the shareholders continue working as salaried employees under the direction of their previous collective manager. These 'new-old' structures retain all the weaknesses and inefficiencies of collectives, although they are often expected to be more efficient because of their modern-sounding names and new charters.

At the opposite extreme are farms that completely dismantle the old structure and physically distribute all land and assets to individuals. The large-scale farm is broken up into small family units, each with its allotment of land and assets (the latter possibly in cash equivalent form). A few hundred private farms are thus created in place of one large collective farm.

A number of farm structures intermediate between the old collective and the new family farm have appeared. In a 'bottom-up' approach, individual private farmers may combine to form an association of agricultural producers. This process is demonstrated in Figure 9.2 by the transition from the collection of independent private farmers at top right to the association structure at bottom right. In some associations, each farm retains its identity, but all farms co-operate in provision of services, where economies of size and the effect of pooling of resources are most pronounced. Co-operative support may include use of transport, input supply, product marketing, purchase and joint use of agricultural machinery, mechanical maintenance, credit, transportation, and even professional services, such as accounting, management consulting, and extension. In other associations, private farmers may pool their small plots into larger allotments and form a partnership for production. Several partnerships or associations in the same region may in turn join to form service co-operatives.

A similar associative structure may evolve 'top-down,' when the old collective, instead of totally dismantling into hundreds of private farms, reorganises internally into a system of relatively small autonomous profit centers. This process initially produces the structure shown schematically at bottom left in Figure 9.2. The land and asset base of each such unit is the sum total of the land and asset shares of its members or 'active investors'. Because of the relative smallness of such units, which are organised as partnerships or small corporations, the active workers maintain the sense of private ownership of production assets. The old management structures of the collective in turn may reorganise to provide the necessary services and support to these autonomous units. Agricultural production is thus concentrated in units that are small enough to maintain a good sense of personal involvement and accountability, while farm support services and the interface with the environment are provided by a relatively large and professionally experienced co-operative structure.

This organisational form has the important advantage of providing a natural solution to the needs of pensioners or other shareholders who do not wish to continue farming. Each production unit, in addition to its active member-owners, may accept land and asset shares from inactive investors, who will be entitled to dividends from the unit's profits. In order to attract land and capital assets from inactive investors, the production units will have to compete among themselves by improving their efficiency and increasing profits.

Initially, the professional co-operative structure representing former management may play a strong guiding role in the new organisation. Gradually, as the autonomous units gain experience and confidence, the roles will be reversed, and the producers will assume the leading role in the associative organisation, while the management 'umbrella' will begin to function as a true service co-operative supporting the producers. This evolution toward more independent subdivisions and a more service-oriented 'core' is the essence of the top-down transition to the associative structure at the bottom right of Figure 9.2. The difference between the two structures depicted at the bottom of Figure 9.2 is effectively in the degree of independence of the member units and the balance between managerial and service functions of the former center. The key difference between this form and the association of private farmers is that the individuals may not have converted their collectively owned land and assets into individual ownership.

What has changed in restructured farms?

The various restructuring modes of collective and state farms observed in the region are summarised in Table 9.6. Experience in the region indicates that in most cases the process of reform does not lead to a fully fragmented farming structure, contrary to initial expectations. The dismantling approach has been implemented only in Armenia, where all collective farms were rapidly disbanded and divided into very small private farms during 1991-1992. In other countries, dismantling is a rare phenomenon. In Russia, among 21,000 farm enterprises that reorganised by January 1993, only 268 broke up completely into private farms (Brooks and Lerman, 1994).

Table 9.6
Restructuring modes for collective and state farms

•	Reconstitution of a collective structure based on individual ownership of land and asset shares
•	Transformation of the collective structure into a joint-stock corporation based on individual shares
•	Division of the collective structure into autonomous profit-oriented entities based on individual investment of land and asset shares and operating within an association or a service co-operative
•	Separation of independent entities from the collective structure (family farms, partnerships, or production co-operatives)
•	Co-operation of independent entities

The new land owners are not particularly willing to leave the supportive umbrella of the collective structure and risk everything in independent farming. The overwhelming majority of farm workers surveyed in Russia and Ukraine prefer to leave their land and asset shares in the collective (Lerman et al., 1995). A recent survey of residents in 49 reorganised enterprises in three Russian provinces showed that the land and asset shares were leased back to the large farm or invested in its equity capital by over 95 percent of shareholders (IFC, 1997).

Farms tend to reorganise as relatively large units, although with some downsizing. The 'stay as is' approach is popular in Russia and other former Soviet republics, where it is referred to as 'changing the sign on the door' (Table 9.7). Fully 35 percent of the reorganised farm enterprises in Russia as of January 1993 decided to keep their old form, without even changing the sign on the door, 12 percent registered as joint-stock societies, and another 40 percent reorganised as limited liability partnerships (Brooks and Lerman, 1995). In Ukraine, 83 percent of the reorganised collective and state farms registered as collective enterprises or partnership, and 3 percent as joint-stock societies (Lerman et al., 1995). An extreme example of 'changing the sign' approach is seen in Turkmenistan, where all large-scale farms were summarily 'reorganised' by a presidential decree of June 1995, which changed their name from *kolkhoz* (collective farm) to 'peasant association'. Further presidential decrees appear to encourage internal restructuring of the large-scale farms through 'intrafarm leasing' of land and assets by groups of workers, with support and control services provided by the former management group.

The statistical data about the actual process of farm restructuring and the evolution of the new organisational forms are unfortunately highly incomplete. It seems, however, that the traditional dual structure of socialist agriculture, where production was concentrated in very large collective and state farms, on the one hand, and in very small household plots, on the other, is becoming less sharp. The average size of large-scale farm enterprises is decreasing. Farm reorganisation results in three Russian provinces show that the average farm size declined from 3,600 ha and 160 workers before restructuring to 1,900 ha and 85 workers after restructuring, while the number of registered entities increased from 170 to 310 (IFC, 1997). Survey data for Russia point to a definite increase in the proportion of farm enterprises in the lower tail of the size distribution between 1990 and 1993 (Figure 9.3), although the smallest farm enterprises are still very large compared to typical farms in market economies. At the other extreme, household plots in all former Soviet republics practically doubled in size to about 1 ha since 1991 through government programmes that distributed land to the rural population, and a new category of private farms with sizes of 10-40 ha has emerged in the distribution between the two extreme modes.

Table 9.7
Restructuring of farm enterprises in Russia and Ukraine
(percent of all farms)

	Russia	*Ukraine*
Not reorganised	22	28
Reorganised	78	72
Keeping old form with new charter	27	9
Collective enterprises and limited-liability partnerships	32	55
Join-stock societies	9	2
Agricultural production co-operatives	6	2
Associations of peasant farms	3	2
Other forms	1	2

Source: Data for Russia as of January 1993, based on Brooks and Lerman (1994); data for Ukraine as of January 1994, based on Lerman et al. (1994).

Despite some evidence of downsizing in restructured farms, internal reorganisation has not produced deep results. In a survey conducted in 1996 in Ukraine, 75 percent of large-farm managers report that the new subdivisions enjoy independence in planning and managing their production activities, but this is practically the limit of their autonomy (Lerman and Csaki, 1997). In less than 5 percent of cases the subdivisions are responsible for input purchasing and product marketing, or have their own administrative staff, and in only 7 percent of farm enterprises the subdivisions have freedom to hire and fire their workers. Practically none of the subdivisions have their own bank account. In line with this extremely limited autonomy of the newly formed subdivisions, virtually all farm enterprises retain collective or central management organs responsible to a different extent for overall production planning and management for the entire enterprise, banking relations and finances, coordination of the production activities of the subdivisions, provision and allocation of commonly owned machinery and equipment to subdivisions, and labour management. The collective framework thus has preserved most of its traditional functions.

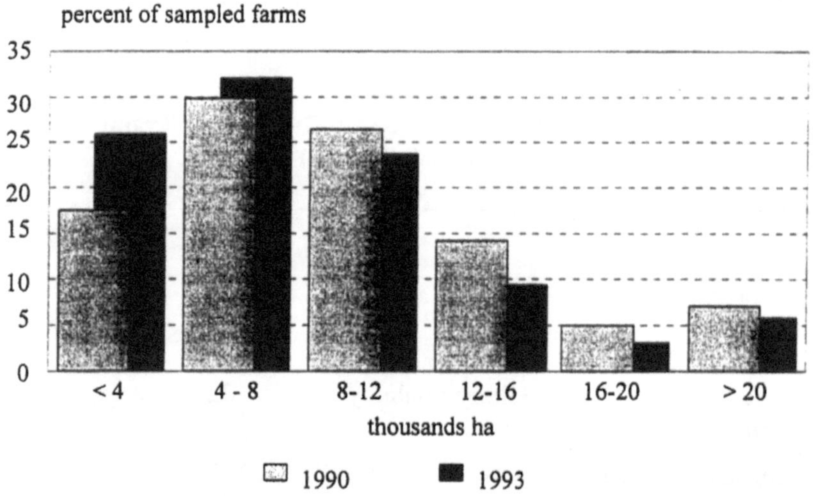

Figure 9.3 Size distribution of farm enterprise, Russia 1990-1993

A basic concept in internal restructuring of large farm enterprises is the exercise of free will by shareholders when forming new functional groupings with their land and asset shares. This principle is strictly observed in the extensive farm restructuring project conducted by the Internal Finance Corporation (IFC) and the UK Know-How Fund since 1993 in several provinces in Russia. Yet outside the experimental project, the principles of voluntary regrouping of shareholders are less apparent. Ukrainian survey results indicate that in over 60 percent of farm enterprises the new units were formed simply on the basis of the old production subdivisions, and in over 80 percent of cases the new units were allocated the land and assets that they had on the old balance sheet (Lerman and Csaki, 1997). The individual shareholders did not exercise their freedom of choice and freedom of association when creating the new units, and the asset base of the new units was not formed by shareholders voluntarily pooling their land and asset shares. Shareholders were simply 'assigned' to their old organisational subdivisions.

It is not surprising that in this situation employees of farm enterprises often fail to discern any significant changes in the way their farms are operated and managed. According to recent surveys in Ukraine and Russia (Lerman and Csaki, 1997; IFC, 1997), about half the employees report that no real change has so far taken place in the reorganised farms. The assessment of reorganisation outcomes by employees of farm enterprises strengthens the feeling based on previously described findings that so far changes in large farms have been largely superficial, and have not touched on the systemic flaws inherent in the socialist system of agriculture.

There have been no radical changes in labour relations either: most managers report that their farm enterprise continues to be committed to a life-time employment policy for its members and do not acknowledge disguised unemployment on their farm (Lerman and Csaki, 1997). Yet there is evidence of employee departures in more than half the reorganised enterprises in the Ukrainian survey (Lerman and Csaki, 1997), and the percentage of Russian farm employees concerned about the possibility of losing their job in reorganised enterprises is higher than in non-reorganised farms (IFC, 1997). Thus, despite the declared commitment of farm mangers to the old socialist ideology of labour, reorganised farms appear to be more sensitive to dangers of labour redundancy.

Another positive result of reorganisation is the marginally higher involvement of employees in decision making at the farm level. Thus, in the Ukrainian survey, 54 percent of respondents report that they now participate in decisions, compared to 48 percent before reorganisation (the difference is statistically significant). Among those who did not participate in decision-making prior to reorganisation fully 20 percent report that they are now involved in farm-level decisions (Lerman and Csaki, 1997).

A definite change is observed in the relations of the farm enterprises with local authorities. Direct dictates from the top have almost ceased completely. There is no pervasive intervention in the production plans of farm enterprises, except in case of strategic commodities, such as cotton in Turkmenistan and Uzbekistan. Farms are allowed considerable independence in their production and marketing decisions. Yet old habits die hard, and there is a strong informal chain of dependency and ongoing consultation between managers and the district bureaucracy. Managers cannot ignore the goals and objectives of district authorities, and in this sense they are not free to adjust their product mix completely in response to market signals. One of the most glaring examples is the relatively slow decrease of livestock production in farm enterprises: although livestock has been unprofitable in recent years, and the new private farmers indeed have changed their orientation to emphasise crop production, managers of large farms cannot afford to ignore the traditional insistence of district authorities on maintaining the herd as 'a source of milk for our school children'.

The continued dominance of large-scale farming structures may be attributed to various reasons. Families apparently feel that a large-scale farm provides a greater measure of safety in the rapidly changing environment than individual farming. The large farms will be unable to provide in the new economic environment the same range of social services to their members as in the past, and yet rural residents seem to believe in the safety of numbers, at least at the present stage. Also large-scale farms, with the benefit of accumulated experience of their professional managers, may be better equipped to operate in an environment without fully functioning market services than small individual operations. In a sense, they may provide a natural transition to service co-operatives of the future. Another factor that must not be ignored is the traditional power of the manager as a representative of the authorities in old collective farms. In many instances, the manager exercises influence to prevent deep restructuring and preserve the large-scale organisation in order to keep his former power. All these factors contribute to the observed inertia and the slow transition to new farming structures despite availability of enabling legislation.

Because the economic environment is still changing and the development of market infrastructure still has a long way to go in the former Soviet Union, options for restructuring should remain open. Thus, recipients of land rights who at this stage choose to remain in collective-type enterprises should retain the right to exit with land and assets at a point in the future. Whether the exit right is protected, or property rights devolve to the enterprise when a shareholding firm is created depends on how the relevant laws are written and how the by-laws of the enterprise treat the issue of withdrawal. A number of laws and decrees in former Soviet republics present severe obstacles to exit from shareholding enterprises with land and assets. When barriers to exit are high, the likelihood is great that the farm structure will be frozen in the form of corporatised large farms created in the first stages of restructuring. These are not likely to be the forms of farm organisation that will allow agriculture in the region to become competitive in relatively open market economies. Flexible exit mechanisms, on the other hand, will allow the development of a multiplicity of farm structures and enable the mechanism of evolutionary selection to take its course.

Conclusions

After five years of reforms, it is clear that the emerging structure of agriculture in the FSU is different from the initial expectations. There has been no wholesale fragmentation of large-scale farm enterprises into family farms, except in Armenia and Georgia, where the respective governments actively pursued a policy of fragmentation. Instead, the transition agriculture in the FSU is characterised by three groups of producers: large-scale shareholder farms managing over a 1000 ha with over 100 workers; household plots of 0.2-1 ha cultivated by rural families; and independent private farms cultivating from 10 ha to 100 ha outside the collectivist framework. The third group is a new product of agrarian reforms since 1991. The first two groups are the traditional components of the dual socialist agriculture that has existed during decades. Yet they also have changed in the process of reform: the large-scale farms have become on average smaller, while the household plots practically doubled in size. The large-scale farms are changing from state-controlled monoliths to agglomerations of individual stakeholders, who in varying degrees are beginning to understand and exercise their rights.

Yet, despite all the changes, the restructuring of the collective sector has been slow and relatively superficial. Most land still remains in collective (although 'privatised') ownership. Legal and political uncertainty, inadequate information, and lack of market services are serious obstacles to deeper restructuring of the old organisations. In the face of high risks individuals are reluctant to assume greater responsibility for private production and at this stage prefer to keep their land and assets in reconstituted collectives. Economic and legal stability, reliable law-enforcement mechanisms, clear, unrestricted, easily transferable property rights, and establishment of competitive, non-monopolistic market services are essential measures for further progress with agrarian reforms in the former Soviet Union.

References

Brooks, K. (1990), 'Soviet Agriculture's Halting Reform,' *Problems of Communism*, Vol. 39, pp. 29-41, March-April.

Brooks, K. and Lerman, Z. (1994), 'Land Reform and Farm Restructuring in Russia', *World Bank Discussion Paper* 233, The World Bank: Washington DC.

Brooks, K. and Lerman, Z. (1995), 'Restructuring of Traditional Farms and New Land Relations in Russia,' *Agricultural Economics*, Vol. 13, pp. 11-25.

Brooks, K., Krylatykh, E. Lerman, Z., Petrikov, A. and Uzun, V. (1996), 'Agricultural Reform in Russia: A View from the Farm Level', *World Bank Discussion Paper* 327, The World Bank: Washington, DC.

CIS (1994), *Strany-Chleny SNG v 1993 g. Statisticheskii Yezhegodnik* (Member-Countries of the CIS in 1993. Statistical Yearbook), Statistical Committee of the CIS: Moscow.

European Commission (1995), *Agricultural Situation and Prospects in the Central and Eastern European Countries*, Summary Report and Country Reports, Directorate-General for Agriculture: Brussels.

Euroconsult (1994), 'Farm Restructuring and Land Tenure in Reforming Socialist Economies: A Comparative Analysis of Eastern and Central Europe', *World Bank Discussion Paper* 268, The World Bank: Washington DC.

FAO (1994), *Production Yearbook*, FAO: Rome.

IFC (1997), *Monitoring Russian Reorganized Farms: An Integrated Analysis of Economic and Social Change*, IFC Land Unit Monitoring Team: Moscow.

Lerman, Z., Brooks, K., and Csaki, C. (1994), 'Land Reform and Farm Restructuring in Ukraine', *World Bank Discussion Paper* 270, The World Bank: Washington DC.

Lerman, Z., Brooks, K. and Csaki, C. (1995), 'Restructuring of Traditional Farms and New Land Relations in Ukraine,' *Agricultural Economics*, Vol. 13, pp. 27-37.

Lerman Z. and Brooks, K. (1996), 'Russia's Legal Framework for Land Reform and Farm Restructuring,' *Problems of Post-Communism*, December.

Lerman Z. and Csaki, C. (1997), 'Land Reform in Ukraine: The First Five Years', *World Bank Discussion Paper* (forthcoming), The World Bank: Washington DC.

Medvedev, Zh.A. (1987), *Soviet Agriculture*, Norton: New York.

Nikonov, A.A. (1995), *The Spiral of a Centuries-Long Drama: Agrarian Science and Politics in Russia from the 18th through the 20th Centuries* [in Russian], Entsiklopediya Rossiiskikh Dereven': Moscow.

Wädekin, K.-E. (1973), *The Private Sector in Soviet Agriculture*, University of California Press: Berkeley.

10 Agricultural privatisation, land reform and farm restructuring in Central and Eastern Europe: A comparative analysis

Johan F.M. Swinnen and Erik Mathijs[1]

Introduction

This concluding chapter presents a comparative analysis of agricultural privatisation and transformation policies, land reform and resulting changes in farming structures in Central and Eastern European countries (CEECs) based on the country studies in this volume, the discussion at two workshops and on additional information from a related study on the political economy of agricultural privatisation and farm restructuring (Swinnen, 1997a). Our comparative analysis yields some general patterns in both the process of agricultural privatisation and land reform, the resulting farm structure, and their determinants. Some important observations are the following:

1 Most CEECs have chosen to restitute collective farm land to former owners. Moreover, former owners who kept legal rights to their land were restituted property rights on their land without exception. Former owners who lost their legal ownership title were restituted land only in Albania, Slovenia and the Baltic States. In Albania the majority of land is distributed to farm workers.

2 In the majority of CEECs, state farm land is leased, pending sale of the land. For example, in Eastern Germany nationalised land is managed by the Land Utilisation and Administration Company and leased to former owners who lost their ownership titles and to legal entities. The main exceptions are the restitution of state farm land in Slovenia, and the distribution to farm workers in Albania.

3 Non-land assets have typically been privatised through other procedures than those used for land. In many cases they are privatised using vouchers

which can be turned into capital shares in the new co-operative farm or used for purchasing non-land assets for private use.

4 Privatisation does not necessarily lead to full transfer of all property rights to new (private) owners. In other words, the post-reform effective property rights distribution is only partially determined by the land privatisation legislation.

5 Following the enactment of the necessary legislation, state and collective farms have been transformed into a wide variety of farm organisations, such as producer co-operatives, joint stock companies, limited liability companies, partnerships and individual farms.

6 Large-scale production organisations still dominate agricultural production in several CEECs. Many new land owners lease their land to the large-scale successor organisations of the collective and state farms. In 1994, they cultivated more than two-thirds of the total agricultural area in Bulgaria, Hungary, the Czech Republic and Slovakia. In several other CEECs individual farming has become the dominant farm organisation. This includes Poland and Slovenia, where small-scale farming dominated under the Communist period, and Albania, Latvia and Romania, where spontaneous privatisation led to the widespread break-up of collective farms.

This chapter presents some explanations for these observations on the remarkable variation in privatisation and land reform procedures on the one hand and their impact on property rights and farm restructuring on the other hand. Three arguments will be presented.

1 The choice of the privatisation and land reform process is determined mostly by legal and historical characteristics of agricultural assets, the pre-collectivisation asset ownership distribution, the nature of post-collectivisation asset ownership and the ethnicity of the former owners. These characteristics affect the distributional effects of various privatisation options and the political consequences of the privatisation choice.

2 The resulting property rights distribution is affected by the choice of the privatisation process. The effective property rights distribution is also, and importantly, influenced (a) by technical factors and transaction costs in the reform implementation, (b) by the extent to which interest groups could influence the reform implementation, and (c) by the detailed implementing

regulations, in turn affected by political influence of interest groups and by political coalitions.

3 Privatisation and transformation policies have influenced farm restructuring. More specifically, land restitution has had a conservative influence on farm restructuring. However, the most important determinants of farm restructuring are economic factors, including relative productivity of existing human and physical capital in different organisations, risk, terms of trade, transition-specific distortions, and access to production factors.

The chapter starts with observations on privatisation and land reform and discusses their explanations. Following this is a summary of data on CEEC farm restructuring. This compares the observations with expectations based on theory. We then explain the apparent divergence, and identify the factors which will affect future restructuring.

The reader should be careful in interpreting the data and the concepts used in the statistics provided by CEEC authorities. For example, whether 'private' refers to ownership rights or all property rights (see further) is often uncertain; also the organisational structure of 'private co-operatives', 'joint stock companies' or 'limited liability companies' and their effective operation is unclear in many cases. It tends to distort the conclusion of outside observers in two ways. The statistics suggest more progress in the reforms than has actually taken place, based on the distribution of final land titles and may substantially overestimate the extent of effective restructuring of property rights. However, it also may lead to the opposite reaction: as much *ad hoc* evidence suggests that in many cases effective property rights have not (yet) been fully distributed and statistics confirm that farm restructuring often seems to be legal window-dressing and less restructuring has occurred in reality, it might induce an underestimation of effective changes because of altered incentives in the emerging farm structures and of unregistered, informal contracting of land, labour and other assets.

Agricultural privatisation and land reform

CEEC agricultural privatisation policy choices

CEEC governments have chosen different ways to privatise agricultural assets. In Bulgaria and in former Czechoslovakia, land is restituted to former owners, in historical boundaries if possible. Poland is selling some of its state land, while Hungary used a voucher scheme for privatising a major part of the land

and Albania distributed most land property rights on an equal per capita basis. However, with the exception of Poland and Albania (and also the former Soviet Union) an important share of the land is restituted to its former owners in all CEECs (see Table 10.1). Former owners are restituted the land in historical boundaries if possible. Otherwise they receive property rights on a plot of land of comparable size and quality. This forms the main process of land privatisation in CEECs.

Exceptions to land restitution are the voucher privatisation and distribution to collective farm employees of two-thirds of the land used by Hungarian co-operatives, the sale of Polish state farm land, and land distribution to the rural population in Albania and former Soviet Union. Other cases where land is not restituted to its former owners include land formerly owned by the Catholic Church in the Czech Republic, land formerly owned by Italian citizens in Slovenia, and former owners who lost their land prior to 1947 in Romania and prior to February 25, 1948 in former Czechoslovakia.

Whereas land is mostly restituted, this is not the general rule for other assets. Non-land assets are restituted in some countries, but in many cases they are privatised using vouchers which can be turned into capital shares in the new co-operative farm or used for purchasing non-land assets for private use. The distribution of the vouchers was the subject of much debate. In general, the principle was that members (or their heirs) who had contributed land, labour or other assets to the co-operatives over the 45 years of their existence should receive some share of the remaining assets. To implement this required an inventory of the assets of the co-operative, their value, and a formula to determine the shares of each contributor. Both aspects posed great difficulties. The solutions chosen varied substantially between CEECs as they were also driven by political considerations, and often became very complicated.

Table 10.2 shows how active members receive only 20 percent of non-land assets in Slovenia, versus 40 percent in Hungary and 100 percent in Lithuania. In Lithuania farm equipment can be purchased with investment vouchers (which everybody received), money, or 'green vouchers' which were only distributed among agricultural workers, based on the number of years they worked in the collective. The consequence is that rural pensioners obtained a large share of the non-land assets. In former Czechoslovakia, vouchers were distributed according to a 50/30/20 scheme: 50 percent for former land owners, 30 percent for former non-land assets owners, and 20 percent for 'labour contributions'. In contrast, Bulgarian and Romanian 'labour contributions' received 50 percent and 60 percent, respectively, of the non-land assets.

Table 10.1
Most important land reform procedures in CEECs

	Collective farm land	*State farm land* [a]
Albania	Distribution (physical)	Distribution (physical)[b]
Bulgaria	Restitution	Miscellaneous[d]
Czech Republic	Restitution	Sale (leasing[c])
East Germany	Restitution	Sale (leasing[c])
Hungary	Restitution + distribution (phys.) + sale for compensation bonds[f]	Sale for compensation bonds + sale (leasing[c])
Latvia	Restitution	Restitution
Lithuania	Restitution	Restitution
Poland	-	Sale (leasing[c])
Romania	Restitution + distribution (phys.)	Undecided[e] + Restitution
Russia	Distribution in shares[g,h]	Distribution in shares[g]
Slovakia	Restitution	Sale (leasing[c])
Slovenia	-	Restitution
Ukraine	Distribution in shares[g]	Distribution in shares[g]

[a] Excluding research farms which are nowhere privatised.

[b] Farm workers received vouchers in newly established joint ventures. However, as most of these joint ventures failed, farm workers received first user rights and eventually full property rights.

[c] Land is leased to individuals or entities pending sale.

[d] In Bulgaria, the distinction between state and collective farms is more complicated than in other CEECs because the creation and later abolishment of the so-called Agro-Industrial Complexes. Part of the land classified under 'state farm land' is restituted, because it was initially collective farmland and has a similar status; another part will not be privatised, and another part is the land on which large pig and poultry enterprises are built and which will be privatised separately.

[e] The Romanian government has not decided how to privatise the state farms, including the land, on two-thirds of the state farm land.

[f] Each of the land reform procedures applies to approximately one-third of the collective farmland.

[g] Distribution of collective and state farmland equally per capita among collective farm members or state farm employees in the form of paper shares or certificates. Outsiders who are not entitled to land shares can receive land for private farming from a special state reserve established for this purpose (15-20% of TAL).

[h] Private ownership is prohibited in 10 ethnic republics of the Russian Federation.

Table 10.2
Rules for distribution of non-land assets*

	Latvia	*Lithuania*	*Bulgaria*	*Czech & Slovak*	*Hungary*	*Slovenia*
Active members	(yes)	100 %	(yes)	(yes)	40%	20%
Of which to:						
- Labour	n.a.	n.a.	50%	20%	n.a.	100%
- Land	n.a.	n.a.	50%	50%	n.a.	0
- Capital	n.a.	n.a.	0	30%	n.a.	0
Former owners	n.a.	0	(yes)	(yes)	20%	10%
Pensioners	n.a.	0	0	0	40%	10%
State fund	n.a.	0	0	0	0	20%
Sale	n.a.	0	0	25%	0	40%

* '(yes)' refers to 'eligible for allocation of non-land assets, but detailed information on the share is unavailable'.

Source: Country studies in this volume

In *Eastern Germany*, the contributions members made to the co-operative's assets during collectivisation had to be returned. If the equity capital was larger than the members' contributions to assets, members also receive the rents and interest rates for their contribution. If the equity capital was higher than the sum of the members' contributions and the rents and interests thereon, additional claims could be submitted and calculated based on the members' labour input to the co-operative farm (Beckmann and Hagedorn, chapter 4).

In *Bulgaria*, the agreed formula was partly conditioned by the need to placate the ethnic Turkish population who, for the most part, only contributed labour to the co-operatives. The formula agreed upon was essentially to weight labour and land shares equally at 50 percent. The land shares were based on the claims of the members of the co-operative adjusted in three ways: a reduction factor to allow for land which moved out of agricultural production; a quality adjustment coefficient to account for changes in the distribution of land by quality grade; and a factor to adjust for the period of time the landowner was a member of the co-operative. The overall share of each member was applied separately for each of the categories of assets. It was expected that coupons would then be traded for the assets at a grand auction or exchange in order to complete the distribution. In this way, individuals exchanged their coupons for different categories of assets or combined their coupons to bid for larger items (Buckwell et al., 1994).

In *former Czechoslovakia*, 25 percent of the assets of the co-operatives was sold to provide some capital for the transformed co-operative. The remaining assets (if there were any left) were divided in kind or in cash as follows: 50 percent according to the contribution of land, 20 percent according the years worked in the co-operative, and 30 percent according the amount of non-land assets used by the co-operative. The new asset owners used their property shares: (1) as starting capital for new private farms, (2) as a membership investment in a transformed co-operative, or (3) as 'dead capital', if neither of the first two options were chosen. In this case repayment will only happen after seven years following the beginning of the transformation (Lindemans, 1997).

In *Hungary*, all the collectively owned non-land assets were valued as they appeared in the accounts and divided as business shares. People eligible to get these business shares were existing members, people who had been members for at least five years before the Transition Law came into force, or their heirs. Business shares could also be given to employees and family of co-operative members by decision of the General Assembly of the transformed co-operative. Criteria used by the General Assembly were the length of membership, the level of salary and the value of any collectivised property. These business shares could then be exchanged for physical assets by consensus or by auction (Mathijs and Mészáros, chapter 5). As a result of the distribution, 41.5 percent was allocated to active members, 38.7 percent to retired members and 19.8 percent to outsiders (Fehér, 1995).

The *Romanian* situation was to a great extent characterised by illegal seizures and destruction of property. Hence, the equipment of co-operatives was distributed and carried away outside the legislative or regulatory framework. Gavrilescu (1994) reports that serious disorders in the application of the 1991 Land Law led to the rapid decline of transferable assets. As a result, the privatisation of non-land assets has proceeded much slower than that of land and livestock (Sarris and Gavrilescu, chapter 6).

In *Slovenia*, non-land assets of agricultural farms had to be divided as follows. Employees receive 20 percent and State Funds 40 percent: 10 percent of shares are transferred to the Pension Fund, 10 percent of shares are transferred to the Compensation Fund to compensate former owners of nationalised property, and 20 percent of shares is transferred to the Development Fund and reserved for distribution through authorised investment funds. The remaining 40 percent could be privatised through several methods of privatisation. Company managers and employees are allowed to choose the methods of privatisation and the Company Privatisation Agency must agree with the proposal. It is possible to offer to managers and employees an internal buy-out for which payment at a discount is required or to sell to domestic and foreign investors (Bojnec and Swinnen, chapter 8).

Determinants of agricultural privatisation and land reform policies

Three legal and historical factors seem to be key determinants of the choice of the privatisation and land reform policies in CEECs. They are (in declining order of influence) (1) the post-collectivisation asset ownership status, (2) the ethnicity of the pre-collectivisation asset ownership, and (3) the equality of pre-collectivisation asset distribution (Swinnen, 1997b).

The single most important factor determining the privatisation policy choice is the legal ownership status of the asset at the outset of the reforms. All agricultural assets which were still legally privately owned in 1989 have been restituted in all CEECs. This factor is the main reason why land is generally treated differently than non-land assets in privatisation. During collectivisation, private farmers and land owners were forced to bring their assets into a collectively organised production unit. Many of the assets that the collective used were initially privately owned, but gradually, most assets became collective property when the original assets (buildings, animals, machinery) were physically replaced. The main exception is land. Land could be changed quite dramatically in qualitative terms, because the collectivisation process resulted in a consolidation of fields, in a new infrastructure, destruction of old buildings and roads and construction of new ones. Plots of land which prior to collectivisation were adjacent to a road, might now be in the middle of a vast grain field. Plots which were previously located close to villages may now be far away. Or plots that had little infrastructure might now have drainage and irrigation facilities, or even buildings on them.[2] However, land as a defined physical location remained legally privately owned throughout the Communist regime in many CEECs. Only in the Baltic countries, as well as in the rest of the former Soviet Union (FSU), and in Albania land was legally nationalised.

As a consequence, CEEC governments had little choice but to restitute this land unless they were prepared first to take away the legal ownership rights from the 'former', but also 'formal' or 'legal', owners. Indeed, some economists (see, for example Davidova et al., chapter 2) even argue that one should not refer to land restitution as 'privatisation', because the land was always legally privately owned.[3] One can imagine that in this situation the political costs of not restituting the land are considerably higher than when legal land ownership was not private. First, opposition from the legal owners is bound to be stronger precisely because of the legal status of the land. Second, such a policy would be inconsistent with the political and economic reforms. It would be ironic if private ownership rights, after surviving 40 years of collectivisation and state control, were confiscated by democratically elected governments supporting a market economy!

The principle that agricultural assets which were formally still privately owned are restituted to their formal owners explains several observations.

1 Differences between CEECs: where land is not restituted to former owners (such as Albania, FSU and Poland), the land was state owned.

2 Differences within CEECs: where part of the land is restituted (e.g., Hungary), the difference is due to legal ownership.

3 State farm land is often not restituted when collective farm land is (as state farm land was mostly state owned);

4 In several CEECs the base for land restitution is the ownership distribution after the land reform(s) implemented under Communist dominated governments after World War II.

5 The difference between the treatment of land and non-land assets.

The second determinant of land privatisation method is ethnicity. The privatisation policy choice affects the distribution of asset ownership between ethnic groups. A general observation is that agricultural assets are not restituted to former owners who are, or have become, foreigners. This factor is important in explaining the difference between the privatisation of state farm land in Poland (sales and leasing) and in Slovenia (restitution), with both countries having a very similar pre-1989 agricultural structure. In Slovenia, most of the former owners of the state farm land were Slovenian, but Polish state farm land was mostly owned by Germans who emigrated after the Second World War. Ethnic arguments also played a role in the choice of the restitution date in the Czech Republic. Had the chosen reference date been 1945 rather than 1948, i.e. before the Communist land reforms, this would have implied restituting land to Sudeten Germans who emigrated from the Czech borderlands after World War II.

The choice of the privatisation process also affects the distribution of (agricultural) assets between ethnic groups within the country. Intuitively, one would expect that this motivation would lead to the choice of a privatisation policy which transfers property rights mostly to the ethnic majority for the same reason that foreigners do not receive land. However, this was not always the case, because it depends on the influence of ethnic minorities on the political economic equilibrium. An example where the privatisation choice benefits ethnic minorities is found in Bulgaria where the allocation of a high share of non-land assets to 'labour contributions' was partly to placate the Turkish ethnic minority who was largely employed in agriculture and not eligible for much land, while being an important political force (Buckwell et al., 1994). An example of where the privatisation choice was used against ethnic minorities is privatisation in the Baltics, where restitution of land was chosen to allocate assets to native citizens in the presence of large ethnic minorities.

A third influencing factor, but less important than legal ownership and ethnicity, is the equality of pre-collectivisation asset ownership. This factor determines the potential conflict between the objectives of 'equity' and 'historical justice'. Theory suggests both positive and negative effects of an egalitarian pre-collectivisation land distribution on the likelihood that restitution will be chosen as privatisation policy. Empirical observations suggest that when social equity conflicts with historical justice, as was the case in Albania, equity (and originally efficiency) prevailed in the government's choice. With more than half of the Albanian population active in agriculture and a highly unequal pre-reform land distribution, land restitution would have transferred most of the fertile land into the hands of a small group of families, leaving the bulk of rural households without land. The government approved distribution of land to rural households instead of land restitution to former owners, because the prime importance if the land issue for the large number of rural voters was more important than the large vested interests and lower collective action costs of fewer former landlords (Cungu and Swinnen, chapter 1).

Results of privatisation and land reform processes

Care is necessary when interpreting the data on 'private' assets in CEEC agriculture since the transition. In some cases, neither legal (ownership) rights, nor property rights are private; in some other cases, land titles have not yet been distributed, but new owners are already determining what is being done with the assets and have acquired (some) property rights. In other cases, ownership rights have been transferred to individuals, but their private property rights are incomplete as they have only limited power over the use and transfer of the assets.

Land restitution programmes typically have five steps: (1) claims are submitted; (2) ownership certificates are issued, indicating the rightful claim of the person or family to an asset; (3) the precise value of the asset is calculated and/or land parcels are redrawn and the precise land parcel identified and registered; (4) surveys are completed; and (5) land titles are issued. Uncertainty on property rights is only removed by finalising the fifth step: issuing the land titles.

By 1994 land reform was most advanced in the Czech Republic where nearly 80 percent of total agricultural land (TAL) was restituted. Most of this land was leased to 'successor' organisations, i.e. transformed co-operatives or state farms. Only 700,000 hectares, or 16.4 percent of TAL, was physically withdrawn. As a result, individual farms cultivated 23 percent of TAL. State farm land, representing 18 percent of TAL, was leased to 'expected' owners. In

Slovakia, only 123,988 former owners claimed their land to set up a new farm (about 8.5 percent of TAL). Most 'new' land owners (44.1 percent of TAL) decided to lease their land to a co-operative or a company. State farms had not yet been privatised and still had 15 percent of TAL. In Hungary, the final auctions of the compensation process were still not accomplished in 1996. Also the return of land titles on restituted land was not finished. Moreover, most land in state farms had not been privatised. In Bulgaria, state and collective farms were converted into 'organisations under liquidation' managed by liquidation councils. The privatisation of assets, including land, supervised by these councils was completed in 1995. In Romania, 70 percent of TAL was reported as 'acknowledged' property, i.e. property designated to be privatised. Individuals have 'possession certificates' on only 50 percent of all acknowledged land and full 'ownership titles' on only 20 percent. State farms have not been de-collectivised: they were renamed as 'commercial companies' in which private individuals own shares. In Eastern Germany, collectivised land has been restituted, but many who benefited leased their land to the new co-operatives. State farm land has been 'pre-privatised' by leasing it to private individuals with and without restitution claims. Finally, restitution in Slovenia is progressing only slowly and property rights have been affected by the introduction of the institution of co-ownership between private individuals and state farms using the land.

Privatisation, land reform and effective property rights

The different privatisation schemes will not necessary restore full property rights. Barzel (1989) defines property rights of individuals over assets as the rights, or the powers, to consume, to obtain income from, and to alienate these assets. Legal rights, as a rule, enhance economic rights, but the former are neither necessary nor sufficient for the existence of the latter.[4] Property rights are fully restored only if individuals obtain all rights (user, income, and alienation rights). Not all privatisation schemes fulfill these criteria, however. On farms created on a usufruct basis in the Baltic countries, the operators had no right to sell the land. Another example of incomplete property rights is the land restitution implementation in several CEECs which specifies that an 'equivalent' piece of land should be given when an individual wants to retrieve his land from its use by the new co-operative. However, there are important transaction costs in this process of being assigned an 'equivalent' or 'comparable' plot of land. The plot has to be assigned by the management of the reformed farm. The claimant can go to court if he disagrees with the assignment. Therefore, a priori property rights that are restituted are clearly incomplete in this process due to high transaction costs.

The latter reflects the importance of the reform implementation and the key role which local institutions and agents play in the implementation. The most radical reform policies on paper may result in little effective change unless they are fully and well implemented. While the government has control over the choice (the design) of the policy, it may have less control over the implementation of the design. There are several problems.

First, there are technical problems. Trying to value assets, to assess the validity of land and asset claims, to assign land parcels to claimants or to develop voucher schemes and auctions has proven a very difficult process. None of the countries (and few of the external 'experts' involved) had much experience to bring to the process. Because of these inherent difficulties, the process simply takes a lot of time. This in turn has caused severe pressure on agricultural producers and food consumers and has translated into intense political fighting.

Second, reform implementation requires shifting powers to the relevant local institutions, which may create principal-agent problems in the presence of imperfect information. In all CEECs commissions and agencies have been created to handle the agricultural privatisation and land reform programme.[5] Those institutions are often influenced by local preferences and biased towards management and employee interests, who are closely involved in the decision-making and the activities of the institutions.[6] Management and workers of the collective and state farms have important advantages in information and monitoring compared to dispersed (often urban) asset owners. This will affect their ability to manipulate the process. The potential for the existing management (whether *nomenklatura* or not) to manipulate the reform implementation and resulting property rights distribution depends strongly on the transaction costs for the various partners (legal owners, potential users) in a potential exchange of the assets

In particular, in a situation where legislation is unclear, or even inconsistent, the ability of the administration to act according to its own interests or preferences increases. Given the asymmetry of information, with the local administration being much better informed, it is often very difficult for a government to implement its reforms. Lack of information makes it hard to intervene for a government as it suspects that the reforms are sabotaged because the delays could also be due to the complexity of the process and actual difficulties of implementation.

Hayami (1991) argues that (1) using simple, transparent and uniform rules and (2) limiting the scope for discretionary government involvement in implementation are two key variables in a successful land reform. Both seem important in CEECs. Little progress was achieved in some programmes because the rules were unclear, insufficient incentives for change were provided by a little committed government and further blocked by collective farm management.[7]

A third problem is timing and government credibility. The privatisation and transformation processes require much time, even when all parties support the reforms. Therefore, there exists a real chance that before the reforms are finished, or even well under way, the government is replaced by another which may be less supportive of the reforms. While the new government may not choose to change the privatisation process it may very well introduce a number of regulations and policies which have an important impact on the speed and the extent of the reform implementation. As a consequence, local administrations and executing institutions may feel less inclined to follow government regulations which they privately oppose, because enforcement procedures may not come into effect if the government might be replaced.

Finally, some of the governments have introduced amendments to the privatisation and land reform laws to limit the distribution of property rights on agricultural production factors towards new (former) owners. The most striking examples of such measures is the introduction of the 'good landlord' and 'co-ownership' rules in Slovenia which allocate part of the decision-making rights on the use of the agricultural land to be restituted to the state farms which are using the land over an extended period, as long as the state farms behave as 'good landlords'. Similar legal initiatives which allocate part of the effective property rights to the users of the land (the former state and collective farms) have also been introduced in other CEECs, e.g. in Bulgaria, often as a result of the return to power of ex-Communist parties.

Farm restructuring

Farm restructuring is part of the general agricultural transformation process. Transformation refers to the conversion and restructuring of the collective and state enterprises into enterprises compatible with a market economy. This is, in part, a legal conversion and, in part, an organisational restructuring of the farms into viable business units. The available options are specified in laws and typically include conversion of collective and state farms into joint stock companies, limited liabilities companies or producer co-operatives, reforming the management of these farms in order to increase efficiency (for instance by creating profit-oriented subdivisions), transforming the farms through enlargement of subsidiary plots, separation of individual farms from collective farms or a complete dismantling of large farms (Csaki and Lerman, 1994).[8]

Particularly, the break-up of state and collective into individual farms implies a tremendous change of production technology and a major restructuring of human capital. A modern farmer in a Western country combines many skills while members of collectives farms often possess very narrow range of skills. Thus, newly established farmers must learn a wide variety of skills and knowledge. Some of these skills cannot be learned quickly from books but emerge through a long accumulation of experience. Therefore, important incentives for individuals are necessary to stimulate them to set up their individual farm (Rabinowicz and Swinnen, 1997).

The outcome of the first five years

In general, the data suggest a wide variety of farm structures emerging in many CEECs. Often it is not fully clear what the various names, such as 'joint stock companies' and 'private co-operatives' mean. Therefore, we distinguish between five production structures depending on ownership and use of assets. A first group of production structures are *public or state farms* which are entirely owned by the state. They have not yet been privatised or will not be privatised, e.g. because, as in the West, they remain active in public research. *Co-operatives* are still large-scale farms ranging on average from 475 hectares in Romania to 1,461 hectares in Eastern Germany. All kinds of enterprise structures (limited liability companies, legal entities, etc.) owned by shareholders constitute a third group which we call *companies*. They are characterised by some form of joint ownership and joint operation of the means of production. They are typically large-scale farms, their average ranging from 281 hectares in Bulgaria to 823 hectares in Eastern Germany. Also *partnerships* are characterised by joint operation and ownership, but at a smaller scale than companies or co-operatives. Moreover, partners are often family members. They are medium-sized farms of 100 to 500 hectares. Finally, *individual farms* are farms managed by a single individual or household. These farms typically include both family farms and household plots and are generally characterised by their small scale: from 1 to 50 hectares. In Eastern Germany, only the 'single proprietorships' are classified as individual farms. A summary of the resulting structure is given in Table 10.3. The category 'Other' refers to non-registered land or land not belonging to any statistical categorisation.

The data suggest that in 1994 large-scale farms still dominated in several CEECs. Co-operatives, partnerships and companies use 66 to 80 percent of TAL in Eastern Germany, the Czech Republic, Slovakia, Hungary and Bulgaria. In the Czech Republic most restituted land was leased to transformed co-operatives or to state farms. Only 700,000 hectares, or 16.4 percent of TAL, was physically withdrawn from these structures. As a result, individual farms cultivated 23 percent of TAL. State farm land, representing 18 percent of TAL, was leased to 'expected' owners. In Slovakia, only 123,988 former owners claimed their land to set up a new farm (about 8.5 percent of TAL). Most 'new' land owners (44.1 percent of TAL) decided to lease their land to a co-operative or a company. State farms had not yet been privatised and still had 15 percent of TAL. In Hungary, 22.3 percent of total land was cultivated by individual farms. In Bulgaria, organisations under liquidation still represented 18.1 percent of arable land, but 43.9 percent was reported to be farmed by individuals. In Eastern Germany, collectivised land has been restituted, but many who benefited leased their land to the new co-operatives. State farm land has been 'pre-privatised' by leasing it to private individuals with and without restitution claims. Individual farms cultivated one-fifth of TAL in 1994.

The main exceptions are Romania where individual farms occupy 61 percent of TAL and Albania where virtually all the land (95 percent) is used by small private farmers. Finally, Slovenia (84 percent) and Poland have a majority of individual farms, which already existed under the Communist period.

The data indicate considerable differences in the average size of an individual farm. Individual farms are larger in Eastern Germany (48 ha) and the Czech Republic (14 ha). They are around 1 ha in Hungary, Bulgaria and Romania. This is partly a statistical reason: it is impossible from the available data to distinguish between household plots, part-time farms and full-time individual farms. Because of this, the data *underestimate* the average size of part-time and (even more) of full-time individual farms.

There are some important conclusions: (1) the importance of individual farming varies strongly between CEECs; (2) in several CEECs very large-scale agricultural production organisations continued to play an important role in 1994; (3) in general, state farms had not been privatised and land reform was not completed. We will now turn to the reasons behind these observations.

Individual farming and fragmentation

The break-up of large-scale agricultural production units into individually operated farms differs considerably for the various CEECs. Our calculations give an individual farming index (IFI) – based on the percentage of agricultural land used by individual farms but corrected for the initial situation – that varies between 5 and 95 percent in the different countries of Central and Eastern Europe (Table 10.3). The IFI is low in countries where large-scale successor organisations to the former state and collective farms still dominate, such as Slovakia (5 percent), Hungary (13 percent), and the Czech Republic (20 percent). The IFI is highest in Albania (95 percent) and Latvia (80 percent) where a massive break-up of the collective farms resulted in a domination of small-scale production units. Within the CEECs there also exists wide variation in the importance of individual farming of different regions and agricultural subsectors.

Table 10.3
Farm structures in Central Europe in 1994 based on agricultural land

	Albania	*Bulgaria*[a]	*Czech Republic*	*East Germany*	*Hungary*	*Romania*	*Slovakia*	*Slovenia*
State farms								
area in '000 ha	n.a.	328,085	199,022	-	[b]	2,051,000	443,977	84,800
number	13	1,009	239	-	36	550	262	n.a.
average size in ha	150-200	325	498	-	n.a.	3,729	1,695	n.a.
% of total area	n.a.	7.0	2.8	-	[b]	15.5	20.3	10.9
Co-operatives								
area in '000 ha	-	2,260,356[d]	2,043,470	1,952,100	2,306,000	1,779,569	1,531,147	-
number	-	2,373	1,468	1,336	2,048	3,970	961	-
average size in ha	-	953	1,392	1,461	1,126	446	1,563	-
% of total area	-	48.2	47.7	35.9	37.7	13.4	69.9	-
Companies								
area in '000 ha	-	39,946	876,600	1,286,200	1,163,600[b]	-	100,522	-
number	-	142	1,461	1,563	3,311	-	219	-
average size in ha	-	281	600	823	n.a.	-	459	-
% of total area	-	0.9	20.5	23.7	19.0[b]	-	4.6	-
Partnerships								
area in '000 ha	-	-	14,570	1,116,400	[b]	1,536,532	-	-
number	-	-	47	2,379	2,369	13,741	-	-
average size in ha	-	-	310	469	n.a.	112	-	-
% of total area	-	-	0.3	20.5	[b]	11.6	-	-

Individual farms								
area in '000 ha	n.a.	2,058,545	871,669	1,081,700	2,652,400[c]	7,905,372	115,331	696,500
number	490,000	2,010,961	60,666	22,505	1,203,000	3,579,234	7,759	n.a.
average size in ha	1.05	1.0	14	48	n.a.	2.21	15	n.a.
% of total area	n.a.	43.9	20.4	19.9	43.3[c]	59.6	5.3	89.1
Other								
area in '000 ha	-	-	355,669	-	c	-	-	-
% of total area	-	-	8.3	-	c	-	-	-

[a] Arable land.
[b] 'State farms' and 'partnerships' are included in 'companies'.
[c] 'Other' is included in 'individual farms'.
[d] Including organisations under liquidation.

Source: Country studies in this volume

Table 10.4
Individual Farming Index (IFI), 1995[a]

	IFI		*IFI*
Albania	94.2	Georgia	21.2
Armenia	81.8	Hungary	17.3
Latvia[b]	80.2	Belarus	15.2
Lithuania[b]	60.4	Ukraine	14.1
Romania	55.2	Uzbekistan	13.1
Bulgaria	45.4	Russia	12.1
Estonia[b]	37.5	Turkmenistan	7.1
Kyrgyzstan	23.2	Slovakia[b]	3.1
Czech Republic	22.1	Kazakhstan	0.0

[a] The IFI is calculated by dividing the difference between the share of individual farms in total agricultural land in 1995 (IND95) and in 1989 (IND89) by 100 minus the share of individual farms in total agricultural land in 1989: IFI=(IND95-IND89)/(100-IND89)×100. For all FSU countries except the Baltic countries, an initial share of individual farms equal to 1 percent is assumed.
[b] 1994.

Source: Own calculations based on the country studies in this volume

Table 10.5 presents some additional information for 1994 based on other (OECD, EU) sources. These data largely confirm the picture of Table 10.3, but indicate some additional insights. The data in Table 10.5 show that, indeed, most of the 'individual farms' tend to be very small. This can be derived from the fact that in most CEECs the share of TAL of 'individual farms' is close to that of the farms of less than 5 hectares. The main exception is the Czech Republic, where farms less than 5 hectares only cover 1 percent. Table 10.5 also has information on the Baltic states that show some important differences with the other CEECs. The fact that the share of individual farms in TAL is rather high, while the share of farms smaller than five hectares is low, indicates that an important share of individual farms in these countries are larger than five hectares.

Determinants of farm restructuring

The post-transition CEEC farm structure will depend on the pre-reform farm structure, on the *design* of the privatisation and transformation policies (itself influenced by political economy factors), on the *implementation* of the policies, and on a series of additional factors, including the economic environment, affecting the *outcome* of the reform process.

Table 10.5
Individual Farming Index and degree of fragmentation of agricultural production based on land use data, 1994

Country	*Individual farming index*[a]	*Share of individual farms in TAL*[b]	*Share of farms smaller than 5 ha in TAL*
Albania	94	94	95
Bulgaria	36	44	30
Czech Republic	20	20	1
East Germany	11	20	na
Estonia	38	40	25
Hungary	13	22	12
Latvia	80	81	23
Lithuania	60	64	33
Romania	47	61	45
Slovak Republic	5	5	2

[a] The IFI measures the increase of individual farming and is calculated by dividing the difference between the share of individual farms in total agricultural land in 1994 (IND94) and in 1989 (IND89) by 100 minus the share of individual farms in total agricultural land in 1989: IFI=(IND94-IND89)/(100-IND89).
[b] Total agricultural land, 1994.
[c] Own estimate based on the 1994 survey of small farmers.

Source: Own calculations based on the country studies in this volume, European Commission, 1995 and OECD, 1996

Pre-reform production organisation

Pre-transition CEEC agriculture was characterised by a dual production structure. A small-scale private sector based on small private gardens and plots co-existed in a symbiotic relationship with very large collective and state farms. However, the production structure differed considerably between countries and has also varied over time. For example, Juhasz (1991) describes how shifts in the political ideology and authoritarian rule have been reflected in the structure of the farming system in Hungary. During more liberal political periods collective farms were given more autonomy, and private farming was allowed to grow. Conservative crackdowns led to freezes on reform programmes and recentralisation of the production system.

Agricultural production was mostly collectivised, with the exception of Poland and former Yugoslavia where even under the Communist regime the majority of farm production came from the private sector (Table 10.6). The other extremes are Albania and the former USSR where private farming was less than 2 percent. In Bulgaria all collectivised farms were consolidated with food-processing enterprises and the rest of the agro-industry into huge agro-industrial complexes (AICs) in the early 1970s. These structures were abolished in 1986.

Table 10.6
Importance of different organisational forms in CEEC agriculture in 1987 (in percentage of total agricultural land)

	Private	*Collective*	*State*
Albania[a]	0.5	78.4	21.1
Bulgaria	10.0	0.0	90.0[b]
Czechoslovakia	6.1	63.5	30.4
GDR	9.7	82.5	7.7
Hungary	13.7	71.4	14.9
Poland	78.0	3.6	18.4
Romania	15.6	54.7	28.9
USSR	1.8	30.4	67.8
Yugoslavia[b]	82.3	0.9	16.8

[a] 1983; cultivated land.
[b] Pryor has classified the Agro-industrial complexes as state farms.
[c] 1981; total farm holdings.

Source: Pryor, 1992

Originally, farm workers in collective farms (*kolkhozes*) held the title to their land, while the collective farm owned the non-land assets. The farmers were remunerated at the end of the year with a return on their inputs, based on the performance of the collective farm. Collective farms enjoyed somewhat more autonomy than state farms (*sovkhozes*) in their decisions. In state farms all assets, including land, were owned by the state and farm workers were like employees in any other (industrial) firm and received fixed wages and social security benefits. However, over the past decades, collective farm workers increasingly received the same social benefits as state farm workers. By the 1980s, workers in both state farms and co-operatives had a high degree of job and wage security, little responsibility for financial performance of the farm and little incentive to improve productivity. Both types of farms were protected from bankruptcy by a soft budget constraint and ownership titles in collective farms had no influence on production technology or farm decision-making (Brooks et al., 1991). For all practical purposes collective farms were very much like state farms in most countries.

Farm workers were allowed to have a small garden or *private plot* where they could produce for their own consumption. This private farming has provided an increasing share of the fresh food supply (especially vegetables and fruits). In addition, the collective farm management often rented out labour intensive activities, such as parts of the animal husbandry activities and provided inputs and services to private farming. The large farm had a monopoly position in supplying inputs, since markets for inputs and services did not exist. Table 10.7 illustrates the differential share of private farming by product in Hungary. The share of private farming was low for highly mechanised and low labour intensive activities such as grain and fodder production. It was high for fruits and vegetables and intensive livestock production.[9]

Table 10.7
Production structure for different agricultural products in Hungary (% in total output in 1987)

	Large scale farms (Collective & state)	*Small scale farms (Private)*
Cereals and leguminous crops	89.7	10.3
Industrial crops	88.5	11.5
Potatoes	24.3	75.7
Roughage and juicy fodder	93.8	6.2
Vegetables	24.1	75.9
Fruit	36.7	63.3
Grapes	40.3	59.7
Cattle breeding	76.8	23.2
Pig breeding	44.2	55.8
Poultry raising	56.4	43.6

Source: Juhasz, 1991

Evidently this pre-reform structure is affecting the emerging farm structure. The most obvious impact is the prevalence of small-scale individual farms in Poland and Slovenia (Table 10.6). However, also the share of state versus collective farms and the share of private farms in various production activities is likely to affect the future farm structure.

Privatisation and transformation policies are not independent. Both policies have distributional impacts. The distributional impacts of privatisation are straightforward as the distribution of property rights affects future income streams. The transformation policy affects restructuring of farms and therefore influences how former management and employees retain their position in the restructured enterprises.

Both policies also affect the resulting production organisation and allocation of assets. Without transaction costs, it does not matter who owns assets from an efficiency point of view, because they will invariably end up being used where their marginal return is highest.[10] In such a world, the privatisation process, as long as it turns all ownership rights to private individuals, might not affect the farm structure and the transformation process beyond the transformation from a socialist to a capitalist mode of production. However, in a world with transaction costs and imperfect markets the impact of the privatisation process on restructuring will go beyond this effect. In the agricultural transition in Central Europe, information is imperfect and transaction costs are high, especially in the implementation of the policies. Therefore, both privatisation and transformation policies are important in determining the distributional effects and the transformed organisations resulting from the reforms.

The privatisation process allocates property rights on production factors used by the collective and state farms to various individuals. The new farm organisation need to formally contract with individuals for the use of these assets. As the property rights are uncertain during the transition period, this increases the difficulties and uncertainties for the farms in their use and allocation of their production factors.[11] These difficulties and uncertainties increase when information is less available and transaction costs are higher.

Problems arise on both sides of the market. Those wishing to manage or work on farms must contract with asset owners. Also the new (or potential) owners have difficulties selling or renting their restituted land and other assets. Monitoring the restitution process and searching for the best potential use of the land can be quite costly for an individual who is not involved in agriculture.

Some feared that restitution of land to former owners, many of whom are no longer involved in agriculture, will lead to a break-up of the collective farm structure. However, we argue that the opposite may happen. More specifically, we argue that restitution of farm assets to outsiders (non-collective farm workers) will reduce the establishment of individual farm structure and stimulate the conservation of the large-scale successor organisations (LSOs) that have been created as a result of the transformation of the collective and state farms.

Consider two extreme (hypothetical) scenarios: (a) restitution to former owners, none of whom are involved in farming, or (b) distribution of assets to collective farm members. The standard argument that restitution to former owners will break-up the LSOs, is based on the assumption of a strong link between asset ownership and asset use. One cause of this link could be high transaction costs in the asset exchange markets. However, this argument ignores (a) differences in transaction costs between insiders and outsiders, (b) differences in relative incentives for starting up an individual farm, and (c) the existence of important transaction costs in the privatisation process. Taking these factors into account leads to a different conclusion. We summarise here the extensive analysis in Mathijs and Swinnen (1996).

First, high transaction costs in privatisation implies that property rights for outsiders are incomplete at best. This reduces their incentives to use the assets themselves for production. Moreover, because they lack farming skills, former owners have low labour productivity, further reducing the incentives to take the land out of the LSO structure.

Second, restitution to 'outsiders' – non-farming landowners – increases the transaction costs for LSO members or stockholders wanting to use the assets for setting up their individual farm. These members (or shareholders) in any case face transaction costs. They need to monitor and control the privatisation and asset distribution process when assets are distributed to themselves. These transaction costs are lower for insiders than for outsiders. If assets are restituted to former owners, however, members (or shareholders) face even higher transaction costs as they need to control both the privatisation process and the restitution process. Moreover, while in the first case they might develop an informal contract with the collective farm, this is more difficult in the second case because former owners have less insight on the location or condition of the assets.

In addition, when the individual member (shareholder) competes with LSO management in renting land from the former owner, he is at a disadvantaged position. LSO management, which typically plays an important role in the privatisation implementation has greater access to information, and possibly also more resources available for overcoming transaction costs.

All these factors combine to the conclusion that restitution of assets to former owners make it more likely that the pre-reform organisational structures will survive in comparison to a privatisation procedure that allocates assets to the LSO farm members or shareholders.

Is there any empirical evidence for this? Data on the impact of land reform and transformation regulations are presented in Figure 10.1 and Table 10.8. Figure 10.1 displays the relationship between individual farming and land reform and transformation policies, which are characterised by the ratio of land distributed to total land to be privatised, the 'land distribution index'. The figure shows a positive relationship: individual farming is lowest in countries that have allocated land to outsiders and highest where land was distributed to insiders. For example, Albania and Romania have used land reform policies that allocate land to insiders, i.e., former collective farm members or state farm employees. Albania distributed most of the land to farm workers, and Romania used a combination of restitution and distribution. Jackson (1997) indicates that as a result of the close ties of Romanian collective farm workers to the land – unlike Bulgarian collective farm members who were more like wage earners by the 1980s – a property settlement without a gesture towards the peasants' needs for more land was unimaginable. In contrast, individual farming is very low in countries such as the Czech and Slovak Republics, where land was restituted to former owners many of whom were no longer active in agriculture. The IFI for Slovakia and Hungary is even lower than expected from the relationship between land reform policies and the IFI, reflected by the curve in Figure 10.1. One reason is that both countries implemented legislation that increased the costs for leaving the LSO considerably more than in other CEECs (see Table 10.8).[12]

More generally, data presented in Table 10.8 suggest that individual farming is more important where (1) more of the land was distributed to farm workers, (2) the share of agriculture in employment is high, and (3) exit costs are low. It is remarkable to see how the two countries at the extremes of the spectrum are exactly opposite in these three factors. Albania, where individual farming is highest, distributed land, has a high share of agriculture in employment and low exit costs. Slovakia, where individual farming is lowest, restituted land, has a low share of agriculture in employment and high exit costs.

Also Latvia has a very high percentage (80 percent) of TAL used in farms larger than 100 ha. The high share of individual farms of Latvian agriculture may be partially explained by (a) the egalitarian pre-1945 land distribution which implies that restitution of land returns land mostly to 'insiders', i.e. people in agriculture, and (b) because of the active restitution policy of the Latvian government. The latter was inspired by nationalistic motivations as land restitution allocated land to native Latvians in a country with a very high share (46 percent) of ethnic non-Latvians in the population, especially Russians (34 percent).[13]

These observations support our hypothesis that individual farming is more likely when agricultural assets are privatised through allocation to collective farm members than through restitution to outsiders.

Table 10.8

Various indicators of land reform and transformation regulations

	Share of total agricultural land (in %)			*Share of agriculture in total employment (1994)*	*Man/land ratio (workers/10 ha)*	*Exit costs due to government regulations**
	Privatised land by		*State-owned land 1994*			
	restitution	*distribution*				
Albania	-	93	3	53	6.8	1
Latvia	64	30	2	17	0.9	1
Lithuania	96	21	1	23	1.1	2
Romania	43-58	15-30	13	36	2.4	2
Estonia	74	22	-	8	0.6	2
Bulgaria	81	-	7	22	1.1	2
Czech Republic	79	-	13	5	0.6	2
Hungary	62	19	5	9	0.6	3
East Germany	82	-	8	n.a.	n.a.	2
Slovakia	74	-	20	7	0.7	3

* Own estimate of exit costs induced by farm transformation regulations (1=Low, 2=Medium, 3=High), based on case studies in Swinnen (1997a) and the country studies in this volume.

Source: Own calculations based on European Commission (1995), OECD (1996) and the country studies in this volume.

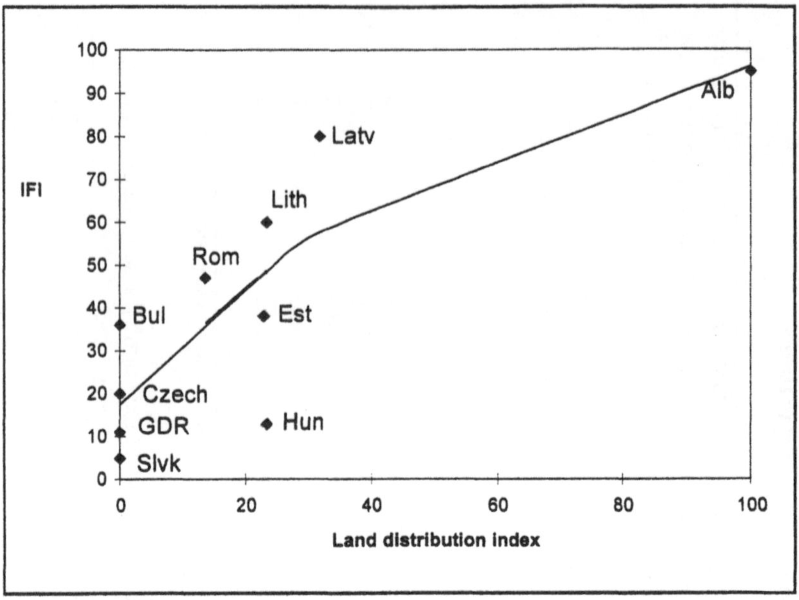

The curve on the graph is based on a least squares regression using a quadratic function. Data for the land distribution index are extracted from the chapters in this volume.

Figure 10.1 Relationship between the 1994 Individual Farming Index (IFI) and the share of distributed land in total land to be privatised (land distribution index)

Transformation policies and reform implementation

Transformation, and in particular the stimulation of individual farms, can be encouraged or discouraged both by the government and by the co-operative management. The process is directly affected by laws, which provide explicit rules for the design of the transformation. Facilitating farmers to withdraw from agricultural production co-operatives can be done through various incentives, e.g. by providing individual private farmers with preferential credit arrangements (e.g., low interest credit and loan guarantees), by canceling their share of the co-operative farm debt, or by the provision of training and extension facilities to support the development of individual farming. Some countries seem to have included a series of regulations that tend to discourage farmers from leaving the co-operatives. We refer to the set of government policies and regulations which attempt to influence the process as the 'transformation policy'.

The choice of the privatisation policy was strongly determined by legal and historical factors, not necessarily reflecting the policy preferences of the government. One objective of the transformation policy was to influence (enhance or mitigate) the impact of the privatisation policy on farm restructuring and on the distribution of effective property rights.[14]

Key characteristics of the transformation policy are (a) the role it allocates for the management of the collective and state farms, and (b) the incentives it provides for leaving the collective farm. For example, the 1992 UDF government in Bulgaria decided to replace the old management of the collective farms by special institutions to liquidate the collective farms, appropriately called 'Liquidation Councils'. Similarly, the 1991 Sajudis government in Lithuania removed the existing management from its controlling positions and created new institutions, the Municipal Agrarian Reform Services, chaired by outsiders. Not surprisingly, in both countries the role and the composition of these institutions was changed when the ex-Communists came back to power.[15]

Other CEECs opted for a less radical approach, emphasising the need to minimise further disruptions. They have done so by giving the members and old management an important role in the transformation of their collective and state farms. For example, in Hungary, Slovakia and the Czech Republic, the former management plays an important role in the transformation of the collective farms: they are the main agents in the 'transformation boards' which have to draw up a plan for transforming the organisation. Individuals are discouraged to leave the collectives e.g. by imposing on them a share of the debt of the former collective and the administration costs of parceling out the land in case they want to leave the transformed collective farm and start up their own individual farm. In both the Czech Republic and in Hungary such policies have been criticised as 'being in favour of the former Communists' by the peasant parties (respectively the Peasant Party in former Czechoslovakia, and the Independent Smallholders' Party in Hungary).

Overall the impact of transformation policies and reform implementation has been a partial allocation of effective land property rights to the users of the land, i.e. the former collective and state farms, and a conservation of the existing large-scale farm structures.

However, in the next section we will argue that the impact of these policies has been less important than the economic environment in which the farms operated prior and during transition.

The economic environment

The economic environment will affect the opportunities and risks for emerging private farmers and, thus, the incentives for individuals to set up their own farms or to stay with a reformed co-operative or state farm. The economic environment since 1989 is characterised by falling terms of trade for agriculture, depressed farm incomes, high inflation, lack of inputs (especially credit), inappropriate technology and human capital, production declines and widespread uncertainty. These factors reduce the incentives for LSO members or shareholders to start up their own private farm. However, only to the extent that large-scale farms can manage risk better than individuals will it have a negative impact on the establishment of individual farms (Deininger, 1995; Mathijs and Swinnen, 1996).

The attitude of the LSO members/shareholders towards individual farming is affected by the lack of a credible legal framework. Investment in agricultural production implies, by its very nature, an important lag between the period of investment and costs and the time of harvest and revenues. It is not hard to understand that individuals are not inclined to start investing unless the legal framework is sufficiently stable. The most important issue is the uncertainty of property rights. As long as property rights are uncertain, decapitalisation of agriculture continues. However, the establishment of strong and stable property rights and institutions takes time. A consequence is that informal contracts have partially replaced formal contracts in agriculture during transition.

Furthermore, after the price and trade liberalisations terms of trade have dramatically turned against agriculture in all CEECs. While substantial improvements have occurred since, agricultural profitability has varied substantially between subsectors and over years. All this increased uncertainty and risk for private producers and reduced incentives to leave the shelter of farming in association. Would-be private farmers face unfavourable market conditions and problems with access to input supplies, processing and marketing facilities. These problems can come from the hostile attitude of old-style management of agro-industrial enterprises or because of physical shortages.

A special case is the capital market. Investment requires capital, which in turn requires credits and loans. If credit is available at all for private farmers, they often do not possess the collateral needed to obtain loans. With low overall profitability in farming, the poorly- developed banking system refuses to lend to agricultural producers unless a large and well defined collateral can be provided (Petranov and Roussinov, 1994). As land titles have not been assigned, producers cannot obtain credit for investment or for purchasing inputs. Credit subsidy programmes in different countries have not succeeded in overcoming this obstacle. The crucial point here is that this problem will only be solved by the full allocation of property rights to (new) producers. This can only be done by handing out the land titles.[16]

There are no consistent data to calculate the impact of prices, risk and legal uncertainty on farm restructuring. We do observe though that the negative development in agricultural terms of trade in 1989-1991 throughout CEECs has stabilised and that the situation has improved substantially (at least for some commodities) since then. We can also conclude that price variation has reduced substantially since 1992, as has the understanding of the emerging market economy by agricultural producers (Jackson and Swinnen, 1995). Both factors can be expected to reduce overall production risk – especially the 'transition-related' risk. Both this risk and negative terms of trade were hypothesised as having a negative impact on individual incentives to leave the LSO, which is largely consistent with an increase in the share of individual farms over the 1990-1994 period.

Informal institutions and human capital

The pre-reform, large farm, production system is not only sustained by formal institutions and organisational arrangements, such as the imposed hierarchy and official rules regulating ownership of assets and decision-making power. The system is further sustained by an underlying network of informal institutions that form the 'glue' of the formal institutional structure. Formal institutions can be suddenly changed in a process of revolution. However, informal institutions such as customs and norms do not change so easily (North, 1990; Eggertson, 1990). While privatisation and land reform effectively destroy and break up the previous formal arrangements, this is not the case with the informal institutions and the existing human capital.

If some non-human capital can be made available for smaller scale production activities, this might not be the case for human capital. Knowledge of the production process and the inherent skills are highly separated in CEEC agricultural production systems and not concentrated in one person. Entrepreneurship among farm workers is negatively influenced by the absence of capabilities and norms which are adjusted to the peculiarities and the heterogeneity of decision-making problems in individual farm operations (Hagedorn, 1992). Human capital cannot be reorganised like other assets and may therefore be a 'conservative' factor in the restructuring process. Furthermore, many of the people employed on collective and state farms are elderly people, who are likely to be more risk-averse in their decision on whether to stay or to start up an individual farm.

Both the lack of (entrepreneurial or general) human capital investment and the existing informal institutions are related to the so-called lack of entrepreneurship and enthusiasm by farm workers for a return to an independent farmer-status, which is widely argued to be a major problem of agricultural privatisation (Brooks et al., 1991; Pryor, 1992). Most pre-1994 surveys indeed indicate that a large share of former collective farm workers prefer to remain in the privatised co-operative farm organisations (Kopeva and Michev, 1993; Kraus et al., 1994; Gavrilescu, 1994).

Both the uncertain and negative economic environment and the lack of human capital and the informal institutions present a very pessimistic picture on farm restructuring and the emergence of new farms. This is inconsistent with the data, as presented in Table 10.3. While successor organisations to the collective and state farms do still dominate agricultural production in several CEECs, we also notice that individual farms have become quite important in others, and that in some CEECs they are already using most of the land. Which factors do account for this?

Agricultural productivity

A key factor explaining the break-up of the former collective (and state) farms is the relative productivity of the former collective farm versus the collective farm members' productivity in individual farms. If the average productivity in the collective farm was relatively high before reform, members have less incentive to leave the collective farm than in case the initial average productivity was relatively low (Mathijs and Swinnen, 1996).

However, the relation between productivity on the collective farm and productivity on an individual farm is not straightforward. It depends on the skills and human capital of collective farm workers, but also on e.g. the technology and the degree of labour specialisation in the collective farm before the reform. Consider two collective farms, one in which the members are highly specialised in performing a specific task and another where everyone performs the same tasks. As a result of specialisation, the individuals' productivity in the first collective farm will be higher than in the second farm. In the first collective farm, average productivity is high, but members are specialised in a specific activity and their relative productivity in a private farm might be low. In contrast, members on another, low productive, collective farm might have a lower absolute productivity, but might have a higher relative productivity when they work on their own farm.

To analyse the relation between individual farming and average collective farm productivity, Figure 10.2 shows the relation between individual farming and value added per capita (as a proxy for productivity). The figure shows a clearly negative relationship for nine CEECs. Countries with low agricultural productivity, such as Albania, have a significantly higher degree of individual farming than CEECs where collective farm productivity was higher, such as Hungary.

There also exists a positive relationship between the individual farming index and the share of food in total consumer expenditure (Figure 10.3). In those countries where productivity on collective farms was too low to provide the basic needs of the members, they will leave. The issue of relative productivity is related with consumption risk, or more generally, food security, reflected in the share of the budget spent on food. In countries characterised by a large share of the budget spent on food, food security is more critical. This factor appears to have been a key factor for countries such as Albania and Romania, where extremely low productivity of collective farms, in combination with government policy favouring state farms and taxing collective farms, caused very low incomes on the collective farms. In the case of Albania the situation was so extreme that food shortages and hunger resulted, causing a massive break-up of the collective farms after 1991 (Cungu and Swinnen, chapter 1).

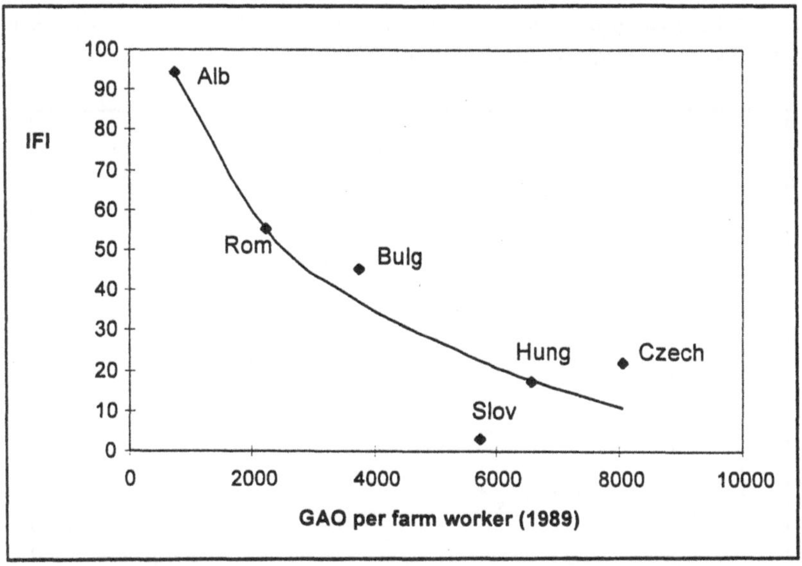

GAO is calculated using data from FAO, *Production Yearbook 1990* (FAO: Rome, 1991).

Figure 10.2 Relationship between IFI in 1995 and initial collective farm productivity in 1989 for six CEECs

Technology and capital intensity in production

The existing capital stock (machinery, etc.) is linked to the previous large-scale production process. The impact of the existing capital stock is hard to assess ex ante, because it is influenced by a number of other factors. First, it depends on the availability of alternative (smaller scale) machinery and other capital assets. This is not only determined by the situation and the restructuring in the down- and upstream industry, but also on the extent to which farmers have access to credit for purchasing new machinery. Second, it depends on the nature of the machinery and of the privatisation process. An example may help to clarify this point. If machinery is not 'too' large and if privatisation allows for indivisibility of sales, this might stimulate the creation of institutions as custom workers or working stations. Examples from Western European agriculture indicate that organisational innovations, such as service co-operatives (Germany and France), custom workers (The Netherlands and Belgium) and farm service contractors (UK) in production, and the existence of co-operatives for processing and input purchasing, allow to realisation of important scale effects where they exist, without having to move to a large-scale *production* system. The development of such institutions surrounding the production system depends not only on the privatisation of non-land assets but also on the restructuring of the up- and downstream industry. A key issue is whether LSOs can contribute to an efficient transition by gradually transforming themselves into service co-operatives (see Deininger, 1995).

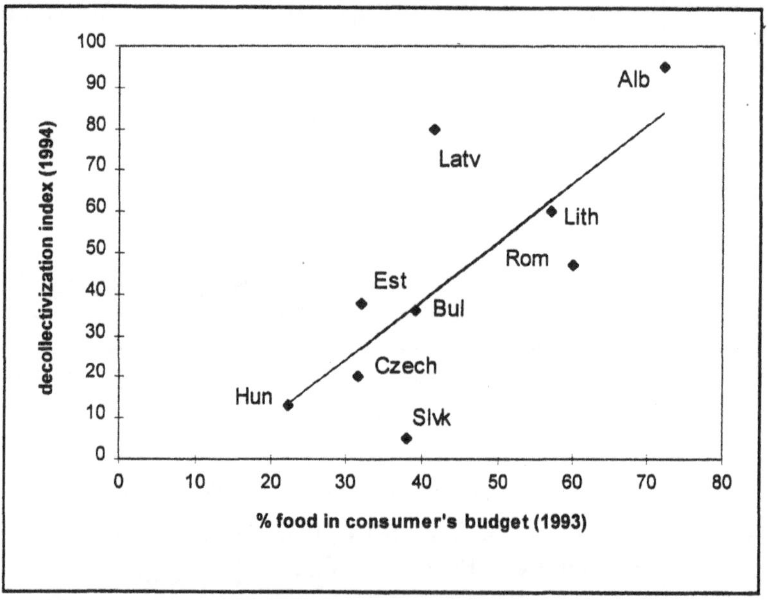

The curve on the graph is based on a least squares regression. Data on the share of food in consumer's budget are from OECD (1996).

Figure 10.3 Relationship between the Individual Farming Index in 1994 and the share of food in consumer's expenditure in 1993

Mathijs and Swinnen (1996) model the impact of land and capital assets on the creation of individual farms through their impact on the *exit costs* of farm members wanting to leave the collective farm. In general, exit costs reduce incentives for members to leave the collective farm, and a higher level of technology and of capital use increases those exit costs. They argue that one should expect the share of individual farms to be highest in production activities with the highest labour intensity in production. Empirical evidence is consistent with this hypothesis. In countries with a high man/land-ratio, such as

Albania and Romania, there was a high share of individual farms compared to countries with a low man/land-ratio, such as the Czech and the Slovak Republic (see Table 10.8).

Figure 10.4 shows the relationship between individual farming and the share of agriculture in total employment. Two factors can be the cause of this relationship. First, the share of agriculture in total employment is typically negatively correlated with the level of development and the use of technology in agriculture. A country with a high share of agriculture in employment is typically less developed and uses inferior technology compared to a country with a low share of agriculture where there was a substitution of labour by capital as a result of the technological progress. Second, the share of agriculture in employment also captures the number of outsiders versus insiders. A high share of agriculture in total employment is an indication that relatively few people have left agriculture and thus that the ratio outsiders to insiders is relatively low. Consequently, a low share of agriculture in total employment means that relatively many farmers left agriculture resulting in a high ratio of outsiders to insiders. World Bank surveys suggest that the overwhelming majority of individual farmers are former co-operative members or employees of state farms (Csaki and Lerman, 1994). Since, in general, outsiders lack the appropriate skills to start up a private farm, the number of farm workers, i.e., insiders, is a good reflection of the number of people actually interested in individual farming.

Figure 10.4 shows a positive correlation between the 1993 share of agriculture in the economy and the IFI. CEECs with more than 15 percent of active people employed in agriculture (Albania, Bulgaria, Latvia, Lithuania, and Romania) show a higher share of individual farms compared with countries where agricultural employment is less than 10 percent of the work force (Czech and Slovak Republics, Hungary). Again the figure indicates some additional regional biases: the Baltic countries all lie above the curve, while the southern CEECs lie on or below the curve. This observation suggests that additional factors have stimulated the establishment of individual farms in the Baltic countries and especially in Latvia.

The impact of reform on fragmentation

While the break-up of state and collective farms into individual farms clearly causes fragmentation in some cases (e.g. Albania), this is not the case in general. In most of the CEECs we study, many individual farms cover 100 hectares and more. The Latvian case in particular shows that a high IFI (80 percent) does not necessarily imply a high degree of fragmentation: only 23 percent of TAL is cultivated by farms smaller than five hectares. Most of the land is farmed by farms between 5 and 100 hectares.

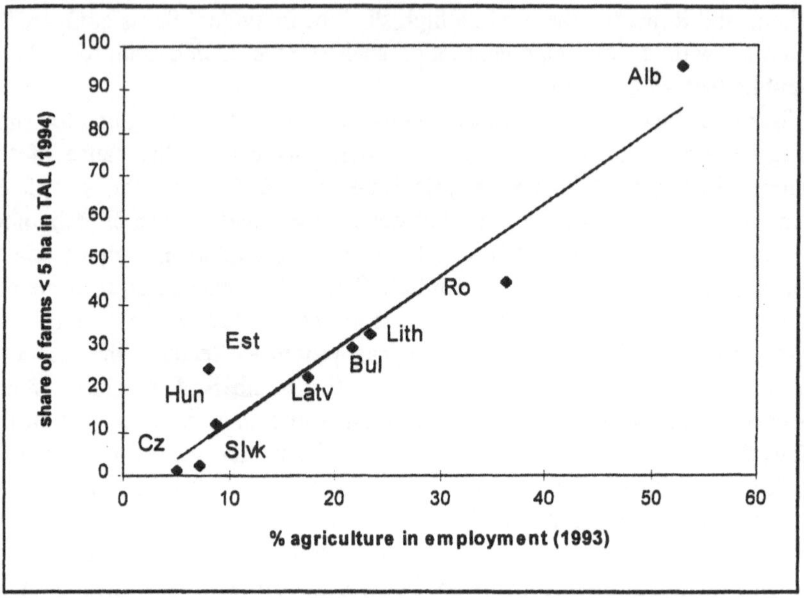

The curve on the graph is based on a least squares regression. Data on the share of agriculture in total employment and on farms smaller than 5 hectares are from OECD (1996).

Figure 10.5 Relationship between land fragmentation in 1994 and the share of agriculture in total employment in 1993

The most important factor explaining fragmentation is the size of the rural work force. A high share of agriculture in employment implies a strong demand for small household plots, thus contributing to the fragmentation of land. Figure 10.5 shows an almost perfect linear correlation (regression $R^2=0.92$) between the share in TAL of small farms as a measure for fragmentation and the share of agriculture in the economy. Additional factors have apparently stimulated the start-up of larger individual farms. This can be derived by comparing Figures 10.4 and 10.5. The comparison with Figure 10.4 suggests that farms larger than five hectares have emerged more strongly in the Baltic countries, and especially in Latvia, than in other CEECs with a similar agricultural share in employment. Possible explanatory factors are the pre-1945 land distribution and the active restitution and radical transformation policies of the Latvian government.

Conclusion

While the available empirical evidence is imperfect, it supports our hypotheses (a) that transition-related risk and adverse terms of trade have negatively affected the creation of individual farms; (b) that the productivity of the new large-scale farm has an important negative impact on the creation of individual farms; (c) that with extremely low collective farm productivity, food security concerns induce individuals to start up small individual farms; (d) that individual farming is less in capital and land intensive production activities; and (e) that privatisation policies affect the extent of individual farming. More specifically, restitution of production factors to outsiders (former owners) leads to less individual farming than asset distribution among farm workers and members.

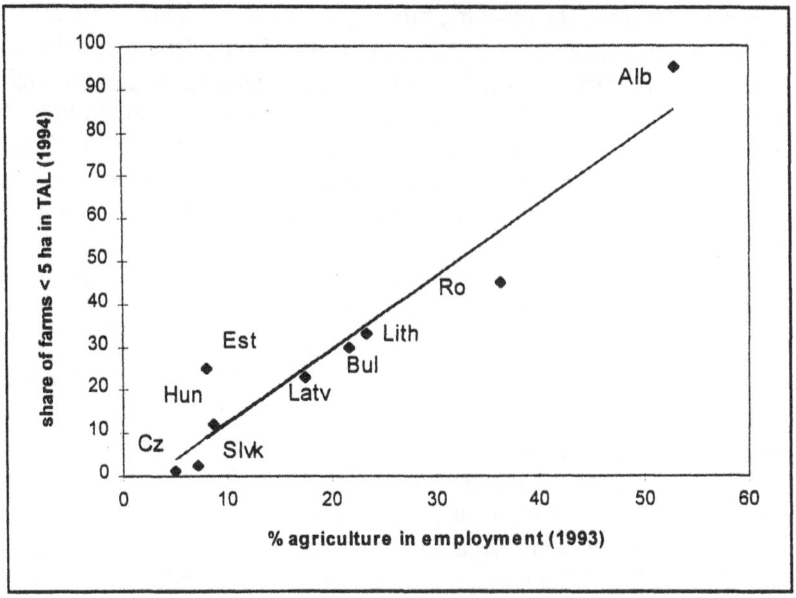

The curve on the graph is based on a least squares regression. Data on the share of agriculture in total employment and on farms smaller than 5 hectares are from OECD (1996).

Figure 10.5 Relationship between land fragmentation in 1994 and the share of agriculture in total employment in 1993

Concluding remarks

This chapter has presented insights on agricultural privatisation and transformation policies, land reform and resulting changes in farming structures in Central and Eastern European countries based on a comparative analysis of the reform policies, processes and their determinants. The comparative analysis indicates both some general patterns and a wide variation in reform processes and results between CEECs.

Most CEECs have restituted land to former owners. Moreover, former owners who kept legal rights to their land were restituted property rights on their land without exception. Non-land assets have typically been privatised through other procedures than land. The choice of the privatisation and land reform process is determined mostly by legal and historical characteristics of agricultural assets, i.e. the nature of post-collectivisation asset ownership, the pre-collectivisation asset ownership distribution and the ethnicity of the former owners.

Privatisation does not necessarily leads to full transfer of property rights to new (private) owners. The resulting property rights distribution is affected by the choice of the privatisation process, more specifically land restitution has had a conservative impact on the farm structure. In addition, the property rights distribution is influenced (a) by technical factors and transaction costs in the reform implementation, (b) by how affected agents could influence the reform implementation, and (c) by additional government regulations, in turn determined by political influence of the affected agents and by political coalitions.

After transformation legislation was enacted, state and collective farms have been transformed into a wide variety of farm organisations, such as producer co-operatives, joint stock companies, limited liability companies and individual farms. Large-scale production organisations still dominate agricultural production in Bulgaria, Hungary, the Czech Republic and Slovakia. However, in Albania, Romania and the Baltic states, individual family farms dominate. Important determinants of farm restructuring are economic factors, including relative productivity of existing human and physical capital in different organisations, risk, terms of trade, transition-specific distortions and factor intensity of agricultural production.

Notes

1 The authors acknowledge financial support from the Belgian Fund for Scientific Research (NFWO) and helpful comments from participants in the COST network and in the Sinaia seminar.

2 In general, these changes are more important when the land is situated in better agricultural areas. Plots in mountainous and agriculturally disadvantaged rural areas often have undergone less change. This change in quality and use of land affects the administrative difficulties in restituting land to previous owners. It also induces compensation claims. Non-land owning members and management of collective farms will claim that they should be compensated if the quality of the land has been improved and the land is restituted to previous owners.

3 The formal owners were deprived of virtually any authority on the use of the land and thus from income produced on their land and their rights of alienation of the land. This illustrates that legal ownership rights are not a sufficient condition for the existence of full, effective property rights. Privatisation in this case has to do more with the transfer of property rights back into private hands, rather than (legal) ownership rights. In other cases it concerns the allocation of both legal ownership rights and effective property rights to individuals or private institutions.

4 The concept of property rights is closely related to that of transaction costs. If transaction costs are associated with the transfer, capture, and protection of rights and if each of these costs is positive and rising with the actions, rights to assets cannot be perfectly delineated. Assets have many attributes which are typically not well known to at least one of the partners in a potential exchange. The information is usually asymmetrical, as present owners or users are better informed. The transfer of assets entails costs, among others resulting from both parties' attempts to determine what the valued attributes of these assets are. Exchanges that otherwise would be attractive may be forsaken because of such exchange costs. Furthermore, the rights people have over assets are not constant. They are a function of their own direct efforts at protection, of other people's capture attempts, and of government protection (Barzel, 1989).

5 In former Czechoslovakia each co-operative and state farm has elected a Transformation Board to handle the privatisation of the farm. The management of Bulgarian collective farms has been replaced with a Liquidation Council, responsible for the privatisation of the farm assets and the management of the farmers during the transition. The claims of land ownership are handled by Municipal Land Councils. Romania has similar commissions in each village to redraw the maps and allocate the land among claimants. In Poland, the central Agricultural Property Agency manages the privatisation of state farms (and their management during the transition period). See Beckmann and Hagedorn (chapter 4) for theoretical issues on optimal agency design in agricultural transformation.

6 One option for the government to break party and officials' resistance is to place the cadres in a position to enrich themselves in the decollectivised system. This happened in the reform process in China (Lin, 1988).

7 In contrast, the 'radical' amendments to the Bulgarian land law of March 1992 specified explicitly and in detail how restitution and liquidation was to take place, and specified that collective management was to be replaced by a 'liquidation council' which was to be filled with individuals more loyal to the reforms. When the Bulgarian Socialist Party came back to power in 1993, they replaced most heads of the liquidation councils (LCs) with individuals less supportive of the reforms. The consequence is that the

LCs have effectively taken the place of the former management in running the collectives.

8 Pryor (1992) defines the latter as decollectivisation: 'the break-up of large-scale farms, organised either as co-operatives or state enterprises, into individually operated farms and their creation as autonomous production units independent of the government. A change of ownership of these farms, e.g. a conversion into a co-operation in which the workers or others hold stock but the essential farming operations remain roughly the same, is quite a different matter and can occur with much greater ease'. We have also used this definition in Mathijs and Swinnen (1996).

9 A double paradox on the role of private farming under central planning is hypothesised by Juhasz (1991). He argues that, first, the existence of the small private farms has made it possible to introduce the system of large-scale farming originally, since they bound the peasants to the land. Second, the output of the private plots has supplied the rural areas with food at a time when the production of large-scale farms ran terribly low. This 'subsistence' level of food supply basically reduced political pressure and, therefore, has postponed fundamental changes in the agricultural system.

10 The original point (while not presented as a theorem) was made in Coase (1960). For a discussion on the (mostly unintended and, according to Coase, mostly irrelevant and incorrect) implications of the discussion: see Coase (1989).

11 This also affects the maintenance and protection of the existing capital stock against destruction during the transition. The uncertainty on future ownership reduces the incentives for the current managers for maintaining the capital stock. In addition, in some CEECs the property of collective and state farms is put under a temporary management for some time. All this encourages bad maintenance or even outright theft.

12 See Swinnen (1997a,b) for an extensive discussion of privatisation policies.

13 See Rabinowicz (1997) for an extensive discussion and Swinnen (1997b) for an analysis of ethnic impacts on privatisation choice.

14 The choice of the transformation policy has to be interpreted as a choice between efficiency and redistribution. Most CEEC governments have *not* chosen for a transformation policy which would induce more disruptions beyond the disruptions already caused by the privatisation policies. Those reform-minded governments which chose for radical and more disruptive transformation policies did so because they calculated that the long run political gains from creating a rural social structure more supportive of democratic and market reforms, outweighed short run political costs of more disapproval because of increased distortions. Their intention was to break-up the existing social and economic organisation of the rural areas to

change the preferences and political structures in the countryside in order to create a long term political basis for securing political and economic reforms. The political gains outweighed the political costs of this strategy only when (a) the 'security' of the overall reforms was low (reflected in continued political strength of the CP and a small political center); (b) the rural area political preferences could significantly alter the power balance; and (c) if the radical reformers had sufficient votes in parliament to push through the policy (Swinnen, 1997b).

15 The 1993-1994 Bulgarian coalition government, which relied on socialist party (BSP) support, replaced 75 percent of the Liquidation Councils' chairpersons. After the 1994 elections in which the BSP won the majority, the BSP government abolished the Liquidation Councils.

16 Recent evidence indicates that even with land titles, banks do not accept agricultural land as a collateral because of the absence of a land market, or because the value is too low (Swinnen and Gow, 1997).

References

Barzel, Y. (1989), *Economic Analysis of Property Rights*, Cambridge University Press: Cambridge.

Beckmann, V. and Hagedorn, K. (1997), chapter 4 of this volume.

Bojnec, S. and Swinnen, J.F.M. (1997), chapter 8 of this volume.

Brooks, K., Guasch, J.L., Braverman, A. and Csaki, C. (1991), 'Agriculture and the transition to the market', *Journal of Economic Perspectives*, Vol. 5, No. 4, pp. 149-161.

Buckwell, A., Davidova, S. and Trendafilov, R. (1994), 'Land Reform: How Will the Future Look?', in Schmitz, A., Moulton, K., Buckwell, A. and Davidova, S. (eds), *Privatization of Agriculture in New Market Economies: Lessons from Bulgaria*, Kluwer Academic Publishers: Norwell.

Coase, R.H. (1960), 'The Problem of Social Cost', *Journal of Law and Economics*, Vol. 3, No. 1, pp. 1-44.

Coase, R.H. (1989), *The Firm, The Market and The Law*, University of Chicago Press: Chicago.

Csaki, C. and Lerman, Z. (1994), 'Land Reform and Farm Sector Restructuring in the Former Socialist Countries in Europe', *European Review of Agricultural Economics*, Vol. 21, No. 3-4, pp. 553-576.

Cungu, A. and Swinnen, J.F.M. (1997), chapter 1 of this volume.

Davidova, S., Buckwell, A. and Kopeva, D. (1997), chapter 2 of this volume.

Deininger, K. (1995), 'Collective Agricultural Production: A Solution for Transition Economies?', *World Development*, Vol. 23, No. 8, pp. 1317-1334.

Eggertson, T. (1990), *Economic Behavior and Institutions*, Cambridge University Press: Cambridge.

European Commission (1995), *Agricultural Situation and Prospects in the Central and Eastern European Countries*, Directorate-General for Agriculture: Brussels.

Fehér, I. (1995), 'Co-operatives in the Changing Hungarian Agriculture Structure', Proceedings of the 41st EAAE Seminar on Challenge and Strategies for Re-establishing East-Central European Agricultures, Gödöllö, Hungary.

Gavrilescu, D. (1994), 'Agricultural Reform in Romania: Between Market Priority and the Strategies for Food Security', in Swinnen J. (ed.), *Policy and Institutional Change in Central European Agriculture*, Avebury: Aldershot.

Hagedorn, K. (1992), 'Transformation of Socialist Agricultural Systems', *Journal of International and Comparative Economics*, Vol. 1, pp. 103-124.

Hayami, Y. (1991), 'Land Reform', in Meier, G. (ed.), *Politics and Policy Making in Developing Countries. Perspectives on the New Political Economy*, ICS Press: San Francisco.

Jackson, M. (1997), 'Political Economy of Agricultural Reform in Romania' in Swinnen, J.F.M. (ed.), *Political Economy of Agrarian Reform in Central and Eastern Europe*, Ashgate: Avebury.

Jackson, M. and J. Swinnen (1995), *A Statistical Analysis and Survey of the Current Situation of Agriculture in Central and Eastern European Countries*, Report to EU Commission DG-I, The Leuven Institute for Central and East European Studies: Leuven.

Juhasz, J. (1991), 'Large-scale and Small-scale Farming in Hungarian Agriculture: Present Situation and Future Prospects', *European Review of Agricultural Economics*, Vol. 18, No. 3, pp. 399-416.

Kopeva, D. and Michev, P. (1993), *Summary of Land Reform*. APAU, Ministry of Agricultural Development: Sofia.

Kraus, J., Doucha, T., Sokol, Z. and Prouza, B. (1994), 'Agricultural Reform and Transformation in the Czech Republic' in Swinnen J. (ed.), *Policy and Institutional Change in Central European Agriculture*, Avebury: Aldershot.

Lin, J.Y. (1988), 'The Household Responsibility in China's Agricultural Reform: A Theoretical and Empirical Study', *Economic Development and Cultural Change*, Vol. 36, No. 3, pp. 199-225.

Lindemans, I. (1997), 'Process and Politics of Agricultural Privatization in the Czech and Slovak Republics', in Swinnen, J.F.M. (ed.), *Political Economy of Agrarian Reform in Central and Eastern Europe*, Ashgate: Aldershot.

Mathijs, E. and Mészáros, S. (1997), chapter 5 of this volume.

Mathijs, E. and Swinnen, J.F.M. (1996), 'The Economics of Agricultural Decollectivization in Central and Eastern Europe', Working Paper Series of

the joint Research Project: *Agricultural Implications of CEEC Accession to the EU*, Working Paper No. 3/1, Department of Agricultural Economics, K.U.Leuven.
North, D. (1990), *Institutions, Institutional Change and Economic Performance*, Cambridge University Press: Cambridge.
OECD (1996), *Agricultural Policies, Markets and Trade in Transition Economies: Monitoring and Evaluation 1996*, Paris: OECD.
Petranov, S. and Roussinov, K. (1994), 'Credit for Agriculture During the Transition', in Schmitz, A., Moulton, K., Buckwell, A. and Davidova, S. (eds), *Privatization of Agriculture in New Market Economies*, Kluwer Academic Publishers: Norwell.
Pryor, F. L. (1992), *The Red and the Green. The Rise and Fall of Collectivized Agriculture in Marxist Regimes*, Princeton University Press: Princeton.
Putterman, L. (1985), 'Extrinsic Versus Intrinsic Problems of Agricultural Co-operation: Anti-incentivism in Tanzania and China', *Journal of Development Studies*, Vol. 21, No. 2, pp. 175-204.
Rabinowicz, E. (1997), 'Political Economy of Agricultural Privatization in the Baltic Countries', in Swinnen, J.F.M. (ed.), *Political Economy of Agrarian Reform in Central and Eastern Europe*, Ashgate: Aldershot.
Rabinowicz, E. and J.F.M. Swinnen (1997), 'Political Economy of Privatization and Decollectivization of Central and East European Agriculture: Definitions, Issues and Methodology', in Swinnen, J.F.M. (ed.), *Political Economy of Agrarian Reform in Central and Eastern Europe*, Ashgate: Aldershot.
Sarris, A.H. and Gavrilescu, D. (1997), chapter 6 of this volume.
Swinnen, J.F.M. (1997a), *Political Economy of Agrarian Reform in Central and Eastern Europe*, Ashgate: Aldershot.
Swinnen, J.F.M. (1997b), 'The Choice of Privatization and Decollectivization policies in Central and East European Agriculture: Observations and Political Economy Hypotheses', in Swinnen, J.F.M. (ed.), *Political Economy of Agrarian Reform in Central and Eastern Europe*, Ashgate: Aldershot.
Swinnen, J.F.M. and H.R. Gow (1997), 'Agricultural Credit Problems and Policies during the Transition to a Market Economy in Central and Eastern Europe', *Policy Research Group Working Paper*, Nr 6, Department of Agricultural Economics, K.U.Leuven.